Global Impact of Ergot Alkaloids

Global Impact of Ergot Alkaloids

Editor

James L. Klotz

MDPI • Basel • Beijing • Wuhan • Barcelona • Belgrade • Manchester • Tokyo • Cluj • Tianjin

Editor
James L. Klotz
United States Department of
Agriculture, Agricultural
Research Service
Forage-Animal Production
Research Unit
Lexington
United States

Editorial Office
MDPI
St. Alban-Anlage 66
4052 Basel, Switzerland

This is a reprint of articles from the Special Issue published online in the open access journal *Toxins* (ISSN 2072-6651) (available at: www.mdpi.com/journal/toxins/special_issues/ergot_alkaloid).

For citation purposes, cite each article independently as indicated on the article page online and as indicated below:

LastName, A.A.; LastName, B.B.; LastName, C.C. Article Title. *Journal Name* **Year**, *Volume Number*, Page Range.

ISBN 978-3-0365-3780-1 (Hbk)
ISBN 978-3-0365-3779-5 (PDF)

Contents

About the Editor

James L. Klotz

James L. Klotz is a research animal scientist with the USDA-ARS, Forage-Animal Production Unit. Dr. Klotz's research has focused on the influence that ergot alkaloids have on grazing livestock physiology. He has collaborated with scientists around the world and across many disciplines to improve the understanding of how ergot alkaloids interact with animal and plant systems.

toxins

MDPI

ditorial

Global Impact of Ergot Alkaloids

ames L. Klotz

United States Department of Agriculture-Agricultural Research Service, Forage-Animal Production Research Unit, Lexington, KY 40546, USA; james.klotz@usda.gov; Tel.: +1-859-257-1647; Fax: +1-859-257-3334

For many years, ergot alkaloids have been considered both a problem to be mitigated and a potential medical cure. These compounds have been primarily studied in the medical/pharmaceutical [1] and agricultural fields [2]. Depending on one's perspective, the impact that ergot alkaloids have had on the progress of human medicine and livestock production can be either positive or negative. The dose or concentration of ergot alkaloid exposure is paramount. This can determine whether these compounds are implicated in the morbidity and mortality of individuals with St. Anthony's Fire, or whether they can be used to treat migraines and post-partum bleeding; it can determine whether they are used to maximize plant resistance and persistence, or whether they constitute an animal welfare concern for grazing livestock [3,4]. The ethics of ergot alkaloid use is debated to this day, but there is no debating the impact of these compounds. Many of the positive and negative issues associated with ergot alkaloids have specific conditions with regional implications, but that does not diminish the magnitude of impact that these compounds have had on humans, livestock, and plants globally.

Research evaluating ergot alkaloids can be both basic and applied. Many types of research perspectives are necessary in understanding ergot alkaloids' functions and how these findings might be applied. The focus of this Special Issue concerns original research and review articles that highlight benefits and detriments, and successes and failures involving ergot alkloids around the world with deference to regional distinctions. Research models range from fungus, to plant, to mammal; and the ergot alkaloids produced by both *Claviceps* and *Epichloë* spp. of fungi are included in this Special Issue. All submissions focus on ergot alkaloids' effects (positive or negative) in different contexts. There is a benefit to this shared interest, even if the issues with ergot alkaloids do not directly overlap.

Significant advancements in the manipulation of plant–endophyte symbioses have been made in recent years to optimize the profile and concentration of the secondary compounds produced [5]. Eady [6] has reviewed the complexities plant breeders encounter when selecting a desired plant–endophyte symbiont, with New Zealand ryegrass as a model. This is a balance between selecting a source of ergot alkaloids that permit greater plant persistence, and inhibiting ergot alkaloid production that results in mycotoxicosis in grazing livestock in combination with desired plant traits. This makes the understanding of ergot alkaloid production paramount. The potential of using various-omics technologies to study ergot alkaloid production has been demonstrated in this Special Issue. In addition to traditional selection processes, Florea et al. [7] demonstrate the use of CRISPR technology to create a non-transgenic strain of *Epichloë* fungus without the genes necessary to produce ergot alkaloids. Fungi that produce ergot alkaloids can be endophytic and parasitic. Ergot contamination of cereal crops in Canadian provinces has become an issue of increasing concern. Hicks et al. [8] evaluate diversity in genes related to ergot alkaloid production in Canadian strains of the parasitic *Claviceps purpurea* to better characterize and understand the variation of ergot alkaloid content. Also looking at Canadian strains of *C. purpurea*, Liu et al. [9] evaluated the evolution patterns of gene clusters associated with different classes of ergot alkaloid production. Work of this caliber is critical to better understand this evolving issue.

Citation: Klotz, J.L. Global Impact of Ergot Alkaloids. *Toxins* **2022**, *14*, 186. https://doi.org/10.3390/toxins14030186

Received: 18 February 2022
Accepted: 1 March 2022
Published: 3 March 2022

Historically, human interactions with ergot alkaloids have been defined by large-scale poisonings through the consumption of contaminated grains [1]. Incidents of human ergot poisonings are increasingly rare due to improvements in crop management, grain screening and cleaning [10], and the regulation of safe quantities in food and feed [11]. However, there are still areas in the world where this can be an issue [12], and there is also still interest in the pharmaceutical potential of ergot alkaloids. Their most prominent use has been the treatment of migraines and controlling post-partum bleeding in the 18th and 19th centuries. In a current review of the past gynecological and obstetric uses of ergot alkaloids, Smakosz et al. [13] defined a potential role for the application of ergot alkaloids in modern obstetrics. In addition to clinical uses of ergot alkaloids, research assessing the sustainable production of ergot alkaloids in desirable formulations is needed. Shahid et al. [14] have developed a response surface methodology to select strains of *Penicillium citrinum* for their ability to produce ergot alkaloids in culture. Many researchers that study ergot alkaloids can relate to the challenges associated with obtaining purified forms of desired ergot alkaloids in any quantity.

Although medical applications focus on ergot alkaloids' positive effects in humans, animal agriculture has historically and consistently viewed ergot alkaloids as a problem to be solved. Further, changing environmental conditions cause the ever-changing fungal production of ergot alkaloid profiles and concentrations. This necessitates routine surveys of grains and grasses. In this Special Issue, these are exemplified by the on-farm monitoring of ergot alkaloid levels in Kentucky horse pastures described by Lea and Smith [15], as well as the ergot alkaloids found in Slovenian feed grains, as described by Babic et al. [16]. Research of this nature is ongoing globally and contributes greatly to the mitigation of large-scale problems as well as the identification of future areas in need of research.

The variation of the content and concentration of ergot alkaloids is further complicated by livestock exposed to ergot alkaloids that demonstrate varied responses to the toxins. Poole et al. [17] and Wilbanks et al. [18] have studied various aspects, including genetics, that may make cattle more resistant to consumed ergot alkaloids. Ault-Seay et al. [19] used advanced-omics technologies to look at the rumen microbial and host metabolomes to provide a whole-animal characterization of impacts of ergot alkaloids. Mote and Filipov [20] reviewed the use of interactomics to provide a systemic understanding of the pathologies caused by ergot alkaloids that cause fescue toxicosis. A very specific pathology associated with ergot alkaloids and ergotism is a chronic vasoconstriction. Yonpaim et al. [21] looked at the acute exposure of ergot alkaloids on vasoactivity in ovine vasculature, and Valente et al. [22] evaluated prolonged ergot alkaloid exposure on the vasoactivity of bovine vasculature. Both studies [21,22] respectively evaluated aspects related to the ability of ergot alkaloids to interact with adrenergic and serotonergic receptors [23], and both papers concluded that receptor-mediated treatments for ergot alkaloid-induced vasoconstriction could be explored as potential therapies. From a systemic evaluation of ergot alkaloids' impact on the whole animal or microbiome, to the study of a specific symptom, there is much yet to be learned about how ergot alkaloids disrupt mammalian physiology.

The collection of papers in the Global Impact of Ergot Alkaloids (https://www.mdpi.com/journal/toxins/special_issues/ergot_alkaloid) (accessed on 17 February 2022) Special Issue highlights the rich diversity of research and the complexity of the problems centered around ergot alkaloids. Although many specific issues related to accidental or intentional consumption of ergot alkaloids can be localized to a certain geographic region, the problems, challenges, and fascination with ergot alkaloids is global.

Funding: This research received no funding.

Acknowledgments: This editor is grateful to all the contributing authors. Their expertise and high-quality research made this special issue possible. Also, special thanks are extended to all the reviewers who provided rigorous reviews that have improved the content of this special issue. Lastly, I would like to thank the staff of MDPI Toxins journal for their patience and steadfast organization.

Conflicts of Interest: The author declares no conflict of interest.

eferences

Schiff, P.L. Ergot and its alkaloids. *Am. J. Pharm. Educ.* **2006**, *70*, 98. [CrossRef] [PubMed]

Klotz, J.L. Activities and Effects of Ergot Alkaloids on Livestock Physiology and Production. *Toxins* **2015**, *7*, 2801–2821. [CrossRef] [PubMed]

. Strickland, J.R.; Looper, M.L.; Matthews, J.C.; Rosenkrans, C.F., Jr.; Flythe, M.D.; Brown, K.R. Board-invited review: St. Anthony's Fire in livestock: Causes, mechanisms, and potential solutions. *J. Anim. Sci.* **2011**, *89*, 1603–1626. [CrossRef] [PubMed]

. Klotz, J.L.; Nicol, A.M. Ergovaline, an endophytic alkaloid. 1. Animal physiology and metabolism. *Anim. Prod. Sci.* **2016**, *56*, 1761. [CrossRef]

. Johnson, L.J.; de Bonth, A.C.M.; Briggs, L.R.; Caradus, J.R.; Finch, S.C.; Fleetwood, D.J.; Fletcher, L.R.; Hume, D.E.; Johnson, R.D.; Popay, A.J.; et al. The exploitation of epichloae endophytes for agricultural benefit. *Fungal Divers.* **2013**, *60*, 171–188. [CrossRef]

. Eady, C. The Impact of Alkaloid-Producing Epichloe Endophyte on Forage Ryegrass Breeding: A New Zealand Perspective. *Toxins* **2021**, *13*, 158. [CrossRef]

'. Florea, S.; Jaromczyk, J.; Schardl, C.L. Non-Transgenic CRISPR-Mediated Knockout of Entire Ergot Alkaloid Gene Clusters in Slow-Growing Asexual Polyploid Fungi. *Toxins* **2021**, *13*, 153. [CrossRef]

;. Hicks, C.; Witte, T.E.; Sproule, A.; Lee, T.; Shoukouhi, P.; Popovic, Z.; Menzies, J.G.; Boddy, C.N.; Liu, M.; Overy, D.P. Evolution of the Ergot Alkaloid Biosynthetic Gene Cluster Results in Divergent Mycotoxin Profiles in *Claviceps purpurea* Sclerotia. *Toxins* **2021**, *13*, 861. [CrossRef]

). Liu, M.; Findlay, W.; Dettman, J.; Wyka, S.A.; Broders, K.; Shoukouhi, P.; Dadej, K.; Kolarik, M.; Basnyat, A.; Menzies, J.G. Mining Indole Alkaloid Synthesis Gene Clusters from Genomes of 53 Claviceps Strains Revealed Redundant Gene Copies and an Approximate Evolutionary Hourglass Model. *Toxins* **2021**, *13*, 799. [CrossRef]

10. Flieger, M.; Wurst, M.; Shelby, R. Ergot alkaloids—Sources, structures and analytical methods. *Folia Microbiol.* **1997**, *42*, 3–30. [CrossRef]

11. EFSA. Scientific Opinion on Ergot Alkaloids in Food and Feed. *EFSA J.* **2012**, *10*, 158. [CrossRef]

12. Urga, K.; Debella, A.; W/Medihn, Y.; Agata, N.; Bayu, A.; Zewdie, W. Laboratory studies on the outbreak of Gangrenous Ergotism associated with consumption of contaminated barley in Arsi, Ethiopia. *Ethiop. J. Health Dev.* **2002**, *16*, 317–323. [CrossRef]

13. Smakosz, A.; Kurzyna, W.; Rudko, M.; Dasal, M. The Usage of Ergot (*Claviceps purpurea* (fr.) Tul.) in Obstetrics and Gynecology: A Historical Perspective. *Toxins* **2021**, *13*, 492. [CrossRef]

14. Shahid, M.G.; Nadeem, M.; Gulzar, A.; Saleem, M.; Rehman, H.U.; Ghafoor, G.Z.; Hayyat, M.U.; Shahzad, L.; Arif, R.; Nelofer, R. Novel Ergot Alkaloids Production from *Penicillium citrinum* Employing Response Surface Methodology Technique. *Toxins* **2020**, *12*, 427. [CrossRef] [PubMed]

15. Lea, K.M.; Smith, S.R. Using On-Farm Monitoring of Ergovaline and Tall Fescue Composition for Horse Pasture Management. *Toxins* **2021**, *13*, 683. [CrossRef] [PubMed]

16. Babic, J.; Tavcar-Kalcher, G.; Celar, F.A.; Kos, K.; Cervek, M.; Jakovac-Strajn, B. Ergot and Ergot Alkaloids in Cereal Grains Intended for Animal Feeding Collected in Slovenia: Occurrence, Pattern and Correlations. *Toxins* **2020**, *12*, 730. [CrossRef] [PubMed]

17. Poole, D.H.; Mayberry, K.J.; Newsome, M.; Poole, R.K.; Galliou, J.M.; Khanal, P.; Poore, M.H.; Serao, N.V.L. Evaluation of Resistance to Fescue Toxicosis in Purebred Angus Cattle Utilizing Animal Performance and Cytokine Response. *Toxins* **2020**, *12*, 796. [CrossRef]

18. Wilbanks, S.A.; Justice, S.M.; West, T.; Klotz, J.L.; Andrae, J.G.; Duckett, S.K. Effects of Tall Fescue Endophyte Type and Dopamine Receptor D2 Genotype on Cow-Calf Performance during Late Gestation and Early Lactation. *Toxins* **2021**, *13*, 195. [CrossRef]

19. Ault-Seay, T.B.; Melchior-Tiffany, E.A.; Clemmons, B.A.; Cordero, J.F.; Bates, G.E.; Flythe, M.D.; Klotz, J.L.; Ji, H.; Goodman, J.P.; McLean, K.J.; et al. Rumen and Serum Metabolomes in Response to Endophyte-Infected Tall Fescue Seed and Isoflavone Supplementation in Beef Steers. *Toxins* **2020**, *12*, 744. [CrossRef]

20. Mote, R.S.; Filipov, N.M. Use of Integrative Interactomics for Improvement of Farm Animal Health and Welfare: An Example with Fescue Toxicosis. *Toxins* **2020**, *12*, 633. [CrossRef]

21. Yonpiam, R.; Gobbet, J.; Jadhav, A.; Desai, K.; Blakley, B.; Al-Dissi, A. Vasoactive Effects of Acute Ergot Exposure in Sheep. *Toxins* **2021**, *13*, 291. [CrossRef] [PubMed]

22. Valente, E.E.L.; Harmon, D.L.; Klotz, J.L. Influence of Prolonged Serotonin and Ergovaline Pre-Exposure on Vasoconstriction Ex Vivo. *Toxins* **2021**, *14*, 9. [CrossRef] [PubMed]

23. Berde, B.; Stürmer, E. *Introduction to the Pharmacology of Ergot Alkaloids and Related Compounds as a Basis of Their Therapeutic Action*; Springer: Berlin/Heidelberg, Germany, 1978; Volume 49.

toxins

Review

The Impact of Alkaloid-Producing *Epichloë* Endophyte on Forage Ryegrass Breeding: A New Zealand Perspective

Colin Eady

Barenbrug, New Zealand Ltd., 2547 Old West Coast Road, Courtenay, Christchurch 7671, New Zealand; ceady@barenbrug.co.nz; Tel.: +64-27-717440

Abstract: For 30 years, forage ryegrass breeding has known that the germplasm may contain a maternally inherited symbiotic *Epichloë* endophyte. These endophytes produce a suite of secondary alkaloid compounds, dependent upon strain. Many produce ergot and other alkaloids, which are associated with both insect deterrence and livestock health issues. The levels of alkaloids and other endophyte characteristics are influenced by strain, host germplasm, and environmental conditions. Some strains in the right host germplasm can confer an advantage over biotic and abiotic stressors, thus acting as a maternally inherited desirable 'trait'. Through seed production, these mutualistic endophytes do not transmit into 100% of the crop seed and are less vigorous than the grass seed itself. This causes stability and longevity issues for seed production and storage should the 'trait' be desired in the germplasm. This makes understanding the precise nature of the relationship vitally important to the plant breeder. These *Epichloë* endophytes cannot be 'bred' in the conventional sense, as they are asexual. Instead, the breeder may modulate endophyte characteristics through selection of host germplasm, a sort of breeding by proxy. This article explores, from a forage seed company perspective, the issues that endophyte characteristics and breeding them by proxy have on ryegrass breeding, and outlines the methods used to assess the 'trait', and the application of these through the breeding, production, and deployment processes. Finally, this article investigates opportunities for enhancing the utilisation of alkaloid-producing endophytes within pastures, with a focus on balancing alkaloid levels to further enhance pest deterrence and improving livestock outcomes.

Keywords: endophyte transmission; livestock safety; insect testing; quality control; alkaloid profile

Key Contribution: This manuscript details the history and commercial exploitation of alkaloid-producing endophyte in New Zealand, highlighting the issues faced with ergot alkaloid-producing strains, and the future opportunities and risks of deploying alkaloids via a plant host.

Citation: Eady, C. The Impact of Alkaloid-Producing *Epichloë* Endophyte on Forage Ryegrass Breeding: A New Zealand Perspective. *Toxins* **2021**, *13*, 158. https://doi.org/10.3390/toxins13020158

Received: 30 November 2020
Accepted: 6 February 2021
Published: 18 February 2021

1. Introduction

Efficiently producing animal-sourced food (ASF) from pasture using 'low-cost livestock' (in situ grazing of forage) is a key requirement in order meet land-use sustainability criteria [1]. New Zealand pastoral farming systems are some of the most efficient 'low-cost livestock' systems in the world, producing ASF economically and with a relatively small environmental footprint [2–4], and recently https://www.dairynz.co.nz/news/research-shows-nz-dairy-the-world-s-most-emissions-efficient/ accessed on: 11 February 2021. This is principally because the NZ climate is conducive to high year-round biomass production of high metabolizable energy, palatable temperate pasture. *Lolium perenne*, ryegrass, although an introduced species, is currently the principle grass of choice in NZ because cultivars have been bred to perform within NZ climatic ranges [5], with a cardinal temperature that enables good growth [6], along with architecture and physiology to provide high year-round photosynthetic conversion efficiency [7]. To complement these plant characteristics, management requires the addition of low-cost nutrients provided in part by addition of a legume, white clover, nitrogen and phosphate [8]. Persistence (biomass production/time) of such pastures is achieved through a combination of good stock and pasture management, breeding of the ryegrass germplasm, e.g., for rust resistance, seasonal growth, and the use of *Epichloë* endophytes to relieve biotic and abiotic stressors. These endophytes are asexual, maternally inherited and produce characteristic alkaloid profiles [9]. The compatibility of endophytes with the grass host, the effect they have on the host phenotype, and the temporal, spatial alkaloid profile produced are key components that need to be understood before deployment decisions can be made. The successful use of *Epichloë* endophytes has been achieved despite numerous application problems, and no single perfect endophyte exists on the market. Biologically, this is not surprising as the asexual *Epichloë* is trapped in the vegetative tissue or seed of a host ecotype. Variants are essentially limited to somatic mutations as recombination and reassortment of chromosomes do not occur. There are exceptions to this and hybridisations between endophytes have occurred throughout their evolution [10,11] to produce diploid and triploid interspecific hybrids. For more on the evolution of endophytes, refer to the works of Schardl and Hettiarachchige [9,10,12]. These biological characteristics favour the development of local variants evolved to live within a geographically constrained grass ecotype with specific growth characteristics, e.g., architecture, flowering dates, dormancy, and under local biotic, and abiotic stressors. Endophyte discovery projects [13] have found localised variants such as AR37, nea2, nea6, and CM142. Transfer of endophytes between ecotypes or species can cause gross changes to the symbiosis [13]. Elite forage ryegrass has been bred, through sexual recombination, primarily for biomass production and performance within an environment. This may rapidly change architecture, heading date, vernalisation requirement, etc., of the plant, which a recent study has suggested may limit the symbiotic association [14]. More studies are required to fully understand what dictates the temporal and spatial growth of the endophyte in the ryegrass and particularly the related alkaloid expression profiles. Many studies have been undertaken that demonstrate host-induced differences in *Epichloë* traits, aside from transmission and viability that are crucial production traits. The same *Epichloë* in a different ryegrass host can demonstrate an >10-fold alkaloid expression range [15] and/or have altered infection characteristics. Previous research has primarily been one-off studies, comparing one combination with another [15]. Trying to exactly copy biotic conditions, sampling, analysis, etc., to compare between manuscripts is difficult, thus making it difficult to home in on the causation(s) of any differences. One theme that comes through is that tetraploid ryegrass tends to have lower alkaloid concentrations than diploid, perhaps not surprising as the larger cell size of the tetraploids would provide a larger plant cell volume: endophyte ratio, thus diluting any endophyte contribution. A better understanding of the interplay between host and endophyte is required and until then, each ryegrass cultivar–*Epichloë* strain relationship must be assessed on its own merits.

This manuscript provides a brief overview of endophyte strains used in New Zealand and the ergot profiles produced, and then focuses on the practical breeding of ryegrass containing *Epichloë* endophyte and outlines some of the key challenges facing a grass seed company's day to day breeding, production, and distribution. This manuscript will highlight quality control (QC) testing requirements and their difficulties, and functionality testing requirements of the industry, including shortfalls in alkaloid knowledge. This manuscript then investigates opportunities for breeding the host to better accept an endophyte as well as advances in endophyte selection, and deployment systems for use in future pasture systems. These highlights will be made whilst focused on *Epichloë*–ryegrass relationships in NZ, the issues and opportunities outlined are likely applicable to endophytes of other crop and forage species.

2. Brief History of Endophyte Strains Used in New Zealand Grass Breeding

Endophytic fungi have been known about for a long time [16] but the connection to animal livestock health and insect deterrence took some time to discover. It was not until the late 1970s in the USA that an endophytic fungus in tall fescue (*Festuca arundinacea*) was shown to cause fescue toxicosis in cattle [17] and then in New Zealand in the early 1980s the closely related endophyte in perennial ryegrass was shown to cause ryegrass staggers in sheep [18]. In the subsequent discovery period and following a series of name changes, they eventually became known as *Epichloë festucae* var. *lolii* and the tall fescue endophyte *Epichloë coenophiala* [19]. The original New Zealand var. *lolii* strain that produces alkaloids peramine, lolitrem B and ergovaline was termed Standard Endophyte (SE), wild type or Common Toxic to distinguish it from the strains discovered from there on. Endosafe™ was the first commercial ryegrass endophyte, which was released in 1990. From the literature, it appears (but is difficult to confirm) that Endosafe™ was originally AR6, which itself was two strains, later identified to be AR5 and AR77, which both produce peramine and ergovaline [20,21]. Initial studies (1991) indicated that Endosafe™ demonstrated good insect deterrence and was safe for animal performance [22]. This claim was quickly questioned as subsequent studies gave animal health issues attributed to the ergovaline [21,23] and it became understood that alkaloid profiles of the same endophyte could differ substantially within different hosts [15]. Endosafe™ in diploid perennial ryegrass cultivar 'Pacific' was withdrawn from the market, but the tetraploid cultivar 'Greenstone' with Endosafe™ continued to be sold. Reselection (possibly from the original Endosafe™) for lower ergovaline identified the strain AR5 which was later marketed as Endo5 and is still sold in Australia today. The known role of ergot alkaloids in fescue toxicosis steered ryegrass endophyte research away from ergot alkaloid-producing *Epichloë* and, shortly after, Endosafe™ strain AR1 was introduced to the New Zealand market in 2001. AR1 produced peramine but no ergovaline or lolitrems and proved to be animal safe and demonstrated good Argentine stem weevil deterrence [24–26], leading to its successful uptake by the NZ pastoral sector. Unfortunately, the single alkaloid profile of AR1 did not deter black beetle and other pests, which led to widespread product failure in the upper North Island of NZ, where these pests are prevalent [27].

AR1 was the first endophyte to obtain plant variety rights (PVR) protection in NZ, which set the precedent for future endophytes (Table 1). After AR1, the NEA2 endophyte in diploid cultivar 'Tolosa' was released. This endophyte was initially not fully characterised (see below) and produced moderate levels of peramine and ergovaline, and low levels of lolitrems. The product was withdrawn due to seed production issues. In 2007, a fourth ryegrass endophyte, AR37, was released to the market. AR37 ryegrass appeared to have similar or better persistence than ryegrass with SE endophyte [28,29]. AR37 still caused occasional outbreaks of ryegrass staggers, which could be severe but generally animal production compared favourably with nil endophyte ryegrass [30]. AR37 produces epoxy-janthitrem alkaloids, via a similar biochemical pathway to that for lolitrem production, but janthitrems have a lower potency than lolitrem B [31,32]. Again, as with Endosafe, it was shown that the individual host–endophyte relationship was important in regulating alkaloid expression with some AR37/ryegrass cultivars causing greater staggers than others (certain AR37 products contain warnings to this effect). In some studies, a decrease in milk solids production in dairy cows has also been observed in pastures containing AR37 [26,33]. License terms for AR37 caused some NZ companies to search for alternative endophytes. This led to additional Novel Endophyte Agriseed (NEA) endophytes being developed. These have primarily originated in Spanish ryegrass germplasm but, as the discovery programme gained momentum endophytes from other regions of the world have been included. Many hundreds of *Epichloë* were screened but present commercial NEA endophytes are derived mainly from three strains—nea2, nea6, and nea3. Note most endophyte strains are denoted by capital letter(s) and a number, for the sake of clarity the 'nea' strains are represented here by lowercase to distinguish them from the commercial 'NEA' products. The nea strains are marketed singularly or in combination as NEA, NEA2, or NEA4; and produce a combination of peramine, ergovaline and a low level of lolitrem B. They have been tested to be animal safe and give good insect deterrence, but like AR37, host–endophyte interactions are important and industry evaluation tables carry caveats for animal performance issues under extreme circumstances (https://www.nzpbra.org/ accessed on: 11 February 2021). Ryegrass containing AR37 or

NEA endophytes have been the principle proprietary perennial or hybrid ryegrasses sold in NZ over the last 10 years. Other companies have followed, either through licensed products or by discovering their own endophytes, although commercial success in the market has been limited. CropMark released U2 a *N*-formylloline-producing meadow fescue endophyte (*Epichloë uncinatum*) and it appears that they have tried to aid stabilisation of this in ryegrass through creation of festulolium hybrids. Recently, they have protected a new endophyte CM142 (NZ PVR website) classed as a novel janthitrem-producing *Epichloë* DLF (DLF.com) market Edge, an *Epichloë festucae* var *lolii* that produces high peramine, low lolitrem B and ergovaline (potentially similar to nea2); and an *Epichloë coenophiala* called Protek that produces low ergovaline and loline, and is derived from (and for) tall fescue (2017 Australian plant breeders rights); and an *Epichloë siegelii* called Happe, a meadow fescue endophyte from Germany that produces lolines and is suitable for use in some ryegrass offering protection against porina [34]. Meanwhile, AgResearch extended their species range through the discovery of AR501, a non-ergovaline-producing tall fescue endophyte, which they superseded with AR542 and AR584 and market as MaxP and MaxQ [35] for use in fescue, and Barenbrug (barusa.com) came out with a similar product E34®. Although a couple of different *Epichloë* species above are being sold in ryegrass (U2 and Happe), it is generally recognised that switching host species for an endophyte is difficult and usually results in gross symbiotic changes that render the relationship unmarketable [13]. Whilst recognising these other host–*Epichloë* relationships, this review focusses on *Epichloë festucae* var *lolii* in ryegrass.

The use of *Epichloë* endophytes as a 'trait' of the plant is an almost unique characteristic of forage grass plant breeding. Its commercial success is demonstrated by expansion of its first designed use in New Zealand to Australia, South America, and South Africa. Development of *Epichloë* for Fescues, for example MaxQ and MaxQII [35], has further extended the use to North American markets. There is also a renewed interest in Europe for the use of endophytes as biological control agents due to increasing constraints on synthetic chemical use. AgResearch has valued the contribution of endophytes to be worth \$200 million NZD (~118M Euro) a year to the NZ economy [36]. The AR37 endophyte patent has been estimated to be worth \$3.6 billion NZD (~2.1B Euro) [37]. Conceptually, using the plant as a solar factory to produce natural biological protectants (via the symbiotic fungus) is an efficient, sustainable, way of delivering protection to broad acre pastures. For reviews on the broader exploitation of *Epichloë* endophytes for agricultural, and future perspectives, the reader is referred to recent reviews [13,35,38]. Use is still mainly limited to temperate grasses, but considerable research is aimed at extending the host range, and endophyte species used, to advance further commercial exploitation [39].

Within commercial ryegrass *Epichloë* associations, there is a dichotomy within the industry around the use of ergot alkaloids. Some groups consider ergot alkaloids to be toxic to livestock and do not support their use in current product lines. This position has arisen from fescue toxicosis observations and early trials in NZ with a high ergovaline-producing *Epichloë* Endosafe™ that did cause health issues [15,38]. Other groups take the position that it is the 'dose that makes the poison'. For this group considerable effort has been undertaken to find products that produce low levels of ergovaline across the pasture, such that it can still confer insect resistance, but intake into the animal is low enough not to cause health issues. A review of the literature [40,41] identified theoretical non-toxic levels but admitted caveats to the research as historical work mainly used SE endophyte and often failed to note pasture alkaloid concentrations for the actual grass consumed. The review work was followed up by a study of NEA2 endophyte demonstrating that in managed pasture situations intake levels by the animal were unlikely to be above detrimental threshold levels [42]. For a review on the use of ergot alkaloid endophytes in New Zealand pastures, the reader is referred to Caradus et al. [15].

Table 1. Lists commercial endophytes, their botanical name, principal host, secondary host in brackets, the PVR approval date and the alkaloid profile, P = peramine, L = lolines, E = ergovaline, Lol = lolitrem, and J = epoxy-janthitrems. * Only basic endpoint compounds identified as data from PVR reports and manuscripts often fail to detail all compounds tested, sampling protocols, temporal and spatial sampling differences, thus making relativities difficult to assess.

Endophyte	Botanical Name	Grass Host	PVR Reg.	Alkaloid Profile *
AR1	Epichloe festucae Leuchtm., Schardl and M. R. Siegel	Perennial Ryegrass	23 April 1996 (expired)	P
AR501	Epichloe coenophiala C.W. Bacon and Schardl	Tall Fescue (Perennial Ryegrass)	23 April 1996 (expired)	P, L
AR542 (MaxP)	Epichloe coenophiala C.W. Bacon and Schardl	Tall Fescue	1 February 1999 (expired)	P. L
UNC1	Epichloe uncinata Leuchtm. and Schardl	Meadow Fescue	14 October 2008	L
nea 2	Epichloe festucae Leuchtm., Schardl and M. R. Siegel	Perennial Ryegrass	25 July 2008	P, E, Lol
AR37	Epichloe festucae Leuchtm., Schardl and M. R. Siegel	Perennial Ryegrass	25 July 2008	J
Happe	Epichloe (Fr.) Tul. and C. Tul. (Epichloë siegelii)	Meadow Fescue (Perennial Ryegrass, Festulolium)	23 June 2010	L
nea 3	Epichloe festucae Leuchtm., Schardl and M. R. Siegel	Perennial Ryegrass	30 June 2009	E, P
U2	Epichloe uncinata Leuchtm. and Schardl	Meadow Fescue	25 July 2008	L
nea 6	Epichloe festucae Leuchtm., Schardl and M. R. Siegel	Perennial Ryegrass	25 July 2008	E, P
AR584 (MaxQ)	Epichloe coenophiala C.W. Bacon and Schardl	Tall Fescue	25 July 2008	L, P
AR95 (Avanex®)	Epichloe festucae Leuchtm., Schardl and M. R. Siegel	Perennial Ryegrass	28 August 2014	E
PTK647	Epichloe coenophiala C.W. Bacon and Schardl	Tall Fescue	18 August 2014	E, L
AR601 (Avanex®)	Epichloe coenophiala C.W. Bacon and Schardl	Tall Fescue	12 May 2010	E, L
AR604	Epichloe coenophiala C.W. Bacon and Schardl	Tall Fescue	12 May 2010	E, L
U12	Epichloe uncinata Leuchtm. and Schardl	Meadow Fescue	28 August 2014	L
AR1006	Epichloe uncinata Leuchtm. and Schardl	Meadow Fescue	21 April 2015	L
E815	Epichloe festucae Leuchtm., Schardl and M. R. Siegel	Perennial Ryegrass	27 August 2014	P, E, Lol
nea10	Epichloe festucae Leuchtm., Schardl and M. R. Siegel	Perennial Ryegrass	29 August 2014	E, P
nea11	Epichloe (Fr.) Tul. and C. Tul.	Perennial Ryegrass	13 August 2014	E, P
nea21	Epichloe (Fr.) Tul. and C. Tul.	Meadow Fescue (Perennial Ryegrass)	29 August 2014	L, P
nea23	Epichloe (Fr.) Tul. and C. Tul.	Meadow Fescue (Perennial Ryegrass)	29 August 2014	L, P

Table 1. *Cont.*

Endophyte	Botanical Name	Grass Host	PVR Reg.	Alkaloid Profile *
U13	Epichloe uncinata Leuchtm. and Schardl	Meadow Fescue	2 September 2016	L
AR1017	Epichloe uncinata Leuchtm. and Schardl	Meadow Fescue	5 October 2016	L
CM142	Epichloe festucae Leuchtm., Schardl and M. R. Siegel	Perennial Ryegrass	17 January 2019	J
nea47	Epichloe festucae Leuchtm., Schardl and M. R. Siegel	Perennial Ryegrass	10 July 2019	E, P

3. Ergot Alkaloid Profiles Produced in Epichloe Endophytes

Deployment of endophytes in grasses has primarily focused on expression of five key alkaloids—peramine, loline, lolitrem, janthitrem, and ergovaline. The myriad of associated precursors and derivatives have largely been ignored. Peramine is produced by a single enzymatic step with no intermediates and is known to be non-toxic to mammals [43]. Loline (*N*-formylloline) is produced via 3 enzymatic steps [43] and has several intermediates but like peramine is not known to be toxic to mammals [43]. Further, it is produced in endophytes primarily used in *Festuca pratensis* type grasses. The indole diterpenes lolitrems and janthitrems are toxic to mammals and have very complex pathways with many intermediates [44], some of which are also toxic. As such, the 'simple' determination of endpoint lolitrems (lolitrem B) and janthitrems (epoxy janthitrem I–IV) may give misleading information regarding animal health effects as intermediate and derivative levels may also have biological impact. Likewise, ergovaline is derived from a complex biochemical pathway that has among its precursors ergoline, clavines, ergoamides and ergopeptines [45,46]. The ergoline ring structure with its similarity to dopamine, serotonin and adrenaline provide ergot structures the basis to act on respective receptors as agonists or antagonists. Thus, producing a multitude of effects depending on the secondary structure. The ergopeptines ergotamine and ergovaline have similar activities in mice models [47] causing vasoconstriction, increase in blood pressure and bradycardiac properties. *Epichloe festuca var lolii* has been studied with respect to its particular ergovaline pathway [48], this analysis was for a standard toxic strain that follows the published pathway to produce a 'standard' level of ergovaline. Although this is often poorly determined [41]. Research in the ergot pathways has been limited to between species or across a few strain(s) within a species, which does not necessarily reflect what the profiles in particular variants might be. For example, nea2 produces a low level of ergovaline and lolitrem B [49,50]. Whilst some progress has been made in identifying the gene clusters and biochemistry present or absent in some of these strains [48,51], little work has been performed to understand differences in levels of ergot intermediates or derivatives in particular strains. For example, ergovaline is seen as a key detrimental compound on livestock health, but intermediate compounds can transfer more readily across the rumen and may in themselves cause animal health effects [52]. Finch et al. 2019 [53] studied mammalian toxicity of chanoclavine and demonstrated this to be safe, but this compound is early in the pathway. A recent study (Barenbrug NZ 2020, unpublished) using a non-lolitrem-producing strain nea3, was shown to produce high levels of paxilline and terpendole C greater than observed in SE strain and under hot dry summer conditions with 'rank' feed this was sufficient to cause ryegrass staggers, even though no lolitrem B was detectable. Analysis of *E. uncinata* haplotypes for loline content identified similar differences in the three forms of loline (NAL, NANL, and NFL) between different haplotypes [54] Analysis of six endophytes by Young 2013 looked at presence vs. absence of the *Eas* gene cluster and whether ergot alkaloids were produced or not [55]. The complex nature of ergot production in the Claviceps [48], between strain variation of the genes, expression profiles, and ergot bioactivity suggest that a greater level of characterisation is needed than is currently undertaken before a new endophyte strain is advanced for commercial exploitation.

4. Strategic Breeding Challenges with Existing Commercial Endophytes

Strategically the requirement for endophyte in a grass breeding programme throws up a conundrum. Does the breeder breed the host to fit the endophyte or breed the best ryegrass and find a compatible endophyte? The first scenario may limit the host grass germplasm to only those genotypes that fit an endophyte, e.g., work of Gagic [56]. The second strategy does not limit the host genotypes but may produce a grass that cannot sufficiently host a suitable endophyte. If several endophyte types are available, then the second strategy is possibly best but if a single endophyte is available then the first strategy is probably advisable. In NZ this has been depicted in the market through the predominant marketing of endophyte AR37, and to a much lesser extent AR1 by PGGW Seeds and Agricom brands (DLF), whilst Barenbrug have marketed NEA, NEA2, NEA4, AR1 and AR37. Most ryegrass breeding pipelines, such as $\frac{1}{2}$sib family selections, and population based family selections have been developed primarily because of basic ryegrass characteristics; outcrossing, S and Z incompatibility, wind pollinated, small male and female organs situated together, and the traits of importance (biomass, persistence, heading date) [57,58]. Research efforts are slowly changing this and developments in genomic selection [58–60], hybrid technology [61], paternity testing [62], self-fertility [63] and biotechnology [64] will likely change the current pipelines. For now though, breeding pipelines generally revolve around 1/2 sib family selection [60] or between- and within-family population breeding approaches [58]. Both these systems essentially identify best families and put together phenotypically similar parent plants to make a new 'synthetic' cultivar. The process of performing this is very different depending on the endophyte strategy taken.

Limiting the host to fit an endophyte is a relatively easy option, and as cycles of the breeding pipeline are completed, more germplasm contains the single endophyte, and subsequent crossing and selection becomes a simple process. Introgression of ecotypes or competitor germplasm may require the elimination of any incumbent endophyte, a relatively simple heat treating of seed and testing of seedlings [65]. Then crossing with the desired endophyte-containing mother line and harvesting from these mother plants will provide introgressed offspring with the endophyte. As this material is further backcrossed checks are required to determine the stability and transmission of the endophyte. Even crosses between two lines that do contain the desired endophyte may create a host with characteristics less conducive to stability and transmission, so testing is always required. Determining compatibility in such a system, using AR37, has been made easier by the development of a genomic estimated breeding value (GEBV) tool for host acceptance of an endophyte [56]. Whilst this is a great advance, care is still required as transmission and stability within a host is not all that needs to be considered. Spatial and temporal growth, and alkaloid profile knowledge is also required, along with potential detrimental effects on the host, e.g., reduced growth. Even with a GEBV for symbiotic potential the strategic decision to use a single endophyte may limit host germplasm, such that some plant characteristics are ultimately compromised.

Breeding the best host and then finding a suitable endophyte has other challenges but does not limit the host germplasm so plant potential is not compromised. As many breeding lines and ecotypes already contain an endophyte, it is first necessary to understand what endophyte is contained and what this contributes, or detracts, to the plant phenotype. Initially, this was problematical as many endophytes were unknown and testing was a costly, skilled exercise. A small set of simple sequence repeat (SSR) were available to detect endophytes [66] with a cost ~23 Euro per test. Therefore, breeders often worked with limited knowledge, using immunological tests for endophyte presence or absence, and only confirmed type when necessary. Breeding a high-performing synthetic ryegrass population would often occur and then a subset of plants would be tested to determine endophyte status. Following this, individual plants could be chosen to constrain endophyte status. Historically at least a couple of early commercial endophyte strains were actually mixtures of two or three endophytes [20,30]. If the correct endophyte was not present, then a sample of seed could be heat cured of undesirable endophyte [65], and seedlings re-inoculated with the endophyte of choice [67]. Following this, further re-characterisation was required potentially putting the breeding programme back three years or more.

Such 'blind' breeding resulted in the release of a ryegrass cultivar 'Trojan^{NEA2}' containing two endophytes, determined by SSR, and PVR protected as nea2 and nea6. This serendipitous mix provided an effective alkaloid profile and saw Trojan^{NEA2} become the top selling ryegrass across NZ between 2015 and 2018. Later, using a broader SSR panel it was shown that the nea6 component was in fact two different endophyte strains [30], with similar alkaloid profiles. So Trojan^{NEA2} in fact consisted of nea2, nea6 and the variant endophyte [30] (later PVR protected as nea47).

Recently (Kompetitive allele specific primer), KASP technology [68] (see molecular testing below) has been applied to typing endophytes [69], which has enabled affordable identification of numerous endophyte strains quickly and easily. The system was developed for quality control determination and is based on sequence information that identifies a unique (single nucleotide polymorphism) SNP for a strain. The test result either concurs with the SNP under test, or identifies it as an 'other' strain, or returns a 'below threshold/negative'. If identified as 'other', then this can be further investigated with different strain-specific KASP primers. KASP works very well when breeding material has known commercial strain(s), providing a simple yes/no test for identification. If unknown ecotypes are to be tested, it is also advisable to check with broad SSR panels or multiplex PCR panels such as those developed by Vikuk et al. [70].

The KASP test has transformed breeding the best host because it allows populations to be easily endophyte typed and so parents, F_1 and F_2 individuals can be combined with a knowledge of the endophyte. This also by default utilises two generations of selection pressure for endophyte transmission and survival such that new synthetics are created with some level of selected compatibility. Further, any plants within the synthetic that have the wrong endophyte, or no endophyte, can be eliminated, or used as pollen donors only. If two endophytes are required (as in Trojan^{NEA2} nea2/6 above), then harvest from equal numbers of mother plants containing nea2 or 6 can favour a balance of seed containing the respective endophytes, which KASP can confirm.

5. Practical Methods for *Epichloë*-Infected *L. perenne* Quality Control

Endophyte transmission is variable and endophyte viability decreases more quickly than the seed that contains it. Thus, moving from 100% infected nucleus seed to breeder's seed, to G1 and G2 production seed results in a loss of endophyte. If this cumulative loss is >30% then the final seed will have insufficient endophyte to be sold as endophyte-containing seed (NZPBRA industry standard). Additionally, ingress of seed with standard toxic endophyte, or packaging, labelling, handling, and storage errors can all affect the endophyte type and percentage in the final product. New host–endophyte combination(s) also must be checked for alkaloid profile and ultimately effectiveness as an insect deterrent, and impact on animal wellbeing. Practically, this means testing for endophyte is required at all stages of the process. Such QC testing must be simple, robust, cost effective, i.e., fit for purpose. The following section provides a brief outline of common protocols used, including potential caveats, and outlines how they are used in each stage of the seed production process. They are not necessarily the latest, or all, research methods but primarily robust and cost effective.

5.1. Endophyte Detection—Microscopy

Tillers—Briefly, a ~1 cm base section of tiller is secured to a microscope slide using glue tape and the outer leaf blade unrolled so that it is flat against the slide. These leaf blade sections are then stained using a drop of 2% analine blue and left for ~30 min before being examined under a compound microscope (×40 magnification). Endophyte hyphae stain blue. For a more detailed description, the reader is referred to Bacon and White [71].

Seed—Briefly, ~250 seeds are placed in a test tube and 15.4 M nitric acid added to double the volume of the seeds. This is then heated at 60 °C for approximately 15 min and then the seeds are rinsed under tap water. Individual seeds are then dissected to leave the caryopsis which is stained as for the tillers above. For more detail, refer to [72].

Caveats—Endophyte is obvious growing alongside the cells of the plant. However, the strain of endophyte cannot be determined and even different endophyte species—for example, *E. occultans, E. uncinata*, or *E. typhina* can easily be mistaken for each other. Low levels of infection may go undetected and, within seeds, it should not be assumed that the endophyte is alive.

5.2. Endophyte Detection—Immunoblot

Briefly, tillers, cut at the base of the plant, are blotted by pressing sap from the freshly cut base onto a nitrocellulose membrane. Once blotted, the membrane is incubated with blocking solution, rinsed, and then incubated with a primary antibody. Following this, the blot is rinsed with blocking and incubated with a secondary antibody. This is rinsed off and a final chromogen mix added to bind to the secondary antibody. Blots containing endophyte protein stain according to the chromogen used. For detailed protocol, refer to [73,74], although most companies have their own custom testing regimes, a kit can be purchased from (Agrinostics, Watkinsville, GA, USA; cat. #ENDO797-3).

Caveats—Immunoblots are not strain specific and can react to different *Epichloë* species. A recent in-house comparison of seed immunoblots with microscopy revealed a small percentage of samples (~2%) were positive on the microscope and negative on the blot or vice versa (Barenbrug NZ, unpublished). Immunoblots are also based on threshold levels of detection and tiller tests use standard 6-week-old tillers, younger tillers can be used but the negative results are less reliable. Additionally, late season tillers with little sap can be difficult to blot, all of which results in semi-quantitative/qualitative data.

5.3. Endophyte Detection—Molecular Testing

Molecular analysis of plant endophyte status was developed with SSRs [75]. This was rapidly developed into a fingerprinting system for endophyte types based on flanking variation around a microsatellite tandem repeat locus B [66] and soon became a routine test in NZ supplied by AgResearch. Similar systems were developed in the USA for fescue endophytes [76]. Testing requires isolation of good quality DNA from tiller or seed samples, usually via freeze dried or fresh samples, followed by multiplex assays using 3 to 5 primer sets to discerning B allele loci. Assays require electrophoretic separation of fragment sizes. Recently, using more complete sequence analysis of different endophyte strains, a catalogue of SNPs has been developed to enable the simple discrimination of known endophytes through KASP [69], this relies on fluorescent (HEX or FAM) primers competing for a binding site, which matches or mismatches the SNP in question. This process can use crudely isolated DNA and provides an immediate digital result (within 2 h, see Figure 1) via real-time PCR, reducing costs and making it suitable for in-house QC testing.

Caveats—Molecular testing requires quality assessment and threshold setting of the data. Is a negative truly negative, or just below the threshold of detection? This is particularly true if performing bulk analysis, although digital droplet PCR and or advancements in next-generation sequencing (NGS) techniques may improve this in the future. In addition, both SSR and KASP do not eliminate the risk that a new but related endophyte is wrongly detected. Using NGS techniques, it is becoming increasingly possible to identify small genetic differences even between identical twins [77], thus molecular differences depend in part on how hard you look for them.

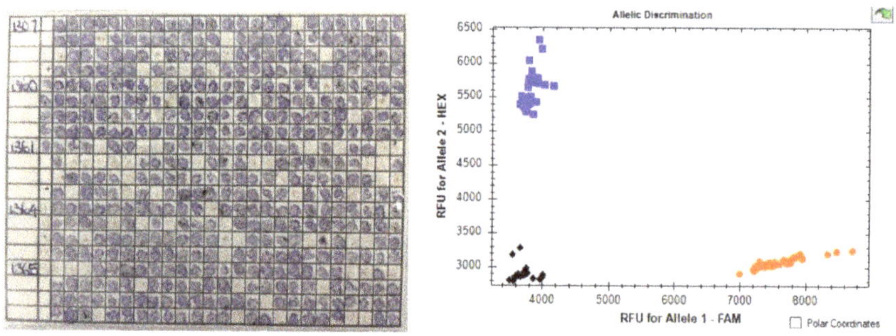

Figure 1. Left: typical immunoblot of tiller sap. Blue staining represents presence of endophyte through the detection of immunoblot antibodies bound to endophyte protein motifs. Clear squares represent ryegrass tillers without endophyte. **Right**: typical KASP assay result—in this case, assaying NEA4. Blue squares represent nea2, orange 'other' (in this case nea3) and black represent below threshold (negative). (Images courtesy of Amanda James.)

5.4. Alkaloid Profiles

Testing in grass breeding situations is usually limited to analysis of leaf material, traditionally, field, or pot grown, material is usually cut to ground level. Samples are placed on ice, then frozen, freeze dried, and ground to <1 mm. This material is then analysed, via HPLC and LC–MS techniques [78–80]. Recently, a higher-throughput system has been developed for ergovaline, lolitrem B and peramine by Agriculture Victoria, Australia [81]. The authors investigated methanol extraction protocols, multiple separate extractions v two extractions combined and concentrated. Matrix effects and recovery efficiency was also investigated. Analysis was undertaken across three analytical platforms, two LC–MS systems, the QE MS and the QQQ MS and ergovaline was also analysed using an established HPLC-FLD method. The epoxy-janthitrem alkaloids have been very difficult to analyse due to instability issues. This has recently been overcome [82] but it is still a complex HPLC protocol requiring acetone extraction in the dark to prevent degradation. For full structural elucidation, nuclear magnetic resonance (NMR) assignments for the four epoxyjanthitrem (I–IV) compounds, and a new epoxyjanthitriol were required. The instability, extraction requirements and specific techniques for detailed analysis highlights the importance of understanding sample preparation right through to a result and highlights the difficulties of comparing data across platforms and between laboratories.

Caveats—Alkaloid analysis requires skilled biochemical knowledge, expensive equipment, and accurate sample preparation. Endophyte alkaloid profiles in a plant are difficult to sample measure and interpret and previous testing methodology and analysis has often raised more questions than it has answered [41]. Cross-laboratory validations are recommended but rarely performed due to costs and some standards such as epoxy-janthitrems are not readily available. Eady et al. [42] compared Australian and New Zealand laboratories for ergovaline and obtained a 95.5% correlation. Although other simpler techniques to detect alkaloids are being investigated such as NIR [83], for now, well-equipped laboratories with specific skills are needed for accurate alkaloid detection. These capabilities are expensive and usually result in essential analysis only being performed. For a semi-quantitative technique that can be used in house, it is possible to quantify ergot alkaloids with ELISA kits (agrinostics.com/shop/), but nothing similar is available for other alkaloids. Most of the reported alkaloid sample testing in ryegrass states 'cut to ground level', but even this depends upon interpretation of ground level and may vary by several cm's (Author's personal observation). Combine this with poor characterisation of the height of the sample, and the knowledge that alkaloid temporal or spatial spread throughout the plant also varies with host, strain and environment, and it becomes almost impossible to conclude anything beyond the individual data set [40,41]. Klotz and Nicol [40] and

Nicol and Klotz [41] highlighted this in a review on animal effects caused by ergovaline in ryegrass and concluded that actual knowledge of alkaloid levels in the grass consumed was a much more relevant measure for animal health studies. It seems that the original purpose of alkaloid testing was to understand levels in plants, not levels ingested by grazing animals. Many animal health studies followed the 'cut to ground level' methodology, causing a reduced ability to deduce meaningful knowledge from, or compare, the results [41]. Since that review Eady et al. [42] estimated animal intake levels against the threshold level deduced by Nicol and Klotz [41]. The data would seem to agree with the threshold. However, more research is required to fully understand the impact on health of different alkaloid levels ingested through 'natural' grazing conditions.

5.5. Livestock Safety Testing

In New Zealand, evaluating novel endophyte ryegrass combinations for animal safety is undertaken voluntarily under agreed industry protocols developed by the Endophyte Technical Committee (ETC), which is part of the New Zealand Plant Breeding Research Association (https://www.nzpbra.org/ accessed on: 11 February 2021). The guidelines set out requirements for trialing including ethical approval and include scores for ryegrass staggers, heat stress, dags, and liveweight gain. Pasture under test must contain >85% viable endophyte, and chemical profiles to ground level determined. For a trial to be valid, a negative health effect must be observed in SE plots. Data are presented to the ETC and, if approved, a star rating for risk of ryegrass staggers and a comment on animal performance is assigned. Each year, industry rated tables are published to allow comparison of the animal safety parameters (https://www.nzpbra.org/ accessed on: 11 February 2021).

Caveats—Animal trialing is difficult and expensive to undertake, and such trials are under increasing pressure regarding animal welfare concerns. Data from the trials are of limited use for several reasons: (1) chemical profiles to ground level of the plant provide little knowledge as to what the actual alkaloid intake by the animal is; (2) environmental conditions vary year on year and influence alkaloid production, with trials often only run for a 4–8 week period within one year; (3) sheep genetics is not controlled and considerable variation is known to exist in regard to alkaloid tolerance; (4) the range between a valid positive control, ~1.1 ppm [84], and an extreme positive control, > 4 ppm [85], for lolitrem B alkaloid level is large; and (5) large differences in pasture quality may exist. This makes comparisons between trials difficult and may underestimate effects by up to 3.5-fold if conditions are such that only the minimum 1.1 ppm threshold is achieved. Thus, the result is open to manipulation by a skilled trial manager and the true potential of an endophyte to cause harm may be missed. The grazing practices of the trial, i.e., 'worst-case scenario', do not represent typical on-farm practice management and so the relevance of the test is also now under question. In addition, presence of other fungi such as *Claviceps purpurea*, or animal diseases may also cause animal fitness issues (though these should be countered by the control plots).

It may be possible with the extensive data set held by the ETC, other published material and on-going experiments, to calculate a relationship between alkaloid level and animal effect. This would allow trial material to be grown under much more defined conditions (e.g., recommended good management practice, vs. extreme, defined hot summer drought conditions) and animal alkaloid intake levels modelled by chemical analysis of the grass profiles. Such an approach would reduce environmental variables and eliminate or reduce the need for animal trials. This concept is not new, as Nicol and Klotz [41] calculated such an effect using published ergovaline data. However, for industry application, more comprehensive models are required.

5.6. Insect Testing

As with the animal testing, insect testing in NZ is via industry-agreed protocols developed by the ETC. Six key pest insects affecting NZ pastures have agreed testing protocols. Testing is customised for each insect but generally involves pot and/or field trials against a susceptible (no endophyte) and control, typically SE plants. Insects under question include Argentine stem weevil (*Listronotus bonariensis*), black beetle (*Heteronychus arator*), grass grub (*Costelytra zealandica*), pasture mealybug (*Balanococcus poae*), porina (*Wiseana spp.*) and root aphid (*Aploneura lentisci*). Other insect pests have also been investigated, and ryegrass *Epichloë* alone have been demonstrated to have activity against over 20 insect species [86]. As needs arise, or commercial advantage is sought, it is likely that additional protocols will be added to the current list of six. Each year, data are submitted to the ETC and industry-approved star rating tables for endophyte insect control are published by the NZPBRA (https://www.nzpbra.org/ accessed on: 11 February 2021).

Caveats—As with livestock testing, insect testing is difficult, with many variables requiring control. Health and life stage of the insect, genetics of the populations, health and maturity stage of the plant, and environmental conditions all impact upon the response. These can cause variability between trials in pots, and in-field trials where infestation, or not, is heavily influenced by geography and season. Trials require entomological expertise, and like many livestock experiments are often contracted out to universities or research organisations at considerable expense.

6. Use of Quality Control Methods within a Ryegrass Seed Company

6.1. Breeding

Once a strategy of how to breed has been decided (see Section 4), then a further strategy on what to test and when must be established. Determining the order of importance of endophyte traits and when and where to test for them through the breeding pipeline is a perplexing task. Any new endophyte discovered, serendipitously, or via screening of seed banks or collections is usually initially phylogenetically identified through SSR, KASP, or NGS based molecular analysis. The new endophyte is then assessed in its natural host and an alkaloid profile deduced (although this is rarely spatially and temporally studied at these early stages). If desirable, transfer to elite host germplasm follows then a repeated round of alkaloid profiling (1–2 years) is undertaken. Ability of the endophyte to be inoculated into a broad range of host germplasm is also desirable but, this is a resource dependent process, as most inoculations are only successful at a low frequency (~4–20%). Stability and transmission tend to be studied in parallel to field evaluations that also investigate the effect of the endophyte on host architecture and performance. Research groups such as those at the Noble Research Institute (www.noble.org accessed on: 11 February 2021), AgResearch (www.agresearch.co.nz accessed on: 11 February 2021), and Agribio (www.agribio.com.au accessed on: 11 February 2021) have research teams dedicated to this process and usually work collaboratively with seed companies.

The success rate of this discovery process has declined with time as it becomes increasingly unlikely to find a novel and useful profile, within the finite species/strains available [87]. Research, e.g., Kaur et al. [88] has looked at many hundreds of *Epichloë* endophytes, of these, few have proceeded beyond their initial screening. The New Zealand PVR database lists 34 endophytes (www.iponz.govt.nz accessed on: 11 February 2021), of these, 3 have expired, 4 have been withdraw, 24 granted, and only 3 are currently filed. Application dates and numbers suggest (given discovery is some time before application) a peak discovery before 2012–2014 for ryegrass endophytes (Table 2).

Table 2. *Epichloë* PVR applications granted in NZ 1997 to 2020.

Year	1997–1999	2000–2002	2003–2005	2006–2008	2009–2011	2012–2014	2015–2017	2018–2020
Epichloë PVR applications	1	2	2	5	4	10	5	2

Inoculation into elite germplasm requires at least 40 individual ryegrass genotypes to be successfully inoculated, otherwise the allele frequency of the host may shift significantly from the original elite germplasm [89]. An alternative way if inoculation is difficult is to inoculate a few plants and then cross onto them with many different elite pollen donor genotypes. If inoculation is not possible, then backcrossing onto the original host, acting as the mother, can also be undertaken. This method may help by introgression of host genes, required for endophyte stability, into the elite germplasm, but requires at least four rounds of backcrossing to get a genotype similar to the original elite germplasm.

In mature breeding pipelines regardless of strategy, existing elite germplasm already contains known functional endophyte(s). With such material the process is easier, and by selective harvesting from mother plants endophyte status can be controlled and new synthetics created. Whatever method is used, each stage of the process needs to be monitored by tiller immunoblots to confirm inoculation, and viable transmission. Microscopy or immunoblots of seeds for presence of the endophyte in the seed is also required. Subsamples of plants, or seed need to be checked by molecular analysis to confirm strain and detect any contamination. Information on transmission, stability and effect on host phenotype is gathered, as resource allows, during the breeding of the new 'synthetic' line. Failure of any of these characteristics can prevent the new endophyte/host symbiont from progressing.

6.2. Agronomy

Once a new ryegrass synthetic line x endophyte strain(s) combination has been chosen for advancement, it requires agronomic assessment. Initially, confirmation of the alkaloid profile, temporally and spatially, is undertaken, along with nationwide trials to understand the host performance. If chosen for advancement, then animal and insect trials, following ETC testing protocols, can be undertaken. For all these trials, seed batches and tiller samples again need to be tested for endophyte presence and subtested for endophyte strain to ensure QC through the process.

6.3. Production

Seedlings of the new line (putative cultivar) are tested for endophyte using immunoblots, and a subset for molecular profiling, any negatives are removed, and seed is harvested from 100% infected plants. This is then multiplied up by seed production specialists within the companies. Some marketed endophytes are a mixture of two different endophytes, e.g., NEA2. Initially, this was serendipitously multiplied up as a mixture but now at each stage of multiplication the ratio of the strains is monitored through molecular testing. Alternatively mixtures can be made by combining individual host–endophyte combinations [90] at various stages of the multiplication process. For single strains, testing is simpler but still required to ensure that no contamination or mislabelling has occurred. Additionally, because endophyte transmission is rarely 100%, endophyte seed percentage, and tiller viability assessments are required at each stage of multiplication from breeders, to pre-basic, to certified seed generation 1 and 2. The final seed for sale requires an endophyte viability of above 70%, otherwise it has to be sold as a low-endophyte product. Despite the importance of transmission and viability, it still varies through management and climatic conditions. Research in this area [35,74,91–93], and considerable in-house research, has made some advancement but endophyte status in production crops is still a major production risk. In addition, crop treatments such as pesticides (especially fungicides) have to be evaluated for potential negative effects on endophyte status [92]. Loss of endophyte status can require the production manager to rework the cultivar starting with a new 100% infected batch of nucleus plants. For new host–endophyte combinations, specific testing of fungicide applications, plant growth regulators, and seed treatments, on the transmission and viability of the endophyte may be necessary.

6.4. Storage and Sale

Endophyte viability in the seed declines quicker than the host seed does [94–97] and this rate varies depending upon the relationship and the storage environment [98]. Storage at low temperature and humidity is required to maintain endophyte viability and insure delivery of good quality product to the farmer [99]. Most seed companies involved in endophyte have strict harvest and storage conditions to quickly cool the seed and maintain low humidity, required for endophyte longevity. This includes strict timelines from harvest to cool storage. Within the NZ industry, freshly harvest seed is given a 6 month grace period, whereby machine dressed endophyte level (seed squash or immune test) is accepted for viability. Following that seed lots must be tested every 6 months usually via a 6 week tiller grow-out test to ensure viability of the endophyte. This is a logistical problem as failure to maintain testing can see lines requested for immediate sale having to wait 6 weeks for a test result. Farmers may go elsewhere for product under such a scenario. Thus, efforts to reduce the time required to perform seed viability tests are important (see Section 5.2 above). Sale to farmers is usually via a third-party store and 'just in time' delivery to these stores reduces the likelihood of poor storage between leaving the production company and sale to the farmer. Advice on on-farm seed storage is given, but there is little monitoring, and once the grass is sown in the paddock little or no monitoring is undertaken. Occasionally, if a poor paddock is produced checks on endophyte status can be made using molecular profiling. This has resulted in the identification of wrong product, or products seemingly mixed with other products (Author, unpublished). Buying from a reputable dealer can help avoid this. Using endophyte status as a QC or product marker is another useful attribute that endophyte can confer to a ryegrass product.

7. Opportunities and Risks for the Future

7.1. Host Breeding

As mentioned above, ryegrass breeding methods are improving, new phenotyping [64] and genotyping techniques [57] are starting to deliver cost effective genomic selection [58,60,100] and these are being applied to improve endophyte stability and transmission [56]. Addition of hybrid breeding [61], self-fertility [101], along with gene editing technologies [102] will likely reduce variation between genotypes within a host germplasm (cultivar or F1 hybrid). This reduced G $\times$ G variation would be expected to result in a more consistent endophyte profile within the product. Still factors affecting the symbioses are many and complex. As an example, Rinklake et al. [14] recently proposed that the process of vernalisation maybe an important transmission factor. Endophyte has a different cardinal temperature to ryegrass and the range of cold required for ryegrass vernalisation may affect endophyte growth and triggers for establishment of reproductive tillers differentially. This could lead to reduced endophyte entering the newly formed tiller, where, through intercalary growth [103], it eventually enters the developing infloresences [103]. Reduced initial colonisation of the shoot apical meristem may therefore reduce the number of infloresences infected. Knowing what traits are important in stability and transmission, are key to future breeding efforts. In a completely different breeding approach, Spangenberg (European Patent Office EP2521442A1) has proposed growing and selecting populations of ryegrass containing multiple different endophyte strains in vivo and simply selecting for the best plants to create a host population containing a plurality of endophytes [104].

7.2. Endophyte Discovery and Manipulation

Discovery and exploitation have become increasingly sophisticated with the use of NGS [12] and high-throughput biochemical analysis techniques [81]. There is no doubt that discovery programmes will continue but as stated this will require smarter or greater efforts to discover new, useful, endophytes. Increasing ease of certain techniques has widened the scope for who can look, and existing collections may still hold interesting endophytes. A recent NZ PVR for CM142 demonstrates this, as it was found in an existing collection from the Margot Forde Germplasm Centre [105]. Use of molecular QC analysis is now routine for many companies and provides opportunities to detect 'unknown' endophytes. This expands discovery capability that was previously limited to a few research laboratories around the world. Aside from discovery it is possible to screen endophyte/host cultivar populations and identify individuals that express different levels of alkaloid. This can be used to select for a particular profile such as lower ergovaline levels. Ryegrass cultivar

Reward containing Endo5 endophyte is an example of such a product being a subselection of AR5, itself a subselection of Endosafe™, with evidence for this shown by their interconnectedness in the paper by Thom et al. [20]).

Genetic modification (GM) and/or gene editing techniques can be used to manipulate the asexual endophyte. With no sexual recombination breeders are reliant on somatic mutations within the endophyte. Evolutionarily hybridisations have arisen but attempts to achieve this in vitro have been unsuccessful to date and use of mutagens invokes Muller's ratchet making accumulation of deleterious mutations far more likely than a specific beneficial one [106]. GM research has been demonstrated in *Epichloë* for a considerable time [107,108] and some site-specific mutagenesis using NHEJ [109], and recently the use of CRISPR/Cas9 gene editing has been demonstrated [110]. Due to regulatory hurdles, GM research is limited to laboratory research activities, but gene editing, having been deregulated in many countries, offers up some unique opportunities. Precise editing of toxin profiles is an obvious target and repair of a perA gene within a LPTG-3 janthitrem-producing endophyte, as currently achieved through GM [107], a potential candidate. If GM becomes acceptable, then adding more complex multigene alkaloid pathways to make a customised profiles might be possible.

7.3. Deployment Systems

Beyond genetic manipulation of endophyte and host, through conventional breeding, mutagenesis or biotechnology, there is also the opportunity to improve endophyte functionality in the pasture through creative deployment systems. One method is simply to dilute the intake by the animal through growing multiple species within the pasture. This is currently a topical approach as regenerative agricultural (RA) proponents advocate multiple species pastures. However, NZ soils already have high, stable carbon reserves [111], thus negating a major proposed benefit of RA. Additionally, ryegrass has an annual biomass, ME, and intake preference advantage in much of NZ [112] so RA would likely dilute production gains achieved by current proprietary ryegrass/endophyte pastures. If diverse species and a more 'natural' habitat is the goal of RA, then perhaps sustainable intensification and set aside land with native species maybe a better approach. Diluting alkaloid levels without losing production could be overcome by blending with a non-endophyte-containing ryegrass, and such products are currently sold as LE (low-endophyte) ryegrass in NZ. However, under high insect pressure without-endophyte grass might be preferentially predated leaving the high alkaloid ryegrass. In reality, it seems that if this occurs, then weeds fill the space and maintain the alkaloid dilution [42]. However, weeds are defined by their poor feed value or growth, so animal performance is again potentially is lost. A second and more sophisticated strategy would be to design novel endophyte combinations within the sward. NEA2 contains nea2 strain, a low ergovaline producer and nea6 strain, a moderate ergovaline producer, resulting in a sward with an ergovaline level sufficient to confer good insect protection but low enough to be unlikely to cause animal health issues under normal grazing conditions [42]. Alternatively, a farmer can purchase two different endophyte-containing cultivars—one with ergovaline and peramine and another producing janthitrem—this could provide a broad spectrum of excellent insect protection properties and reduce the individual alkaloid levels within the pasture. As janthitrem and ergovaline have different effects on livestock, neurological (janthitrem) vs. physiological (ergovaline), the dilution helps prevent either reaching a threshold that will cause animal health issues. Designer mixtures within the best-performing cultivar has been proposed [90] and offers the additional opportunity of maintaining optimal pasture yield. Currently in NZ nitrogen leaching issues are encouraging farmers to reduce synthetic fertiliser application and increase clover and herb (plantain or chicory) percentage within the pasture. This science-based approach results in a diverse pasture blend (ryegrass, clover, herb, endophyte, rhizobia) with effective alkaloid levels, whilst maintaining a productive and sustainable pasture.

7.4. Risks

The success of ergovaline- and non-ergovaline-producing endophytes in New Zealand ryegrass is well documented and without endophytes insect pests can devastate pastures especially in the North Island. However, this situation is unique, and uptake and deployment in Australia, South America, Africa, Europe and North America has been much less. Fescue toxicosis, ryegrass staggers and ergot poisoning are key contributors behind this restraint and represent real issues with historically severe outbreaks reported in many countries [113] that have led to numerous animal deaths. In addition, many other regions of the world are not suited, or reliant, on the almost exclusive temperate ryegrass growth that is practiced in New Zealand. Different grass species require different endophytes, different insects require different alkaloid deterrents, greater reliance on other feed detracts from the value of the endophyte, annual production over perennial, and other factors all change the commercial equation that is made in estimating the cost to benefit ratio, the risk, of using endophyte. Climate change will alter insect pests within regions of the globe and what pasture species are grown within specific regions. Such changes are both risk and opportunity for endophyte-based alkaloid deployment but given the long lead in time (6 to 16 years) to incorporate new endophytes into ryegrass cultivars, these decisions also represent a risk to the breeding companies concerned.

8. Conclusions

Plant breeding of ryegrass would be much simpler without endophytes. Most breeding data management software have issues coping with a maternally inherited trait that can be lost, added, or interchanged with a different 'trait'. For the breeder, following the trait, and all the QC required around this, is an enormous drain on resources. Having to re-test the performance of a novel host germplasm with a new endophyte can take three years from re-inoculation. An inability to routinely produce commercial quantities of seed with sufficient (>70%) viable endophyte can see a huge loss in margin for that crop, and disrupt supply leading to many marketing issues. Likewise, loss of endophyte viability in storage is a logistical challenge requiring constant re-evaluation and re-prioritising of stock for sale. Contamination with wild-type endophyte, or another proprietary endophyte, for example through growing on contaminated land, can see a crop production ruined, and if unknowingly on-sold cost millions in reparation costs. This recently happened with an Australian tall fescue cultivar Barnaby [114]. Sale of the wrong endophyte for consumption by deer or horses, which are more sensitive to alkaloids, can also result in large lawsuits or reparation costs.

Despite the catalogue of negatives, *Epichloë* endophytes have stood the test of time and proved to be a vital component of NZ pasture systems. The obvious natural, sustainable protection that *Epichloë* alkaloids offer to pasture is too valuable to ignore. Farming systems continue to come under ever increasing pressure to 'be natural' and many synthetic pesticides are being removed from the market. This provides unique opportunities that will no doubt see the use of endophytes within and beyond pasture systems expand in the future. In addition, the new knowledge of host–endophyte associations along with better breeding capability for the associations will see improved production and deployment. Finally, the deregulation (in many parts of the world) and use of gene editing, and other biotechnologies, bring an exciting capability to customise alkaloid profiles and or other attributes of the *Epichloë* endophyte to potentially provide new sustainable solutions for the agriculture industry.

Funding: This research received no external funding.

Institutional Review Board Statement: Not applicable.

Informed Consent Statement: Not applicable.

Data Availability Statement: Not applicable.

Acknowledgments: The author wishes to acknowledge Barenbrug for the time to undertake this MS, Courtney Inch, Will Clayton and Graham Kerr for review of the MS and the NZ pasture industry for immersion into the practical world of endophytes.

Conflicts of Interest: The author is employed by Barenbrug NZ Ltd a company that benefits from the use of endophytes.

References

1. Hhe, V.Z.; Herrero, M.; Van Hal, O.; Röös, E.; Muller, M.; Garnett, T.; Gerber, P.; Schader, C.; De Boer, L. Defining a Land Boundary for Sustainable Livestock Consumption. *Glob. Chang. Biol.* **2018**, *24*, 4185–4194.
2. Lorenz, H.; Reinsch, T.; Hess, S.; Taube, F. Is Low-Input Dairy Farming More Climate Friendly? A Meta-Analysis of the Carbon Footprints of Different Production Systems. *J. Clean. Prod.* **2019**, *211*, 161–170. [CrossRef]
3. Chobtang, J.; Ledgard, S.; Mclaren, S.; Donaghy, D.J. Life Cycle Environmental Impacts of High and Low Intensification Pasture-Based Milk Production Systems: A Case Study of the Waikato Region, New Zealand. *J. Clean. Prod.* **2017**, *140*, 664–674. [CrossRef]
4. Case, B.; Ryan, C. An Analysis of Carbon Stocks and Net Carbon Position for New Zealand Sheep and Beef Farmland. 2020. Available online: https://beeflambnz.com/sites/default/files/news-docs/BL_Carbon_report_for_review_final_submit.pdf (accessed on 11 February 2021).
5. Stewart, A. Genetic Origins of Perennial Ryegrass (Lolium Perenne) for New Zealand Pastures. *Adv. Plant Breed.* **2006**, *12*, 55–61.
6. Monks, D.P.; Sadat Asilan, K.; Moot, D.J. Cardinal Temperatures and Thermal Time Requirements for Germination of Annual and Perennial Temperate Pasture Species. *Agron. N. Z.* **2009**, *39*, 95–109.
7. Crush, J.R.; Rowarth, J.S. The Role of C_4 Grasses in New Zealand Pastoral Systems. *N. Z. J. Agric. Res.* **2007**, *50*, 125–137. [CrossRef]
8. Rowarth, J.S.; Pennell, C.G.; Fraser, T.J.; Baird, D.B. Pasture Response to Fertiliser Inputs under Dairy Grazing. *Proc. N. Z. Grassl. Assoc.* **1996**, 123–127. [CrossRef]
9. Schardl, C.L.; Leuchtmann, A.; Spiering, M.J. Symbioses of Grasses with Seedborne Fungal Endophytes. *Annu. Rev. Plant Biol.* **2004**, *55*, 315–340. [CrossRef] [PubMed]
10. Schardl, C.L.; Collett, M.A.; Watt, D.; Scott, D.B. Origin of a Fungal Symbiont of Perennial Ryegrass by Interspecific Hybridization of a Mutualist with the Ryegrass Choke Pathogen. *Genetics* **1994**, *136*, 1307–1317. [CrossRef]
11. Saikkonen, K.; Young, C.A.; Helander, M.; Schardl, C.L. Endophytic Epichloë Species and Their Grass Hosts: From Evolution to Applications. *Plant Mol. Biol.* **2016**, *90*, 665–675. [CrossRef]
12. Hettiarachchige, I.K.; Ekanayake, P.N.; Mann, R.C.; Guthridge, K.M.; Sawbridge, T.I.; Spangenberg, G.C.; Forster, J.W. Phylogenomics of Asexual Epichloë Fungal Endophytes Forming Associations with Perennial Ryegrass. *BMC Evol. Biol.* **2015**, *15*, 1–14. [CrossRef]
13. Johnson, L.J.; Voisey, C.R.; Faville, M.J.; Moon, C.D.; Simpson, W.R.; Johnson, R.D.; Stewart, A.V.; Caradus, J.R.; Hume, D.E. Advances and Perspectives in Breeding for Improved Grass-Endophyte Associations. In Proceedings of the "Improving Sown Grasslands through Breeding and Management" Joint Symposium EFG/Eucarpia, Zurich, Switzerland, 24–27 June 2019; pp. 351–363.
14. Rinklake, I.; James, A.; Brownfield, L.; MacKnight, R.; Eady, C. Relative Epichloë Endophyte Fungal Biomass in Ryegrass Tillers Grown Pre and Post Vernalization. *N. Z. Agron. Soc. J.* **2020**, in press.
15. Caradus, J.R.; Card, S.D.; Finch, S.C.; Hume, D.E.; Johnson, L.J.; Mace, W.J.; Popay, A.J. Ergot Alkaloids in New Zealand Pastures and Their Impact. *N. Z. J. Agric. Res.* **2020**, 1–41. [CrossRef]
16. Neill, J.C. The Endophyte of Rye-Grass (Lolium Perenne). *N. Z. J. Sci. Technol. Sect. A* **1940**, *21*, 280–291.
17. Bacon, C.W.; Porter, J.K.; Robbins, J.D.; Luttrell, E.S. Epichloë Typhina from Toxic Tall Fescue Grasses. *Appl. Environ. Microbiol.* **1977**, *34*, 576–581. [CrossRef]
18. Fletcher, L.R.; Harvey, I.C. An Association of a Lolium Endophyte with Ryegrass Staggers. *N. Z. Vet. J.* **1981**, *29*, 185–186. [CrossRef] [PubMed]
19. Leuchtmann, A.; Bacon, C.W.; Schardl, C.L.; White, J.F.; Tadych, M. Nomenclatural Realignment of *Neotyphodium* Species with Genus Epichloë. *Mycologia* **2014**, *106*, 202–215. [CrossRef]
20. Thom, E.R.; Popay, A.J.; Hume, D.E.; Fletcher, L.R. Evaluating the Performance of Endophytes in Farm Systems to Improve Farmer Outcomes—A Review. *Crop Pasture Sci.* **2012**, *63*, 927. [CrossRef]
21. Sewell, J.C. Recurrent Selection in Perennial Ryegrass (Lolium Perenne L.) for Reduced Levels of Ergovaline with Particular Emphasis on the Effect of Other Ergot Alkaloid Concentrations. Master's Thesis, Lincoln University, Lincoln, New Zealand, 2015.
22. Fletcher, L.R.; Popay, A.J.; Tapper, B.A. Evaluation of Several Lolitrem-Free Endophyte/Perennial Ryegrass Combinations. *Proc. N. Z. Grassl. Assoc.* **1991**, *53*, 215–219. [CrossRef]
23. Latch, G.C.M. Influence of *Acremonium* Endophytes on Perennial Grass Improvement. *N. Z. J. Agric. Res.* **1994**, *37*, 311–318. [CrossRef]
24. Fletcher, L.R.; Sutherland, B.L.; Fletcher, C.G. The Impact of Endophyte on the Health and Productivity of Sheep Grazing Ryegrass-Based Pastures. *Grassl. Res. Pract. Ser.* **1999**, *7*, 11–17.

25. Popay, A.J.; Hume, D.E.; Baltus, J.G.; Latch, G.C.M.; Tapper, B.A.; Lyons, T.B.; Cooper, B.M.; Pennell, C.G.; Eerens, J.P.J.; Marshall, S.L. Field Performance of Perennial Ryegrass (Lolium Perenne) Infected with Toxin-Free Fungal Endophytes (*Neotyphodium Spp.*). *Grassl. Res. Pract. Ser.* **1999**, *7*, 113–122.
26. Thom, E.R.; Waugh, C.D.; Minneé, E.M.K.; Waghorn, G.C. Effects of Novel and Wild-Type Endophytes in Perennial Ryegrass on Cow Health and Production. *N. Z. Vet. J.* **2013**, *61*, 87–97. [CrossRef]
27. Hume, D.E.; Stewart, A.V.; Simpson, W.R.; Johnson, R.D. *Epichloë* Fungal Endophytes Play a Fundamental Role in New Zealand Grasslands. *J. R. Soc. N. Z.* **2020**, *50*, 279–298. [CrossRef]
28. Bultman, T.L. Neotyphodium in Cool-Season Grasses. *Crop Sci.* **2006**, *46*, 493–495. [CrossRef]
29. Hume, D.E.; Cooper, B.M.; Panckhurst, K.A. The Role of Endophyte in Determining the Persistence and Productivity of Ryegrass, Tall Fescue and Meadow Fescue in Northland. *Proc. N. Z. Grassl. Assoc.* **2009**, 145–150. [CrossRef]
30. Fletcher, L.; Finch, S.; Sutherland, B.; deNicolo, G.; Mace, W.; van Koten, C.; Hume, D. The Occurrence of Ryegrass Staggers and Heat Stress in Sheep Grazing Ryegrass-Endophyte Associations with Diverse Alkaloid Profiles. *N. Z. Vet. J.* **2017**, *65*, 232–241. [CrossRef]
31. Finch, S.C.; Thom, E.R.; Babu, J.V.; Hawkes, A.D.; Waugh, C.D. The Evaluation of Fungal Endophyte Toxin Residues in Milk. *N. Z. Vet. J.* **2013**, *61*, 11–17. [CrossRef] [PubMed]
32. Reddy, P.; Guthridge, K.; Vassiliadis, S.; Hemsworth, J.; Hettiarachchige, I.; Spangenberg, G.; Rochfort, S. Tremorgenic Mycotoxins: Structure Diversity and Biological Activity. *Toxins* **2019**, *11*, 302. [CrossRef] [PubMed]
33. Thom, E.R.; Waugh, C.D.; Minnee, E.M.K. A New Generation Ryegrass Endophyte—The First Results from Dairy Cows Fed AR37. In Proceedings of the 6th International Symposium on Fungal Endophytes of Grasses, Christchurch, New Zealand, 25–28 March 2007; pp. 293–296.
34. Kitson, E.R. Viability of Endophytic Fungus in Different Perennial Ryegrass (Lolium Perenne) Varieties Kept in Different Storage Conditions. Master's Thesis, Massey University, Manawatū, New Zealand, 2017.
35. Johnson, L.J.; Caradus, J.R. The Science Required to Deliver *Epichloë* Endophytes to Commerce. In *Endophytes for a Growing World*; Hodkinson, T.R., Doohan, F.M., Saunders, M.J., Murphy, B.R., Eds.; Cambridge University Press: Cambridge, UK, 2019; pp. 343–370, ISBN 978-1-108-60766-7.
36. Use of Grass Fungi Saving NZ Billions Agriculture Science. Available online: https://www.agresearch.co.nz/news/use-of-grass-fungi-saving-nz-billions/ (accessed on 23 September 2020).
37. Rivero, M.J.; Lee, M.R.F.; Cone, J.W. *The Role of Pasture in the Diet of Ruminant Livestock. In Balancing Pasture Productivity with Environmental and Animal Health Requirements*; Burleigh Dodds Science Publishing: Cambridge, UK, 2018; Available online: https://www.perlego.com/book/1435352/balancing-pasture-productivity-with-environmental-and-animal-health-requirements-pdf (accessed on 23 September 2020).
38. Johnson, L.J.; de Bonth, A.C.M.; Briggs, L.R.; Caradus, J.R.; Finch, S.C.; Fleetwood, D.J.; Fletcher, L.R.; Hume, D.E.; Johnson, R.D.; Popay, A.J.; et al. The Exploitation of Epichloae Endophytes for Agricultural Benefit. *Fungal Divers.* **2013**, *60*, 171–188. [CrossRef]
39. Panaccione, D.G.; Beaulieu, W.T.; Cook, D.; Allen, E. Bioactive Alkaloids in Vertically Transmitted Fungal Endophytes. *Funct. Ecol.* **2014**, *28*, 299–314. [CrossRef]
40. Klotz, J.L.; Nicol, A.M. Ergovaline, an Endophytic Alkaloid. 1. Animal Physiology and Metabolism. *Anim. Prod. Sci.* **2016**, *56*, 1761–1774. [CrossRef]
41. Nicol, A.M.; Klotz, J.L. Ergovaline, an Endophytic Alkaloid. 2. Intake and Impact on Animal Production, with Reference to New Zealand. *Anim. Prod. Sci.* **2016**, *56*, 1775. [CrossRef]
42. Eady, C.C.; Corkran, J.R.; Bailey, K.M.; Kerr, G.A.; Nicol, A.M. Estimation of Ergovaline Intake of Cows from Grazed Perennial Ryegrass Containing NEA2 or Standard Endophyte. *J. N. Z. Grassl.* **2017**, *79*, 197–203. [CrossRef]
43. Schardl, C.L.; Grossman, R.B.; Nagabhyru, P.; Faulkner, J.R.; Mallik, U.P. Loline Alkaloids: Currencies of Mutualism. *Phytochemistry* **2007**, *68*, 980–996. [CrossRef] [PubMed]
44. Ludlow, E.J.; Vassiliadis, S.; Ekanayake, P.N.; Hettiarachchige, I.K.; Reddy, P.; Sawbridge, T.I.; Rochfort, S.J.; Spangenberg, G.C.; Guthridge, K.M. Analysis of the Indole Diterpene Gene Cluster for Biosynthesis of the Epoxy-Janthitrems in Epichloë Endophytes. *Microorganisms* **2019**, *7*, 560. [CrossRef]
45. Panaccione, D.G. Origins and Significance of Ergot Alkaloid Diversity in Fungi. *FEMS Microbiol. Lett.* **2005**, *251*, 9–17. [CrossRef]
46. Gerhards, N.; Neubauer, L.; Tudzynski, P.; Li, S.-M. Biosynthetic Pathways of Ergot Alkaloids. *Toxins* **2014**, *6*, 3281–3295. [CrossRef]
47. Reddy, P.; Hemsworth, J.; Guthridge, K.M.; Vinh, A.; Vassiliadis, S.; Ezernieks, V.; Spangenberg, G.C.; Rochfort, S.J. Ergot Alkaloid Mycotoxins: Physiological Effects, Metabolism and Distribution of the Residual Toxin in Mice. *Sci. Rep.* **2020**, *10*. [CrossRef]
48. Florea, S.; Panaccione, D.G.; Schardl, C.L. Ergot Alkaloids of the Family Clavicipitaceae. *Phytopathology* **2017**. [CrossRef] [PubMed]
49. Tian, P.; Le, T.-N.; Ludlow, E.J.; Smith, K.F.; Forster, J.W.; Guthridge, K.M.; Spangenberg, G.C. Characterisation of Novel Perennial Ryegrass Host–Neotyphodium Endophyte Associations. *Crop Pasture Sci.* **2013**, *64*, 716. [CrossRef]
50. Jong, E.V.Z.D.; Dobrowolski, M.P.; Sandford, A.; Smith, K.F.; Willocks, M.J.; Spangenberg, G.C.; Forster, J.W. Detection and Characterisation of Novel Fungal Endophyte Genotypic Variation in Cultivars of Perennial Ryegrass (Lolium Perenne L.). *Aust. J. Agric. Res.* **2008**, *59*, 214–221. [CrossRef]
51. Lorenz, N.; Haarmann, T.; Pažoutová, S.; Jung, M.; Tudzynski, P. The Ergot Alkaloid Gene Cluster: Functional Analyses and Evolutionary Aspects. *Phytochemistry* **2009**, *70*, 1822–1832. [CrossRef]

52. Hill, N.S.; Thompson, F.N.; Stuedemann, J.A.; Rottinghaus, G.W.; Ju, H.J.; Dawe, D.L.; Hiatt, E.E. Ergot Alkaloid Transport across Ruminant Gastric Tissues. *J. Anim. Sci.* **2001**, *79*, 542. [CrossRef]
53. Finch, S.C.; Munday, J.S.; Sprosen, J.M.; Bhattarai, S. Toxicity Studies of Chanoclavine in Mice. *Toxins* **2019**, *11*, 249. [CrossRef] [PubMed]
54. Cagnano, G.; Roulund, N.; Jensen, C.S.; Forte, F.P.; Asp, T.; Leuchtmann, A. Large Scale Screening of Epichloë Endophytes Infecting Schedonorus Pratensis and Other Forage Grasses Reveals a Relation Between Microsatellite-Based Haplotypes and Loline Alkaloid Levels. *Front. Plant Sci.* **2019**, *10*. [CrossRef]
55. Young, C.A.; Hume, D.E.; McCulley, R.L. Forages and Pastures Symposium: Fungal Endophytes of Tall Fescue and Perennial Ryegrass: Pasture Friend or Foe? *J. Anim. Sci.* **2013**, *91*, 2379–2394. [CrossRef] [PubMed]
56. Gagic, M.; Faville, M.J.; Zhang, W.; Forester, N.T.; Rolston, M.P.; Johnson, R.D.; Ganesh, S.; Koolaard, J.P.; Easton, H.S.; Hudson, D.; et al. Seed Transmission of Epichloë Endophytes in Lolium Perenne Is Heavily Influenced by Host Genetics. *Front. Plant Sci.* **2018**, *9*, 1580. [CrossRef]
57. Faville, M.J.; Ganesh, S.; Cao, M.; Jahufer, M.Z.Z.; Bilton, T.P.; Easton, H.S.; Ryan, D.L.; Trethewey, J.A.K.; Rolston, M.P.; Griffiths, A.G.; et al. Predictive Ability of Genomic Selection Models in a Multi-Population Perennial Ryegrass Training Set Using Genotyping-by-Sequencing. *Theor. Appl. Genet.* **2018**, *131*, 703–720. [CrossRef] [PubMed]
58. Pembleton, L.W.; Inch, C.; Baillie, R.C.; Drayton, M.C.; Thakur, P.; Ogaji, Y.O.; Spangenberg, G.C.; Forster, J.W.; Daetwyler, H.D.; Cogan, N.O.I. Exploitation of Data from Breeding Programs Supports Rapid Implementation of Genomic Selection for Key Agronomic Traits in Perennial Ryegrass. *Theor. Appl. Genet.* **2018**, *131*, 1891–1902. [CrossRef] [PubMed]
59. Faville, M.J.; Richardson, K.; Gagic, M.; Mace, W.; Sun, X.Z.; Harrison, S.; Knapp, K.; Jahufer, M.Z.Z.; Palanisamy, R.; Pirlo, S.; et al. Genetic improvement of fibre traits in perennial ryegrass. *Proc. N. Z. Grassl. Assoc.* **2010**, *72*, 71–78.
60. Faville, M.; Cao, M.; Schmidt, J.; Ryan, D.; Ganesh, S.; Jahufer, M.; Hong, S.; George, R.; Barrett, B. Divergent Genomic Selection for Herbage Accumulation and Days-To-Heading in Perennial Ryegrass. *Agronomy* **2020**, *10*, 340. [CrossRef]
61. Pembleton, L.W.; Shinozuka, H.; Wang, J.; Spangenberg, G.C.; Forster, J.W.; Cogan, N.O.I. Design of an F1 Hybrid Breeding Strategy for Ryegrasses Based on Selection of Self-Incompatibility Locus-Specific Alleles. *Front. Plant Sci.* **2015**, *6*. [CrossRef] [PubMed]
62. Riday, H. Paternity Testing: A Non-Linkage Based Marker-Assisted Selection Scheme for Outbred Forage Species. *Crop Sci.* **2011**, *51*, 631–641. [CrossRef]
63. Do Canto, J.; Studer, B.; Frei, U.; Lübberstedt, T. Fine Mapping a Self-Fertility Locus in Perennial Ryegrass. *Theor. Appl. Genet.* **2018**, *131*, 817–827. [CrossRef]
64. Ghamkhar, K.; Irie, K.; Hagedorn, M.; Hsiao, J.; Fourie, J.; Gebbie, S.; Hoyos-Villegas, V.; George, R.; Stewart, A.; Inch, C.; et al. Real-Time, Non-Destructive and in-Field Foliage Yield and Growth Rate Measurement in Perennial Ryegrass (Lolium Perenne L.). *Plant Methods* **2019**, *15*, 72. [CrossRef] [PubMed]
65. Kirkby, K.A.; Hume, D.E.; Pratley, J.E.; Broster, J.C. Effect of Temperature on Endophyte and Plant Growth of Annual Ryegrass, Perennial Ryegrass and Tall Fescue. In Proceedings of the 17th Australasian Weeds Conference, New Frontiers in New Zealand: Together We Can Beat the Weeds, Christchurch, New Zealand, 26–30 September 2010; pp. 56–59.
66. Moon, C.D.; Tapper, B.A.; Scott, B. Identification of Epichloë Endophytes In Planta by a Microsatellite-Based PCR Fingerprinting Assay with Automated Analysis. *Appl. Environ. Microbiol.* **1999**, *65*, 1268–1279. [CrossRef]
67. Latch, G.C.M.; Christensen, M.J. Artificial Infection of Grasses with Endophytes. *Assoc. Appl. Biol.* **1985**, *107*, 17–24. [CrossRef]
68. Semagn, K.; Babu, R.; Hearne, S.; Olsen, M. Single Nucleotide Polymorphism Genotyping Using Kompetitive Allele Specific PCR (KASP): Overview of the Technology and Its Application in Crop Improvement. *Mol. Breed.* **2014**, *33*, 1–14. [CrossRef]
69. Kaur, J.; Gutheridge, K.M.; Sawbridge, T.I.; Mann, R.M.; Forster, J.W.; Spangenberg, G.C. SNP Specific Genotyping of Pasture Grass Endophytes Using KASP (Kompetitive Allele Specific PCR) Assay. In Proceedings of the 9th International Symposium on Fungal Endophytes of Grasses (ISFEG 2015) and 1st International Symposium on Plant Microbiomes (ISPM), Melbourne, Australia, 28 September–1 October 2015; p. 103.
70. Vikuk, V.; Young, C.A.; Lee, S.T.; Nagabhyru, P.; Krischke, M.; Mueller, M.J.; Krauss, J. Infection Rates and Alkaloid Patterns of Different Grass Species with Systemic *Epichloë* Endophytes. *Appl. Environ. Microbiol.* **2019**, *85*, 1–17. [CrossRef] [PubMed]
71. Bacon, C.W.; White, J.F., Jr. Stains media and procedures for analyzing endophytes. In *Biotechnology of Endophytic Fungi of Grasses*; CRC Press: Ann Arbour, MI, USA, 1994; pp. 47–56.
72. White, J.F.; Morgan-Jones, G.; Morrow, A.C. Taxonomy, Life Cycle, Reproduction and Detection of Acremonium Endophytes. *Agric. Ecosyst. Environ.* **1993**, *44*, 13–37. [CrossRef]
73. Simpson, W.R.; Faville, M.J.; Moraga, R.A.; Williams, W.M.; Mcmanus, M.T.; Johnson, R.D. Epichloë Fungal Endophytes and the Formation of Synthetic Symbioses in Hordeeae (=Triticeae) Grasses. *J. Syst. Evol.* **2014**, *52*, 794–806. [CrossRef]
74. Hillis, S.R.J. Transmission and Survival of Perennial Ryegrass Endophyte during Field Based Seed Production. Master's Thesis, Lincoln University, Lincoln, New Zealand, 2019.
75. Groppe, K.; Boller, T. PCR Assay Based on a Microsatellite-Containing Locus for Detection and Quantification of Epichloë Endophytes in Grass Tissue. *Appl. Environ. Microbiol.* **1997**, *63*, 1543–1550. [CrossRef] [PubMed]
76. Young, C.A.; Charlton, N.D.; Takach, J.E.; Swoboda, G.A.; Trammell, M.A.; Huhman, D.V.; Hopkins, A.A. Characterization of Epichloë Coenophiala within the US: Are All Tall Fescue Endophytes Created Equal? *Front. Chem.* **2014**, *2*. [CrossRef] [PubMed]

77. Budowle, B. Molecular Genetic Investigative Leads to Differentiate Monozygotic Twins. *Investig. Genet.* **2014**, *5*, 11. [CrossRef] [PubMed]
78. Hovermale, J.T.; Craig, A.M. Correlation of Ergovaline and Lolitrem B Levels in Endophyte-Infected Perennial Ryegrass (*Lolium Perenne*). *J. Vet. Diagn. Invest.* **2001**, *13*, 323–327. [CrossRef] [PubMed]
79. Baldauf, M.W.; Mace, W.J.; Richmond, D.S. Endophyte-Mediated Resistance to Black Cutworm as a Function of Plant Cultivar and Endophyte Strain in Tall Fescue. *Environ. Entomol.* **2011**, *40*, 639–647. [CrossRef]
80. Spiering, M.J.; Davies, E.; Tapper, B.A.; Schmid, J.; Lane, G.A. Simplified Extraction of Ergovaline and Peramine for Analysis of Tissue Distribution in Endophyte-Infected Grass Tillers. *J. Agric. Food Chem.* **2002**, *50*, 5856–5862. [CrossRef]
81. Vassiliadis, S.; Elkins, A.C.; Reddy, P.; Guthridge, K.M.; Spangenberg, G.C.; Rochfort, S.J. A Simple LC–MS Method for the Quantitation of Alkaloids in Endophyte-Infected Perennial Ryegrass. *Toxins* **2019**, *11*, 649. [CrossRef]
82. Finch, S.C.; Prinsep, M.R.; Popay, A.J.; Wilkins, A.L.; Webb, N.G.; Bhattarai, S.; Jensen, J.G.; Hawkes, A.D.; Babu, J.V.; Tapper, B.A.; et al. Identification and Structure Elucidation of Epoxyjanthitrems from Lolium Perenne Infected with the Endophytic Fungus Epichloë Festucae Var. Lolii and Determination of the Tremorgenic and Anti-Insect Activity of Epoxyjanthitrem I. *Toxins* **2020**, *12*, 526. [CrossRef]
83. Soto-Barajas, M.C.; Zabalgogeazcoa, I.; González-Martin, I.; Vázquez-de-Aldana, B.R. Qualitative and Quantitative Analysis of Endophyte Alkaloids in Perennial Ryegrass Using Near-Infrared Spectroscopy: NIRS to Detect *Epichloë* Alkaloids in Ryegrass. *J. Sci. Food Agric.* **2017**, *97*, 5028–5036. [CrossRef]
84. Logan, C.; Edwards, G.; Kerr, G.; Williams, S. Ryegrass Staggers and Liveweight Gain of Ewe Lambs and Hoggets Grazing Four Combinations of Perennial Ryegrass and Strains of Endophyte. *Proc. N. Z. Soc. Anim. Prod.* **2015**, *75*, 4.
85. Dimenna, M.E.; Mortimer, P.H.; Prestidge, R.A.; Hawkes, A.D.; Sprosen, J.M. Lolitrem B Concentrations, Counts of *Acremonium Lolii* Hyphae, and the Incidence of Ryegrass Staggers in Lambs on Plots of *A. Lolii* -Infected Perennial Ryegrass. *N. Z. J. Agric. Res.* **1992**, *35*, 211–217. [CrossRef]
86. Rowan, D.D.; Garrick, C.M.L. Utilization of endophyte-infected perennial ryegrass for increased insect resistance. In *Biotechnology of Endophytic Fungi of Grasses*, 1st ed.; Bacon, C.W., White, J.F., Eds.; CRC Press: Boca Raton, FL, USA, 2018; pp. 169–184, ISBN 978-1-351-07032-4.
87. Chao, A. On Estimating the Probability of Discovering a New Species. *Ann. Stat.* **1981**, *9*, 1339–1342. [CrossRef]
88. Kaur, J.; Ekanayake, P.N.; Tian, P.; Jong, E.V.Z.D.; Dobrowolski, M.P.; Rochfort, S.J.; Mann, R.C.; Smith, K.F.; Forster, J.W.; Guthridge, K.M.; et al. Discovery and Characterisation of Novel Asexual Epichloë Endophytes from Perennial Ryegrass (Lolium Perenne L.). *Crop Pasture Sci.* **2015**, *66*, 1058–1070. [CrossRef]
89. Kliman, R.; Sheehy, B.; Schulyz, J. Genetic Drift and Effective Population Size | Learn Science at Scitable. *Nat. Educ.* **2008**, *1*, 3.
90. Eady, C.C. Methods of Producing Cultivar Swards with Tailored Properties. U.S. Patent WO/2018/208174, 15 November 2018.
91. Harvey, I.C.; Fletcher, L.R.; Emms, L.M. Effects of Several Fungicides on the Lolium Endophyte in Ryegrass Plants, Seeds, and in Culture. *N. Z. J. Agric. Res.* **1982**, *25*, 601–606. [CrossRef]
92. Rolston, M.P.; Archie, W.J.; Simpson, W.R. Tolerance of AR1 *Neotyphodium* Endophyte to Fungicides Used in Perennial Ryegrass Seed Production. *N. Z. Plant Prot.* **2002**, *55*, 322–326. [CrossRef]
93. Stewart, A.V. Effect on the *Lolium* Endophyte of Nitrogen Applied to Perennial Ryegrass Seed Crops. *N. Z. J. Exp. Agric.* **1986**, *14*, 393–397. [CrossRef]
94. Rolston, M.P.; Hare, M.D.; Moore, K.K.; Christensen, M.J. Viability of *Lolium* Endophyte Fungus in Seed Stored at Different Moisture Contents and Temperatures. *N. Z. J. Exp. Agric.* **1986**, *14*, 297–300. [CrossRef]
95. Hill, N.S.; Roach, P.K. Endophyte Survival during Seed Storage: Endophyte-Host Interactions and Heritability. *Crop Sci.* **2009**, *49*, 1425–1430. [CrossRef]
96. Welty, R.E. Influence of Moisture Content, Temperature, and Length of Storage on Seed Germination and Survival of Endophytic Fungi in Seeds of Tall Fescue and Perennial Ryegrass. *Phytopathology* **1987**, *77*, 893. [CrossRef]
97. Hume, D.E.; Card, S.D.; Rolston, M.P. Effects of Storage Conditions on Endophyte and Seed Viability in Pasture Grasses. In Proceedings of the 22nd International Grasslands Congress, Sydney, Australia, 15–19 September 2013; pp. 405–408.
98. Gundel, P.E.; Garibaldi, L.A.; Martínez-Ghersa, M.A.; Ghersa, C.M. Neotyphodium Endophyte Transmission to Lolium Multiflorum Seeds Depends on the Host Plant Fitness. *Environ. Exp. Bot.* **2011**, S0098847211000256. [CrossRef]
99. Rolston, M.P.; Agee, C. Delivery of quality seed to specification—The USA and NZ novel endophyte experience. In Proceedings of the 6th International Symposium on Fungal Endophytes of Grasses, Christchurch, New Zealand, 25–28 March 2007; pp. 229–231.
100. Arojju, S.K.; Cao, M.; Trolove, M.; Barrett, B.A.; Inch, C.; Eady, C.; Stewart, A.; Faville, M.J. Multi-Trait Genomic Prediction Improves Predictive Ability for Dry Matter Yield and Water-Soluble Carbohydrates in Perennial Ryegrass. *Front. Plant Sci.* **2020**, *11*, 1197. [CrossRef] [PubMed]
101. Do Canto, J.; Studer, B.; Frei, U.; Lübberstedt, T. Pattern of Inheritance of a Self-fertility Gene in an Autotetraploid Perennial Ryegrass (*Lolium Perenne*) Population. *Plant Breed.* **2020**, *139*, 207–213. [CrossRef]
102. Zhang, Y.; Ran, Y.; Nagy, I.; Lenk, I.; Qiu, J.-L.; Asp, T.; Jensen, C.S.; Gao, C. Targeted Mutagenesis in Ryegrass (Lolium Spp.) Using the CRISPR/Cas9 System. *Plant Biotechnol. J.* **2020**, *18*, 1854–1856. [CrossRef]
103. Christensen, M.J.; Bennett, R.J.; Ansari, H.A.; Koga, H.; Johnson, R.D.; Bryan, G.T.; Simpson, W.R.; Koolaard, J.P.; Nickless, E.M.; Voisey, C.R. Epichloë Endophytes Grow by Intercalary Hyphal Extension in Elongating Grass Leaves. *Fungal Genet. Biol.* **2008**, *45*, 84–93. [CrossRef] [PubMed]

104. Spangenberg, G.C.; Guthridge, K.M.; Forster, J.; Sawbridge, T.I.; Ludlow, E.; Kaur, J.; Rochfort, S.J.; Rabinovich, M.; Ekanayake, P.N. Endophytes and Related Methods. U.S. Patent WO2011082455A1, 7 January 2011.
105. Australian Government. Fungal Endophyte (Epichloe Festucae Var. Lolii). *Aust. Gov. Plant Var. J.* **2019**, *32*, 49.
106. Muller, H.J. The Relation of Recombination to Mutational Advance. *Mutat. Res. Mol. Mech. Mutagen.* **1964**, *1*, 2–9. [CrossRef]
107. Hettiarachchige, I.K.; Elkins, A.C.; Reddy, P.; Mann, R.C.; Guthridge, K.M.; Sawbridge, T.I.; Forster, J.W.; Spangenberg, G.C. Genetic Modification of Asexual Epichloë Endophytes with the PerA Gene for Peramine Biosynthesis. *Mol. Genet. Genom.* **2019**, *294*, 315–328. [CrossRef] [PubMed]
108. Panaccione, D.G.; Johnson, R.D.; Wang, J.; Young, C.A.; Damrongkool, P.; Scott, B.; Schardl, C.L. Elimination of Ergovaline from a Grass-Neotyphodium Endophyte Symbiosis by Genetic Modification of the Endophyte. *Proc. Natl. Acad. Sci USA* **2001**, *98*, 12820–12825. [CrossRef]
109. Rahnama, M.; Forester, N.; Ariyawansa, K.G.S.U.; Voisey, C.R.; Johnson, L.J.; Johnson, R.D.; Fleetwood, D.J. Efficient Targeted Mutagenesis in Epichloë Festucae Using a Split Marker System. *J. Microbiol. Methods* **2017**, *134*, 62–65. [CrossRef] [PubMed]
110. Wang, R. Abstract: CRISPR-Cas9 Knockout of the Epichloë Festucae Antifungal Protein Gene (ASA, CSSA and SSSA International Annual Meetings). In Proceedings of the Managing Global Resources for a Secure Future, Tampa, FL, USA, 22–25 October 2017; p. 515.
111. Schipper, L.A.; Mudge, P.L.; Kirschbaum, M.U.F.; Hedley, C.B.; Golubiewski, N.E.; Smaill, S.J.; Kelliher, F.M. A Review of Soil Carbon Change in New Zealand's Grazed Grasslands. *N. Z. J. Agric. Res.* **2017**, *60*, 93–118. [CrossRef]
112. Eady, C.; Courtney, C. A pasture seed production perspective. New Zealand Institute of Agricultural and Horticultural Science AgScience 2021 57 18–19 Regenerative Agriculture Issue. Available online: https://indd.adobe.com/view/693a575a-5482-4df0-bc4d-f986d3bce648 (accessed on 6 January 2021).
113. Krauss, J.; Vikuk, V.; Young, C.A.; Krischke, M.; Mueller, M.J.; Baerenfaller, K. Epichloë Endophyte Infection Rates and Alkaloid Content in Commercially Available Grass Seed Mixtures in Europe. *Microorganisms* **2020**, *8*, 498. [CrossRef] [PubMed]
114. Reading, K. Grass Named after Deputy PM Barnaby Joyce Recalled from Seed Market. Available online: https://www.abc.net.au/news/rural/2016-10-13/deputy-pm-barnaby-joyce-grass-recalled-toxic-fungus/7929322 (accessed on 24 September 2020).

toxins

Article

Non-Transgenic CRISPR-Mediated Knockout of Entire Ergot Alkaloid Gene Clusters in Slow-Growing Asexual Polyploid Fungi

Simona Florea [1], Jolanta Jaromczyk [2] and Christopher L. Schardl [1,*]

1 Department of Plant Pathology, University of Kentucky, Lexington, KY 40546, USA; sflor2@uky.edu
2 Computer Science Department, University of Kentucky, Lexington, KY 40546, USA; jolan-ta.jaromczyk@uky.edu
* Correspondence: schardl@uky.edu; Tel.: +1-859-218-0730

Abstract: The *Epichloë* species of fungi include seed-borne symbionts (endophytes) of cool-season grasses that enhance plant fitness, although some also produce alkaloids that are toxic to livestock. Selected or mutated toxin-free endophytes can be introduced into forage cultivars for improved livestock performance. Long-read genome sequencing revealed clusters of ergot alkaloid biosynthesis (*EAS*) genes in *Epichloë coenophiala* strain e19 from tall fescue (*Lolium arundinaceum*) and *Epichloë hybrida* Lp1 from perennial ryegrass (*Lolium perenne*). The two homeologous clusters in *E. coenophiala*—a triploid hybrid species—were 196 kb (*EAS1*) and 75 kb (*EAS2*), and the *E. hybrida EAS* cluster was 83 kb. As a CRISPR-based approach to target these clusters, the fungi were transformed with ribonucleoprotein (RNP) complexes of modified Cas9 nuclease (Cas9-2NLS) and pairs of single guide RNAs (sgRNAs), plus a transiently selected plasmid. In *E. coenophiala*, the procedure generated deletions of *EAS1* and *EAS2* separately, as well as both clusters simultaneously. The technique also gave deletions of the *EAS* cluster in *E. hybrida* and of individual alkaloid biosynthesis genes (*dmaW* and *lolC*) that had previously proved difficult to delete in *E. coenophiala*. Thus, this facile CRISPR RNP approach readily generates non-transgenic endophytes without toxin genes for use in research and forage cultivar improvement.

Keywords: CRISPR/Cas9; non-transgenic engineered fungi; genome editing; genome sequencing; MinION; nanopore; secondary metabolites

Key Contribution: A CRISPR-based approach was used to generate non-transgenic deletion mutants of *Epichloë* species that are seed-borne fungal symbionts (endophytes) of important forage grasses. Targeted deletions of several genes and genome regions were demonstrated, including the simultaneous deletion of two large gene clusters.

Citation: Florea, S.; Jaromczyk, J.; Schardl, C.L. Non-Transgenic CRISPR-Mediated Knockout of Entire Ergot Alkaloid Gene Clusters in Slow-Growing Asexual Polyploid Fungi. *Toxins* **2021**, *13*, 153. https://doi.org/10.3390/toxins13020153

Received: 30 October 2020
Accepted: 21 November 2020
Published: 16 February 2021

1. Introduction

A few species of filamentous fungi have been genetic models of choice since the 1950s due to their haploid growth stage, facile sexual cycles, abundant sporulation, rapid growth and, with time, large repertoires of mutants and molecular transformation systems. However, given the importance of fungi in medicine, agriculture, and ecosystems, considerable efforts have been invested over several decades to establish molecular transformation and targeted mutation systems for a much broader range of species. These include the *Epichloë* species (family Clavicipitaceae, order Hypocreales), which are systemic, constitutive, and often seed-transmitted symbionts (endophytes) of cool-season grasses (Poaceae, subfam. Poöideae), and which are capable of producing a panoply of bioprotective alkaloids [1–3]. However, features of important *Epichloë* species that present special difficulties for genetic experimentation include growth rates far slower than model fungi, sparse sporulation, limited availability of selectable markers, and for many, diploid or triploid genomes and the lack of a sexual cycle [2,4].

The development of CRISPR technologies has opened the door to facile gene inactivation, removal, or even replacement for a wide range of organisms, including filamentous fungi [5,6]. Initially, the application of CRISPR in eukaryotes involved transformation or transfection with a gene construct expressing the Cas9 double-strand DNase and another construct to transcribe guide and tracrRNAs to direct the activity of Cas9 to the target sites. In *Aspergillus* species, Nødvig et al. [7] developed a system based on a single plasmid harboring a chimeric RNA guide and the *cas*9 gene under fungal promoters, along with a marker gene required for fungal selection. More recently, Cas9-sgRNA (single guide and tracrRNA) ribonucleoprotein complexes (RNPs) have been employed in a wide range of fungi [5,6].

In an example of the RNP approach, targeted mutations have been introduced into the genome of the fungus *Pyricularia oryzae* [8], which is among the most important fungal pathogens globally impacting cereal grain production. The fungus was transformed simultaneously with the RNPs to mutate the target gene and others to generate mutant genes that are positively selectable. This strategy is based on the presumption that those nuclei in which the genes are converted to their selectable forms are also those most likely to have taken up the RNP and consequently mutate the target gene as well.

We use *Epichloë* spp. as exemplars of particularly difficult but important fungi for targeted genetic modification. *Epichloë coenophiala* is widespread as a seed-borne endophyte of the highly popular pasture, forage, and turf grass, *Lolium arundinaceum* (= *Schedonorus arundinaceus* = *Festuca arundinacea*; tall fescue), although its existence was unsuspected in the first decades of widespread propagation of the grass during the mid-20th century. The fungus provides important benefits that translate to enhanced stand longevity and productivity and improved tolerance of drought and other stresses [9,10], and it is capable of producing up to four different classes of alkaloids that protect the grass hosts against invertebrates [2,11]. Unfortunately, the strains of *E. coenophiala* that have been unwittingly co-propagated with tall fescue, and which remain dominant in much of the cool-season pasturelands, produce ergovaline, which is an ergot alkaloid of the highly toxic ergopeptine type [12–14]. Levels of ergovaline tend to be very low, but they are often sufficient to at least cause reproductive problems and reduce livestock health and productivity. For the same reason, cultivars of *Lolium perenne* (perennial ryegrass) with *Epichloë hybrida* [15] strain Lp1 were pulled from the market in 1992 after it was determined that they had toxic levels of ergovaline [16,17]. A study including deletion of the dimethylallyltryptophan synthase gene (*dmaW*) from *E. hybrida* Lp1 and its subsequent complementation with the ortholog from *Claviceps fusiformis* [18] has demonstrated that *dmaW* is essential for ergot alkaloid biosynthesis [19]. There is a potential to develop and deploy such genetically altered strains of *Epichloë* species in forage cultivars because the fungus can be cultured, manipulated, and reintroduced to produce new, stable symbioses with forage and pasture cultivars of tall fescue. However, it is desirable and perhaps essential that such modifications should not involve integration of any foreign gene in the genome, which is a requirement that makes the CRISPR RNP approach especially attractive.

The endophyte *E. coenophiala* is a particularly difficult system for genetic inquiry because of its slow growth [20] and the fact that it is a triploid interspecific hybrid [21]. Two of its three ancestors were ergovaline producers, so *E. coenophiala* has two homeologous copies of the ergot alkaloid biosynthesis (*EAS*) gene clusters [2]. An effort to delete the key gene *dmaW*2 by marker exchange mutagenesis with a hygromycin B-resistance gene (loxP-flanked *hph*) was a particular *tour de force* in which over 1500 transformants were screened and the frequency of homologous gene replacement was 0.2% [22]. Once a Δ*dmaW*2 mutant was obtained and reintroduced into host plants, there was essentially no effect on ergovaline production because of the presence of its homeologue *dmaW*1. Furthermore, although the selectable marker was readily removed from the Δ*dmaW*2 mutant by transformation with a plasmid for the transient expression of Cre recombinase [22], the resulting marker-free mutants consistently had lost the ability to establish stable symbiosis with tall fescue (unpublished results).

Natural mutants of *Epichloë* species with deleted or inactivated alkaloid biosynthesis genes consistently have inactivated the entire set of genes for downstream steps [23–25]. Whether or not this relates to the host incompatibility of the aforementioned marker-free Δ*dmaW2* mutants, the nature of these natural variants suggests that there is selection against expression of enzymes that, due to loss of upstream genes, no longer have access to their normal substrates. Therefore, we consider the most prudent approach to generating ergot alkaloid-negative mutants to be the deletion of both *EAS* clusters in their entirety.

After genome sequencing revealed the subterminal location of the *EAS1* cluster in *E. coenophiala* isolate e19, we devised a new approach to replace that cluster with a telomere-repeat array [24]. Since the homeologous cluster, *EAS2*, has an inactivating mutation in a late-pathway gene, *lpsB2*, the resulting "*EAS1*-knockoff" mutant produced only two early products of the ergot alkaloid pathway, chanoclavine and ergotryptamine. The technique employed transient expression of antibiotic resistance conferred by an *hph* gene positioned in the vector to be lost subsequently by breakage of the integrated DNA at the introduced telomere repeat array. The success of this approach suggested that transient antibiotic selection could be used in other approaches for mutation. In this study, this strategy is applied to CRISPR-based deletion of both *EAS* clusters as well as individual genes.

Both for research and for the practical aim of completely eliminating production of all ergot alkaloids from an agriculturally important grass symbiont, we have chosen to adapt a Cas9-sgRNA RNP approach to entirely eliminate both the 196-kb *EAS1* cluster and the 75-kb *EAS2* cluster. Here, we describe the success of that effort and follow-up experiments to demonstrate the facile nature of our approach and its broader applicability, opening the door to a wide range of non-transgenic manipulations of even slow-growing, asexual, polyploid fungi.

2. Results

2.1. Assembly of *E. coenophiala* and *E. hybrida* Genome Sequences Including Nanopore Data

The genome of *E. coenophiala* e19 wild-type strain was previously sequenced by a combination of pyrosequencing (Roche) and Sanger sequencing of fosmid-cloned ends [24]. Due to its triploid hybrid nature, the genome is complex and has been difficult to assemble, especially across repetitive regions. To assess the potential of Oxford Nanopore technology to improve the genome data and assembly, the *E. coenophiala* e19 genome was sequenced using the portable DNA sequencer MinION. The MaSuRCA v. 3.4.1 de novo assembly was manually curated to give 216 scaffolds with the contig sizes varying from 1915 to 12,291,650 bp with an N50 of 1,403,312 bp (Supplementary Table S1; GenBank accession number JAFEMN000000000). The estimated genome size was 104.2 Mb. The 196.2 kb complete sequence of the *EAS1* cluster was identified on a 676 kb scaffold that ended with a telomere repeat array, and the 75.2 kb *EAS2* cluster was identified on a 2.6 Mb scaffold sequence where it was flanked by housekeeping genes. Both clusters had the 11 known ergot alkaloid biosynthesis genes similarly arranged and oriented, and the difference in their sizes was due to lengths of AT-rich noncoding regions consisting mainly of repeats.

The *E. hybrida* Lp1 genome was sequenced using several sequencing platforms (Supplementary Figure S1) and the assembled data generated 158 scaffolds with the contig size varying from 9259 to 8,313,425 bp, with an N50 of 2,103,505 bp, and estimated total genome size of 79.9 Mb sequence (GenBank accession number JAFEKR000000000). The 83.2-kb *EAS* cluster was located on a 6.7 Mb scaffold.

2.2. Deletion of Ergot Alkaloid Biosynthesis Gene Clusters from the *E. coenophiala* Genome

After *E. coenophiala* e19 protoplasts were treated simultaneously with the *EAS2*-directed RNPs with specific sgRNAs (Figure 1 and Table 1) and the plasmid pKAES329 with a fungal-active *hph* gene, 115 hygromycin B-resistant colonies were recovered and subsequently single-spore isolated on plates without selection. DNA was extracted and subjected to a series of PCR screens (Figure 2 and Supplementary Figure S1) with primers designed for identification of the expected deletion mutants (Table 2). The first screen with primers specific for *easA*1 and *easA*2 indicated 31 colonies lacking only *easA*1, 14 lacking only *easA*2, and five lacking both *easA*1 and *easA*2. As a further check if both *EAS*1 and *EAS*2 clusters were absent in the last five transformants, they were subjected to a second PCR screen with primers targeting a common region in the two *dmaW* homeologs, and all five tested negative for both (Figure 2 and Supplementary Figure S1).

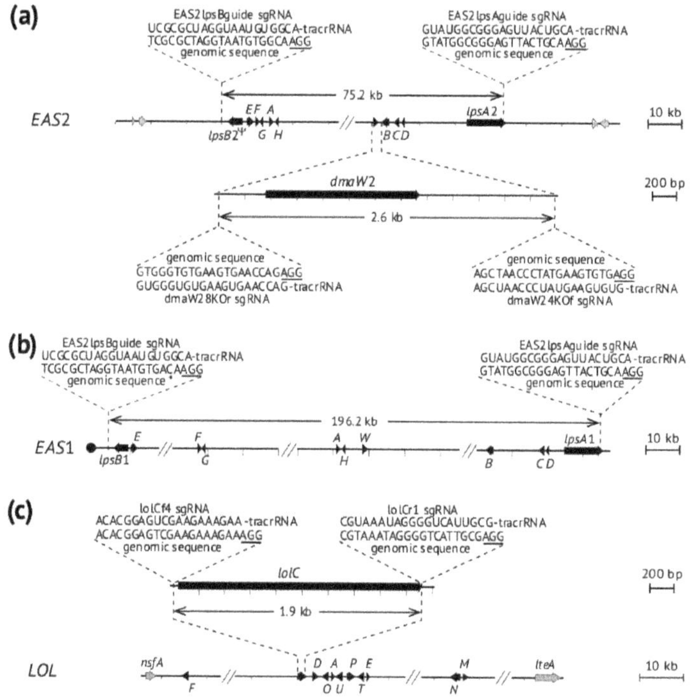

Figure 1. Maps of genes and gene clusters in *E. coenophiala* e19 indicating target locations and sequences of the single guide RNAs (sgRNAs) that direct cleavage by modified Cas9 nuclease. On each map, the AGG (underlined) protospacer adjacent motifs (PAM) required for Cas9 nuclease to generate double-strand break at the target site is 3′-adjacent to the 20-nucleotide target DNA sequence, and the trans-activating crispr RNA (tracrRNA) segment is a 67 nt sequence that interacts with Cas9. Black box-arrows indicate *EAS* or *LOL* genes, which are labeled with the full name or an abbreviation with the last letter of the gene name or with *B* for *cloA*. Hash marks on the *EAS* maps indicate long stretches of noncoding sequences. (**a**) Ergot alkaloid biosynthesis gene cluster *EAS*2 with the *dmaW*2 region magnified. The sgRNA sequences for *EAS*2 deletion match genomic sequences outside but near the 3′ end of the *lpsB*2 pseudogene and within but near the 3′ end of *lpsA*2. The sgRNA sequences for *dmaW*2 deletion match flanking intergenic regions. (**b**) Ergot alkaloid biosynthesis gene cluster *EAS*1. The sgRNA guides are the same as those for *EAS*2, and the sequence flanking *lpsB*1 has a single mismatch to the sgRNA (asterisk). (**c**) Loline alkaloid biosynthesis gene cluster *LOL* with *lolC* magnified.

Table 1. sgRNA guides.

Scheme 1.	Target Site	PAM	Modified sgRNA Sequence [1]
dmaW24KOf	*dmaW*2 3′-flank	AGG	A*G*C*UAACCCUAUGAAGUGUG
dmaW28KOr	*dmaW*2 5′-flank	AGG	G*U*G*GGUGUGAAGUGAACCAG
EAS2lpsAguide	*lpsA* 3′-region	AGG	G*U*A*UGGCGGGAGUUACUGCA
EAS2lpsBguide	*lpsB* 3′-flank	AGG	U*C*G*CGCUAGGUAAUGUGGCA
lolCf4	*lolC* 5′-flank	AGG	A*C*A*CGGAGUCGAAGAAAGAA
lolCr1	*lolC* 3′-flank	AGG	C*G*U*AAAUAGGGGUCAUUGCG

[1] Asterisks indicate 2′ O-methyl analogs with 3′ phosphorothioate linkages.

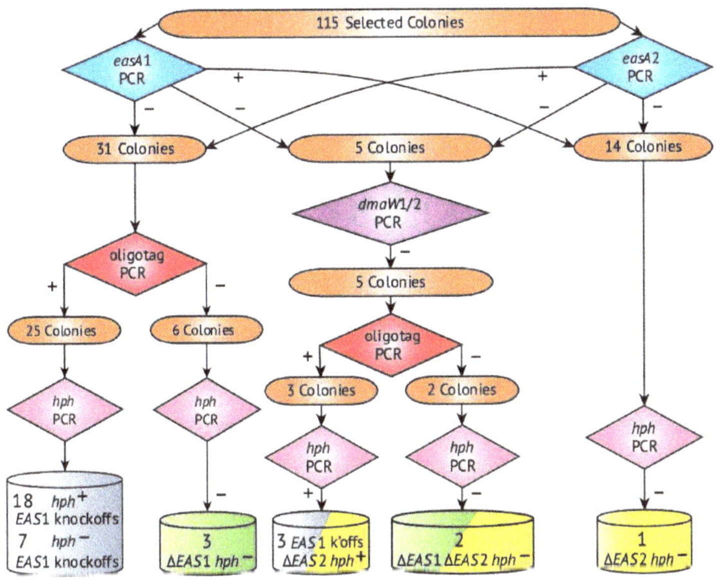

Figure 2. Graphic summary of the screening strategy and results for ergot alkaloid biosynthesis (*EAS*)-cluster deletions in *E. coenophiala* e19. Cylinders represent mutants identified as *EAS*1 knockoffs by chromosome-end deletions [24] (gray), Cas9-mediated *EAS*1 deletions (green; Δ*EAS*1), and Cas9-mediated *EAS*2 deletions (yellow; Δ*EAS*2). Cylinders with two colors represent losses of both *EAS*1 and *EAS*2 gene clusters in the same mutants.

Table 2. Primers used for PCR tests.

Primer Name	Target Gene(s) or Sites	Sequence
dmaW1&2f	*dmaW*1, *dmaW*2	GCAAAGACACTCCACCAGGAAGTT
dmaW1&2r	*dmaW*1, *dmaW*2	AGTTGCGGCGTTAATAGGCTCGTA
hph.4d	*hph*	GACCTGATGCAGCTCTCGGA
hph.3u	*hph*	TCGGCGAGTACTTCTACACA
RTeasA1f	*easA*1	ACAACTTTGGGCGACTGGG
RteasA1r	*easA*1	CCGTTGGTTGCAAGAAGATTGA
RteasA2	*easA*2	GCAGCTTTGGGCGACTGGA
RteasA2r	*easA*2	ACGTTGGTTGCAAGGAGATTGG
lolC-3a	*lolC*	GGTCTAGTATTACGTTGCCAGGG
lolC-5b	*lolC*	TCTAAACTTGACGCAGTTCGGC
oligoscreen(f)	Oligotag	GATGGCCTTTAAAGTCTACGTACTC
lpsAoligoR	*lpsA*1, *lpsA*2	ATATCATGGCAACATTCAGCGCAC

Since the pKAES329 plasmid used to transiently select transformants carried a truncated fragment of *lpsA1*, and there was a single-base mismatch of the EAS2lpsBguide to the *EAS1* locus target site (Figure 1), we expected that the loss of the *EAS1* cluster in some mutants might be due to a homologous recombination of the *lpsA1* sequence followed by chromosome-end knockoff, as was previously accomplished using the same plasmid [24], rather than by Cas9-mediated deletion. To check this possibility, 31 colonies that were negative for *easA1* and another five that were negative for both *easA1* and *easA2* were tested by PCR with primers targeting a 223 bp fragment spanning from the oligotag into the truncated *lpsA1* gene on pKAES329, which was a sequence that was expected to be present in such knockoff mutants [24] but to be absent from mutants induced only by CRISPR. The result was negative for six of the mutants that lacked only *easA1* and two that lacked both *easA* genes, suggesting that the loss of *EAS1* in those mutants was due to Cas9-catalyzed cleavage. The other mutants, which were positive for the oligotag, were not investigated further. The CRISPR-mediated cluster deletions were designated Δ*EAS1*, Δ*EAS2* and Δ*EAS1* Δ*EAS2* (Figure 2).

In addition to generating strains completely lacking ergot alkaloid genes, the goal was to avoid integration of any foreign gene in the genome of the modified strains. A PCR test for *hph* identified two Δ*EAS1* Δ*EAS2* mutants (designated e7801 and e7802) and one Δ*EAS2* mutant (designated e7803) that were marker-free (Figure 2 and Supplementary Figure S1).

2.3. Deletion of *dmaW2*

Epichloë coenophiala e19 is a triploid interspecific hybrid with orthologous *EAS1* and *EAS2* gene clusters that can largely complement each other's ergot alkaloid-biosynthesis gene mutants [22,24]. We previously eliminated the telomere-linked *EAS1* cluster to generate strain e7480 [24]. This "knockoff" mutant retained the *EAS2* cluster with most functional ergot alkaloid biosynthesis genes including *dmaW2*, which encodes the enzyme for the first determinant step in the pathway [19], and which has proven exceptionally difficult to eliminate by marker-exchange homologous recombination [22]. Therefore, *dmaW2* in e7480 was targeted to test if the use of CRISPR RNP technology might provide more efficient editing of the locus. Protoplasts of e7480 were simultaneously treated with the RNP mixture and pKAES329; then, they were regenerated on medium with hygromycin B to obtain a total of 318 selected colonies. Then, these were propagated without selection, and their DNA was screened by PCR for *dmaW2* to identify 54 putative Δ*dmaW2* mutants (Supplementary Figure S1). Screening those for *hph* indicated 50 out of the 54 putative mutants that also lacked the selection marker, which could have occurred either because the marker was expressed without plasmid integration or because of recapitulation of the chromosome-end knockoff [24] whereby the plasmid integrated at the *lpsA1* site and was subsequently lost by breakage within the introduced telomere-repeat array. Positive controls were PCR screens with primers specific for *easA2* and the oligotag and, as expected, all 50 putative marker-free mutants tested positive for both. Moreover, since the *EAS1* cluster is missing in the e7480 genome, the PCR test for *easA1* was negative for all the samples. Two of the Δ*dmaW2* mutants were designated e7799 and e7800 (Table 3).

Table 3. Mutants confirmed by genome sequencing.

Mutant	Parental Strain	Genotype
e7799	e7480	EAS1-knockoff Δ*dmaW2*
e7800	e7480	EAS1-knockoff Δ*dmaW2*
e7801	e19	Δ*EAS1* Δ*EAS2*
e7802	e19	EAS1-knockoff Δ*EAS2*
e7803	e19	Δ*EAS2*
e7804	e19	Δ*lolC*
e7805	e19	Δ*lolC*
e7806	Lp1	Δ*EAS*

2.4. Deletion of the EAS Cluster in E. hybrida Lp1

Epichloë hybrida is a diploid interspecific hybrid that inherited, from its *Epichloë festucae* ancestor, an *EAS* cluster [15] similar to *EAS2* of *E. coenophiala*. The sgRNAs designed for deletion of the *EAS* clusters in e19 were used in an attempt to delete the *EAS* cluster in *E. hybrida* Lp1. The transformation plasmid providing transient hygromycin B resistance was pKAES328, which has the fungal-active gene *hph* but differs from pKAES329 in lacking an *lpsA1* fragment [24]. Following the same PCR procedures as mentioned above for *E. coenophiala* (Supplementary Figure S1), 16 *E. hybrida* colonies were screened, out of which one (designated e7806) was a putative marker-free Δ*EAS* mutant (Table 3).

2.5. Deletion of lolC in E. coenophiala e19

Genomic analysis of *E. coenophiala* e19 indicated a single cluster, designated *LOL*, with the loline alkaloid biosynthesis genes. The *lolC* gene has been suggested to encode the enzyme for the first step in the loline alkaloid pathway and its role has been tested previously by RNA interference (RNAi) in *Epichloë uncinata*, resulting in significantly reduced loline-alkaloid production [26]. To further evaluate the general utility of the CRISPR RNP technology in *E. coenophiala*, sgRNAs were designed and used for the deletion of *lolC*. Following protoplast treatment with the RNP mixture followed by initial selection with hygromycin B, a total of 185 colonies were screened with *lolC*-specific primers to identify 11 that tested negative for the gene (Supplementary Figure S1). These putative Δ*lolC* mutants were further screened with primers for *hph*, identifying two marker-free mutants designated e7804 and e7805 (Table 3).

2.6. Genome Analysis of the CRISPR-Derived Mutants

A de novo genome sequence assembly was performed for each deletion mutant in Table 3. The dataset size (in bases), number of reads (average length 130 bp), genome coverage, and assembly quality metrics for each are presented in Supplementary Table S2. In every case, the sequences were consistent with the PCR results regarding the gene losses due to Cas9 nuclease, absence of the marker gene, and absence of the oligotag sequence. Moreover, to check if any plasmid sequences were present in the deletion mutants, the reads were mapped against the pKAES328 and pKAES329. None of the mutants had sequences from those plasmids.

Interestingly, almost all of the gene deletions resulted from the cleavage and flawless rejoining of the two flanking ends with no other sequence changes (Figure 3). Positions of the cleavage by Cas9 in the e7801 and e7802 Δ*EAS1* Δ*EAS2* mutants varied between the strains but also between the two clusters in the same strain. For instance, in e7802, the cleavage at the *EAS1* cluster near the *lpsB1* locus was inferred to be 3-bp upstream of the PAM site, even though the position had a mismatch between the sgRNA and *EAS1* target site, and 4-bp upstream of the PAM site in *EAS2* where the sgRNA was an exact match to the target sequence. Furthermore, the cleavage near *lpsA1* was 4-bp upstream of the PAM site but 3-bp upstream of the PAM site near *lpsA2*. Among the six sequenced mutants with 15 Cas9-mediated cleavage sites in all, 11 cleavages were 3-bp from the PAM site and four (27%) were 4-bp from the PAM site.

The changes at both *EAS* clusters in e7801 differed from those in e7802 in several interesting ways. In addition to the different cleavage sites mentioned above, e7801 (but not e7802) had undergone a reciprocal recombination event between the two *EAS* clusters, which was likely triggered by the Cas9-induced double-strand breaks in the orthologous positions of the two *lpsA* genes. Furthermore, in e7801, the cleaved end of the *EAS1* cluster was linked to a new telomere repeat array, from which it was separated by only a single base pair. In contrast, both the *EAS1* and the *EAS2* clusters of e7802 were deleted with the flanking sequences joined.

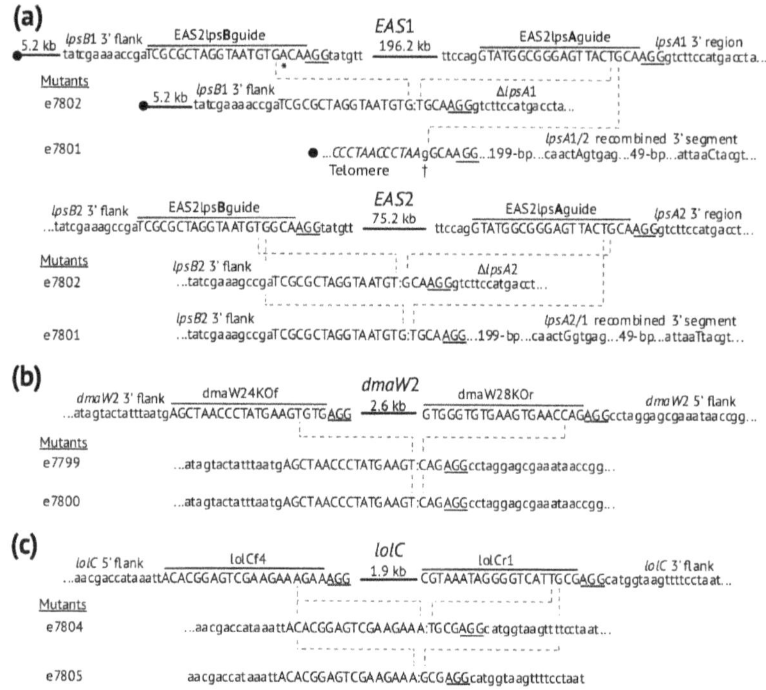

Figure 3. CRISPR-induced deletions in *E. coenophiala* e19 as deduced from genome sequencing. Targets of the sgRNAs (Table 1) are indicated, adjacent PAM sites (AGG in all cases) are underlined, and cleavage sites are assumed to have been 3 bp or 4 bp 5′ of the PAM sites. Colons (:) indicate joined ends. (**a**) Deletions and recombination at *EAS* gene loci. Filled circles indicate telomere repeat arrays of $(CCCTAA)_n$ near *EAS1*, and an asterisk (*) indicates a mismatch between EAS2lpsBguide sgRNA and its target near the *EAS1* gene cluster. In the Δ*EAS1* locus of e7801 a single base-pair addition is indicated (†) just inside the new telomere (italic text and filled circle). In e7801, single-nucleotide polymorphisms within the recombined segments labeled *lpsA1/2* and *lpsA2/1* are indicated as capital letters. (**b**) CRISPR-mediated deletion of the *dmaW2* gene for the first step in ergot alkaloid biosynthesis. (**c**) CRISPR-mediated deletion of the *lolC* gene for the presumed first step in loline alkaloid biosynthesis.

A BLAST (basic local alignment search tool) search against the *E. hybrida* e7806 mutant genome with sequences for the *EAS* cluster and *hph* gene indicated their absence, which was consistent with the previous PCR results. The Cas9-mediated cleavage occurred 3 bp upstream from the PAM site near *lpsB* and 4 bp upstream from the PAM site in *lpsA*. The junction generated by joining the cleaved ends was identical to the junction in the deleted *EAS2* locus of mutant e7801 shown in Figure 3.

The *E. coenophiala* mutants e7999 and e7800 had lost *dmaW2* as indicated by the PCR screen and validated by their sequenced genomes. Since the cleavages induced by Cas9 occurred 3 bp upstream of PAM sequence for the dmaW24KOf and dmaW28KOr sgRNA target sites, both Δ*dmaW2* mutants had identical junctions with no indels (Figure 3).

The genome sequence of *E. coenophiala* Δ*lolC* mutants e7804 and e7805 also confirmed that no other foreign DNA was integrated into the genome. When comparing the *lolC* deletion site in the two mutants, it appeared that Cas9 had generated cleavages 3 bp upstream from PAM site at all sites except for the lolCr1 sgRNA target site of e7404, for which the cleavage occurred 4 bp upstream of the PAM site. The rejoining of the ends was perfect without introduction of indels or any sequence changes (Figure 3).

2.7. Inoculation Efficiency

Endophyte infection determined by tissue-print immunoblot indicated that all CRISPR-mediated deletion mutants established symbiotic relationships with the plant at infection rates similar or higher than those typically seen for wild-type strain inoculations (Table 4).

Table 4. Establishment of symbiosis of mutant endophytes with host plants.

Genotype Inoculated	Parental Strain	Plant Species	Positive/Total	Infection Rate %
Δ*EAS1* Δ*EAS2*	*Epichloë coenophiala* e19	*Lolium arundinaceum*	65/143	45
Δ*EAS2*	*E. coenophiala* e19	*L. arundinaceum*	22/48	46
Δ*EAS*	*Epichloë hybrida* Lp1	*Lolium perenne*	32/46	70
Δ*dmaW2*	*E. coenophiala* e7480	*L. arundinaceum*	125/251	50
Δ*lolC*	*E. coenophiala* e19	*L. arundinaceum*	25/81	31

3. Discussion

We have demonstrated that a CRISPR/Cas9 technology can be applied to polyploid *Epichloë* species for the precise removal of individual genes or gene clusters, and even the simultaneous removal of a pair of large clusters (196 kb and 75 kb). Specifically, the method utilized RNPs—consisting of sgRNAs and Cas9 protein that was translationally fused with two NLRs—which were introduced into the fungus by cotransformation with a transiently selected antibiotic resistance gene. There are two obvious advantages to this approach. One is that more than one gene or genome region can be deleted in a single procedure, and the other is that the procedure leaves no selectable marker or transgenes in the genome. The absence of the selectable marker in the final product allows for its reuse to eliminate additional genes or to reintroduce genes for complementation analysis, and it also addresses regulatory and public concerns about the use of transgenic organisms in applied research and agriculture.

We previously constructed a plasmid (pKAES329) to transiently integrate a selectable antibiotic-resistance gene (*hph*) at a chromosome end [24], and here, we have combined that approach with the RNP approach and found it to be successful and facile. However, we also questioned whether a transient integration of the marker was even necessary. We show here that marker integration is unnecessary by demonstrating the simultaneous elimination of two long gene clusters (*EAS1* and *EAS2*) by treatment with the appropriate RNP mixtures together with a plasmid that, without integrating into the genome, provided expression of the selectable marker.

In this study, we did not test whether even transient marker selection was needed. However, we previously used the transformation of *Epichloë* species without selection to delete a loxP-flanked *hph* gene by transient expression of the Cre-recombinase gene [22]. Although successful, the screen was tedious and time consuming, resulting in a 0.5–2.1% frequency of Cre-mediated deletions among the unselected colonies. In contrast, here, we report that plasmids mixed with the RNPs provided for temporary selection of a limited number of initially antibiotic-resistant transformants. From those, mutants with the target deletions were readily identified by PCR screens and a high proportion were marker free (18–100% depending on the experiment). Therefore, although integration of the plasmid borne selectable marker was not required, our results suggest that that its inclusion in the transformation mixture aided in selection of the RNP-transformed isolates, including those with the desired deletions.

In targeting the *EAS1* gene cluster, a mismatch 3 bp upstream of the PAM site between the EAS2lpsBguide and its target genomic sequence near *lpsB1* did not prevent cleavage at this position. The mismatch at a single-nucleotide polymorphism between the sites near *EAS1* and *EAS2* (the sgRNA sequence was based on the latter) was in the "seed" region but not part of the "core" region for sgRNA-directed Cas9 cleavage [27]. Since deletion of the 196-kb *EAS1* region was achieved with cleavage at the mismatch position, a precise match to the target was evidently not required.

Two additional genome alterations were observed in one of the ΔEAS1 ΔEAS2 mutants. One was the introduction of a telomere at the cut site near *EAS1* in one of the mutants. Interestingly, that cut was repaired very simply with the addition of a single base pair followed by a telomere array. The other alteration in the same mutant was a recombination between the remnants of *lpsA1* and *lpsA2*, presumably moving the genes near *lpsA2* from genomic locations far from a telomere to close proximity to the newly generated telomere. Whether that alteration affects the expression of those genes may be an interesting topic for future inquiry.

Our results demonstrate the high efficiency of the CRISPR/Cas9 technology while not indicating any size limit for genome segments that can be efficiently and precisely deleted. The frequency of *dmaW2* deletion was 17%, compared to only 0.2% previously obtained for marker exchange mutagenesis of the same gene in the same *E. coenophiala* strain [22]. In addition, the frequency of the 75-kb *EAS2* deletion (5.2%) was not dramatically less than that of the 2.6-kb *dmaW2* deletion (17%) and was very close to that of the 1.9-kb *lolC* deletion (5.9%). Conceivably, size limits might be imposed by boundaries of chromatin states, so it is worth considering whether or not the clean deletions of large genome regions are more likely when those genes are part of a coordinately regulated cluster.

Though often characterized as "error-prone", non-homologous end joining (NHEJ) frequently occurs without error. A study in mouse cell lines reported 70% precise NHEJ events after Cas9-mediated cleavage [28]. Many target sites also exhibited approximately 10% 1-bp and 2-bp templated insertions. Insertions of 1 bp following Cas9 cleavage and NHEJ have also been described in *Saccharomyces cerevisiae*, with 74–100% apparently being templated, implying that in those cases, Cas9 cleavage left a staggered (5′-overhanging) end that was filled in by DNA polymerase 4 [29]. In our study, all of the eight sequenced junctions appeared to be precisely joined, but four of those had one or the other end cleaved 4 bp from the PAM site. In those four cases, the result was similar to the common 1-bp templated insertions observed by Lemos et al. [29], so it is reasonable to speculate that they also resulted from staggered cleavage and end-repair.

As commercial sources have recently made Cas9-NLS fusion proteins and synthetic sgRNAs available, a variety of approaches that include Cas9-sgRNA RNPs have been employed in fungi. Most involve the integration of a stable selectable marker [30] or simultaneous generation of a selectable mutation [8]. Khan et al. [31] report using the Cas9-sgRNA RNP system without selection to target the *TOX3* effector gene in *Parastagonospora nodorum*, with all of the six analyzed "transformants" exhibiting mutations at the repair site. This result contrasts with ours in that we found no mutations other than the deletions resulting from rejoining of the cleaved ends and, in approximately half of the junctions, an apparent 1-bp templated insertion. Thus, applications of this technology in different fungal systems may give substantially different outcomes.

4. Conclusions

We report five findings with regard to the application of Cas9-sgRNA RNP technology for deleting DNA segments in *Epichloë* species. First, transient selection for an antibiotic resistance gene included in the transformation mix allowed for the efficient identification of deletion mutants. Second, the deletions were precisely repaired by NHEJ (sometimes with templated 1-bp insertions). Third, even large segments up to 196 kb in our tests were efficiently deleted. Fourth, simultaneous deletions of two large DNA segments with entire biosynthetic gene clusters were obtained. Fifth, the deletion mutants retained their compatibility as grass symbionts. These results highlight the range of non-transgenic manipulations of even slow-growing, asexual, polyploid fungi that can be undertaken with this CRISPR RNP approach. The application to *Epichloë* species provides for tailored genotypes to use in turf and forage, for example to generate elite lines of livestock-friendly cultivars.

5. Materials and Methods

5.1. Biological Materials

The wild-type *Epichloë coenophiala* strain e19 [= American Type Culture Collection (ATCC) 90664] from tall fescue (*Lolium arundinaceum*) cv. Kentucky 31 [32], and the *E. coenophiala EAS1*-knockoff strain e7480 (=ATCC PTA-126679) [24] were maintained in tall fescue elite breeding line KYFA0601 [24]. The wild-type *Epichloë hybrida* Lp1 (=ATCC TSD-66) strain from perennial ryegrass (*Lolium perenne*) [33] was maintained in perennial ryegrass cv. Rosalin [19]. The fungi were isolated from these symbiotic plants and grown as described in Florea et al. [34].

5.2. Miscellaneous Molecular Methods

Plasmid DNA was isolated from bacterial cultures by use of the ZR Plasmid Miniprep-Classic kit (Zymo Research, Irvine, CA, USA). Fungal DNA was isolated from fresh mycelium by use of the DNeasy 96 Plant Kit (Qiagen, Valencia, CA, USA) and a Geno/Grinder 2000 (SPEX CertiPrep, Metuchen, NJ, USA) or by use of the ZR Fungal/Bacterial DNA MiniPrep kit (Zymo Research). The DNA quality and concentrations were assessed by NanoDrop and Qubit (ThermoFisher Scientific, Waltham, MA, USA). Plasmids pKAES328 and pKAES329, with a fungal-active hygromycin B-resistance gene (*hph*), are described in Florea et al. [24].

5.3. Nanopore Single-Molecule Sequencing of Genomic DNA

Genomes of wild-type *E. coenophiala* e19 and *E. hybrida* Lp1 were sequenced by a hybrid approach including Illumina, pyrosequencing, and single-molecule sequencing. For Oxford Nanopore (Oxford, UK) single-molecule sequencing, the fungi were grown on potato dextrose agar (PDA) (BD, Franklin Lakes, NJ, USA) plates topped with cellophane for ease of removal. High molecular weight (> 40 kb) DNA was extracted from young mycelium as described by the method of Al-Samarrai and Schmid [35] and recovered by spooling on glass rods. The Ligation Sequencing Kit 1D (SQK-LSK108) was used for library preparation. For each library, the end-repair and dA-tailing step was performed with the NEBNext End repair/dA-tailing kit (New England Biolabs, Ipswich, MA, USA) by incubating 100 µL Ultra II End-prep reaction buffer containing 3 µg of high molecular weight DNA and 6 µL of Ultra II End-prep enzyme mix for 20 min at 20 °C, then for 20 min at 65 °C. Then, the end-repaired DNA was mixed with 120 µL AMPure XP beads (Beckman Coulter, Pasadena, CA, USA) and incubated 5 min at room temperature. The beads were pelleted on a magnetic rack and the supernatant was discarded; then, they were washed twice with 200 µL of 70% ethanol and allowed to dry. The beads were resuspended in 31 µL of nuclease-free water and incubated for 10 min at room temperature; then, they were pelleted on the magnetic rack until the solution became clear. Then, the DNA solution was transferred into a new 1.5 µL Eppendorf loBind DNA tube, and the DNA concentration was determined using the Qubit. The adapter ligation reaction was performed with 50 µL NEB Blunt/TA Ligase Master Mix and 20 µL adapter mix added to 30 µL containing 2 µg of the End-prep-treated DNA and incubated for 10 min at room temperature. Then, the adapter-ligated DNA was added to 60 µL AMPure XP beads and incubated for 5 min at room temperature; then, it was pelleted on a magnetic rack, and the supernatant was discarded. The beads were resuspended in 500 µL adapter bead binding buffer (ABB), pelleted again on the magnetic rack for the removal of the ABB buffer, allowed to dry, and then resuspended in 15 µL elution buffer. After 10 min incubation at room temperature, the beads were pelleted once more on the magnetic rack to recover the DNA library in the supernatant.

Oxford Nanopore FLO-MIN106 flowcells were primed with a mix consisting of 480 µL RBF (running buffer with fuel mix) and 520 µL nuclease-free water by loading 800 µL of the priming mix into the priming port and incubating at room temperature for 5 min. The flowcell priming was completed by lifting the SpotOn port cover and then loading the remaining 200 µL of priming mix through the priming port. Then, the 75 µL DNA library loading mix consisting of 12 µL DNA library mixed with 35 µL of RBF, 2.5 µL nuclease-free water, and 25.5 µL LLB (library loading beads, EXP-LLB001) was loaded on the flowcell via the SpotON port. Sequence runs were implemented with MinKNOW software version 3.1.0.30 with the NC_48Hr_e19_exp_Run_FLO-MIN106_SQK-LSK108 and NC_48Hr_Lp1_exp_Run_FLO-MIN106_SQK-LSK108 sequencing scripts, and with the live base calling turned off. The MinION FAST5 output files were processed using Guppy basecalling software ver. 2.2.2 supporting GPU processing with NVIDIA-SMI driver version 418.67 and CUDA version 10.1. Guppy was run as a singularity container on the University of Kentucky HPC system.

5.4. Genome Sequence Assembly

Oxford Nanopore and Illumina HiSeq data (described above) were combined with Roche 454 GS FLX+ (454 Life Sciences, Branford, CT, USA) and paired-end Illumina MiSeq data described previously [24] in genome assemblies using MaSuRCA ver. 3.4.1 [36], which is a de novo assembler that combines the benefits of deBruijn graph and overlap-layout-consensus assembly approaches [37]. For Oxford Nanopore data, only the longest reads providing 25-fold genome coverage were included.

5.5. Illumina HiSeq Sequencing of Genomic DNA

DNA libraries were prepared by use of the Nextera DNA library prep and index kits per the manufacturer's reference guide (Epicentre Biotechnologies, Madison, WI, USA). Sequencing was carried out by Novogene Corporation Inc. (Sacramento, CA, USA) on the HiSeq platform with 2 × 150-cycle paired-end reads with 11 barcoded samples multiplexed on one lane (Illumina, San Diego, CA, USA). The data were evaluated for quality using FastQC v. 0.11.9 (https://www.bioinformatics.babraham.ac.uk/projects/fastqc/ (accessed on 20 October 2020)). Based on the FastQC results, the reads were trimmed, and the remnants of the adapters were removed using Trimmomatic v. 0.39 [38]. Duplicate reads were removed using Prinseq-lite v. 0.20.4 [39]. Assembly was performed with MaSuRCA v. 3.4.1.

5.6. Design of sgRNA Molecules and Assembly of RNP Complexes

The sgRNAs were designed using the Benchling platform (https://www.benchling.com/ (accessed on 20 October 2020)), having the genome of *Claviceps purpurea* 20.1 set as reference in the guide parameter design. All the sgRNAs used in this study were designed upstream of an AGG protospacer adjacent motif (PAM) recognition site (Table 1). Cas9 nuclease typically cleaves the DNA 3-bp upstream of the PAM site [40]. The sgRNAs chosen to delete *EAS* clusters were exact matches to target sequences near the ends of the *EAS*2 cluster. One (EAS2lpsBguide) was 163 bp downstream of *lpsB*2, and the other (EAS2lpsAguide) was 718 bp upstream of the stop codon of *lpsA*2. The guide sequence in EAS2lpsAguide was also an exact match to the homologous sequence in *lpsA*1, but EAS2lpsBguide had a single mismatch located 3 bp from the PAM site at the sequence near *lpsB*1 (Figure 1).

The *dmaW*2 sgRNAs were designed to recognize target sequences 431 bp upstream (dmaW28KOr) and 1055 kb downstream (dmaW24KOf) of the *dmaW*2 coding region and were an exact match to the genomic sequence at the *dmaW*2 locus. The *lolC* sgRNAs were designed to target sequences 140 bp upstream (lolCf4) of the start codon and 15 bp downstream (lolCr1) of the stop codon, and they were an exact match to the genomic sequence near the *lolC* gene (Figure 1). The *Streptococcus pyogenes* Cas9 nuclease fused with two nuclear localization signal motifs (Cas9-2NLS), and the sgRNAs were purchased from Synthego Corp. (Redwood City, CA, USA). To form RNP complexes, 180 pmol of each sgRNAs was incubated 10 min at room temperature with 20 pmol Cas9-2NLS nuclease and nuclease-free water in 15 µL volume. For transformation, the RNP complexes were paired and mixed with the plasmid DNA, as indicated in Table 1.

5.7. Fungal Transformation and Selection

Epichloë coenophiala e19, its derivative e7480 [24], and *E. hybrida* Lp1 were transformed with the plasmid–RNP mixtures by a modification of previously described methods [24,41] as follows: Fungus was grown in potato dextrose broth (PDB) (BD, Franklin Lakes, NJ, USA) in an incubator shaker for 5–10 days at 22 °C and 200 rpm. The culture was transferred into 50 mL conical tubes and harvested by centrifugation at 5525× *g* for 20 min at 4 °C. The mycelium was resuspended in 20 mL osmotic medium (1.2 M $MgSO_4$, 10 mM $NaHPO_4$) containing an enzyme mixture consisting of 5 mg/mL Vinoflow FCE (Novozymes, Franklington, NC, USA), 5 mg/mL lysing enzymes from *Trichoderma harzianum*, 5 mg/mL Driselase Basidiomycetes sp., and 3 mg/mL bovine serum albumin (all from Sigma-Aldrich, St. Louis, MO, USA); then, it was incubated for 3 h on a rocker shaker at 30 °C. The remaining mycelial mass was removed from the protoplast suspension by passage through autoclaved Miracloth; then, it was transferred into 30 mL Corex tubes at 10 mL of the suspension in each, which was gently overlayed with 10 mL of ST solution (0.6 M sorbitol, 0.1 M Tris-Cl pH 7.4). To isolate the protoplasts, the tubes were centrifuged at 3329× *g* for 20 min. The protoplasts were rescued from the interface of the two solutions with a pipette and transferred into a tube containing 5 mL of STC solution (1 M sorbitol, 50 mM Tris-Cl pH 7.4, 50 mM $CaCl_2$). The protoplasts in STC were pelleted in a centrifuge at 3329× *g* for 10 min. The supernatant was discarded, and the protoplast pellet was gently resuspended in 5 mL STC and pelleted again, and the supernatant was discarded. Finally, the protoplasts were gently resuspended in a small volume of STC, counted by microscopy with a hemocytometer, and diluted in STC so that 100 μL solution would contain at least 5×10^6 protoplasts.

Protoplasts were transformed using a modification of the polyethyleneglycol (PEG) method as described by Panaccione et al. [41]. The transformation mix was prepared by combining 15 μL of each preassembled RNP complex with 7–10 μg of *Mlu*I-linearized plasmid DNA that was previously incubated 30 min at room temperature with 10 μg of Lipofectin Transfection Reagent (ThermoFisher Scientific, Waltham, MA, USA). The PEG-amendment solution was prepared by mixing two parts of 60% (*w*/*v*) PEG 3350 with one part amendments (1.8 M KCl, 150 mM $CaCl_2$, 150 mM Tris-HCl pH 7.4). The transformation was done in a sterile borosilicate tube by gently mixing 100 μL of protoplast suspension with 50 μL PEG-amendment solution and the DNA mix, and it was incubated on ice for 30 min. Then, 1 mL of PEG-amendment solution was added to the content of the tube and mixed by flicking the tube several times; then, it was incubated for 20 min at room temperature.

The treated protoplasts were plated on complete regeneration medium (CRM) [41], which contained per liter 304 g of sucrose, 1 g of KH_2PO_4, 1 g of NaCl, 0.46 g of $MgSO_4$ $-7H_2O$, 0.13 g of $CaCl_2$-$2H_2O$, 1 g of NH_4NO_3, 1 g of yeast extract, 12 g of PDB powder, 1 g of peptone, 1 g of casein (acid hydrolysate), and 7 g of agarose (Sigma-Aldrich). Plates of 20 mL CRM bottom-agarose contained hygromycin B at concentrations calculated to give a final 50 μg/mL for *E. coenophiala* or 200 μg/mL for *E. hybrida*. Each aliquot (250 μL) of treated protoplast suspension was added to 7 mL of CRM prepared with low melting Sea Plaque Agarose (Lonza, Mapleton, IL, USA) and kept molten at 50 °C; then, it was immediately poured and distributed onto the surface of a CRM plate. The plates were incubated at 21 °C for 3–4 weeks, and then, the fungal colonies were transferred to PDA without hygromycin B for sporulation and single-spore isolation.

5.8. Screening and Analysis of Deletion Mutants

Putative transformants were subjected to three rounds of single-conidiospore isolation on PDA without selection; then, they were grown and maintained on PDA plates. DNA was extracted by use of the DNeasy 96 Plant Kit (Qiagen, Valencia, CA, USA). PCR primers were purchased from Integrated DNA Technologies (Coralville, IA, USA) and are listed in Table 2. PCR was performed in 25 μL reaction mixtures with 5–10 ng DNA template, 200 μM each dNTP, 0.2 μM each primer, 2.5 units AmpliTaq Gold, and AmpliTaq Gold

PCR buffer with $MgCl_2$ at 1.5 mM final conc. (Applied Biosystems, Foster City, CA, USA). Thermocycler conditions were 9 min at 94 °C, 35 cycles of 94 °C for 30 s, annealing temperature 59 °C for 30 s, 72 °C for 1 min, and then a final 7-min incubation at 72 °C.

Selected isolates were subjected to Illumina HiSeq DNA sequencing (see above) to check for the results of deletions and nonhomologous end joining at the target sites. The reads were also screened for sequences from transformation plasmids pKAES328 and pKAES329 by using Deconseq ver. 0.4.3 [42].

5.9. Establishment of Symbiota

To establish symbiotic relationships with host plants, wild-type *E. coenophiala* and mutants were surgically introduced into seedlings of tall fescue elite breeding line KYFA0601, and *E. hybrida* mutants were similarly introduced into seedlings of the perennial ryegrass experimental line GA66 [43] by the procedure described by Latch and Christensen [44] and modified by Chung et al. [45]. After seedlings developed at least three tillers, one tiller each was sacrificed and analyzed for presence of the fungus by tissue-print immunoblot with antiserum raised against *E. coenophiala* protein [46].

6. Patents

Pending: Schardl, C.L.; Florea, S.; Farman, M.L. Fungal chromosome-end knockoff strategy. US 2017 /0349899 Al, 2017.

Supplementary Materials: The following are available online at https://www.mdpi.com/2072-6651/13/2/153/s1, Figure S1: PCR tests of putative CRISPR/Cas9-mediated deletion mutants, Table S1: Genome assembly statistics for wild-type *Epichloë* species, Table S2: Genome assembly statistics for CRISP/Cas9-mediated mutants.

Author Contributions: Conceptualization, C.L.S. and S.F.; methodology, S.F.; formal analysis, J.J.; investigation, S.F.; data curation, J.J.; writing, S.F. and C.L.S.; project administration, C.L.S.; funding acquisition, C.L.S. All authors have read and agreed to the published version of the manuscript.

Funding: This research was funded by the U.S. Department of Agriculture National Institute of Food and Agriculture Hatch project KY012044, and by the Mycological Society of America.

Acknowledgments: The authors thank Lusekelo Nkuwi for maintenance of plants, Alfred D. Byrd for technical support, and Neil Moore (University of Kentucky) for assistance in genomic data analysis.

Conflicts of Interest: The authors declare no conflict of interest.

References

1. Schardl, C.L.; Florea, S.; Pan, J.; Nagabhyru, P.; Bec, S.; Calie, P.J. The epichloae: Alkaloid diversity and roles in symbiosis with grasses. *Curr. Opin. Plant. Biol.* **2013**, *16*, 480–488. [CrossRef]
2. Schardl, C.L.; Young, C.A.; Pan, J.; Florea, S.; Takach, J.E.; Panaccione, D.G.; Farman, M.L.; Webb, J.S.; Jaromczyk, J.; Charlton, N.D.; et al. Currencies of mutualisms: Sources of alkaloid genes in vertically transmitted epichloae. *Toxins* **2013**, *5*, 1064–1088. [CrossRef]
3. Tanaka, A.; Takemoto, D.; Chujo, T.; Scott, B. Fungal endophytes of grasses. *Curr. Opin. Plant. Biol.* **2012**, *15*, 462–468. [CrossRef]
4. Moon, C.D.; Craven, K.D.; Leuchtmann, A.; Clement, S.L.; Schardl, C.L. Prevalence of interspecific hybrids amongst asexual fungal endophytes of grasses. *Mol. Ecol.* **2004**, *13*, 1455–1467. [CrossRef]
5. Schuster, M.; Kahmann, R. CRISPR-Cas9 genome editing approaches in filamentous fungi and oomycetes. *Fungal Genet. Biol.* **2019**, *130*, 43–53. [CrossRef] [PubMed]
6. Song, R.; Zhai, Q.; Sun, L.; Huang, E.; Zhang, Y.; Zhu, Y.; Guo, Q.; Tian, Y.; Zhao, B.; Lu, H. CRISPR/Cas9 genome editing technology in filamentous fungi: Progress and perspective. *Appl. Microbiol. Biotechnol.* **2019**, *103*, 6919–6932. [CrossRef] [PubMed]
7. Nødvig, C.S.; Nielsen, J.B.; Kogle, M.E.; Mortensen, U.H. A CRISPR-Cas9 system for genetic engineering of filamentous fungi. *PLoS ONE* **2015**, *10*, e0133085. [CrossRef] [PubMed]
8. Foster, A.J.; Martin-Urdiroz, M.; Yan, X.; Wright, H.S.; Soanes, D.M.; Talbot, N.J. CRISPR-Cas9 ribonucleoprotein-mediated co-editing and counterselection in the rice blast fungus. *Sci. Rep.* **2018**, *8*, 1–12. [CrossRef]
9. Malinowski, D.P.; Belesky, D.P. Adaptations of endophyte-infected cool-season grasses to environmental stresses: Mechanisms of drought and mineral stress tolerance. *Crop. Sci.* **2000**, *40*, 923–940. [CrossRef]
10. Malinowski, D.P.; Belesky, D.P. Epichloë (formerly Neotyphodium) fungal endophytes increase adaptation of cool-season perennial grasses to environmental stresses. *Acta Agrobot.* **2019**, *72*. [CrossRef]

11. Clay, K.; Schardl, C. Evolutionary origins and ecological consequences of endophyte symbiosis with grasses. *Am. Nat.* **2002**, *160*, S99. [CrossRef] [PubMed]
12. Aiken, G.E.; Klotz, J.L.; Johnson, J.M.; Strickland, J.R.; Schrick, F.N. Postgraze assessment of toxicosis symptoms for steers grazed on toxic endophyte-infected tall fescue pasture. *J. Anim. Sci.* **2013**, *91*, 5878–5884. [CrossRef]
13. Thompson, F.N.; Stuedemann, J.A.; Hill, N.S. Anti-quality factors associated with alkaloids in eastern temperate pasture. *J. Range Manag.* **2001**, *54*, 474. [CrossRef]
14. Klotz, J.L. Activities and effects of ergot alkaloids on livestock physiology and production. *Toxins* **2015**, *7*, 2801–2821. [CrossRef] [PubMed]
15. Campbell, M.A.; Tapper, B.A.; Simpson, W.R.; Johnson, R.D.; Mace, W.J.; Ram, A.; Lukito, Y.; Dupont, P.-Y.; Johnson, L.J.; Scott, B.; et al. *Epichloë hybrida*, sp. nov., an emerging model system for investigating fungal allopolyploidy. *Mycology* **2017**, *109*, 1–15. [CrossRef] [PubMed]
16. Anonymous. New ryegrass withdrawn after side-effects found. *Manawatu Evening Standard*, 16 October 1992; p. 1.
17. Schardl, C.L.; Panaccione, D.G.; Tudzynski, P. Chapter 2 Ergot Alkaloids—Biology and Molecular Biology. In *The Alkaloids: Chemistry and Biology*; Elsevier: Amsterdam, The Netherlands, 2006; Volume 63, pp. 45–86.
18. Tsai, H.F.; Wang, H.; Gebler, J.C.; Poulter, C.D.; Schardl, C.L. The claviceps purpurea gene encoding dimethylallyltryptophan synthase, the committed step for ergot alkaloid biosynthesis. *Biochem. Biophys. Res. Commun.* **1995**, *216*, 119–125. [CrossRef]
19. Wang, J.; Machado, C.; Panaccione, D.G.; Tsai, H.-F.; Schardl, C.L. The determinant step in ergot alkaloid biosynthesis by an endophyte of perennial ryegrass. *Fungal Genet. Biol.* **2004**, *41*, 189–198. [CrossRef]
20. Davis, N.D.; Clark, E.M.; Schrey, K.A.; Diener, U.L. In Vitro Growth of Acremonium coenophialum, an Endophyte of Toxic Tall Fescue Grass. *Appl. Environ. Microbiol.* **1986**, *52*, 888–891. [CrossRef] [PubMed]
21. Tsai, H.F.; Liu, J.S.; Staben, C.; Christensen, M.J.; Latch, G.C.; Siegel, M.R.; Schardl, C.L. Evolutionary diversification of fungal endophytes of tall fescue grass by hybridization with Epichloe species. *Proc. Natl. Acad. Sci. USA* **1994**, *91*, 2542–2546. [CrossRef] [PubMed]
22. Florea, S.; Andreeva, K.; Machado, C.; Mirabito, P.M.; Schardl, C.L. Elimination of marker genes from transformed filamentous fungi by unselected transient transfection with a CRE-expressing plasmid. *Fungal Genet. Biol.* **2009**, *46*, 721–730. [CrossRef]
23. Schardl, C.L.; Young, C.A.; Hesse, U.; Amyotte, S.G.; Andreeva, K.; Calie, P.J.; Fleetwood, D.J.; Haws, D.C.; Moore, N.; Oeser, B.; et al. Plant-Symbiotic Fungi as Chemical Engineers: Multi-Genome Analysis of the Clavicipitaceae Reveals Dynamics of Alkaloid Loci. *PLoS Genet.* **2013**, *9*, e1003323. [CrossRef]
24. Florea, S.; Phillips, T.D.; Panaccione, D.G.; Farman, M.L.; Schardl, C.L. Chromosome-End Knockoff Strategy to Reshape Alkaloid Profiles of a Fungal Endophyte. *G3 (Bethesda)* **2016**, *6*, 2601–2610. [CrossRef]
25. Florea, S.; Panaccione, D.G.; Schardl, C.L. Ergot Alkaloids of the Family Clavicipitaceae. *Phytopathology* **2017**, *107*, 504–518. [CrossRef] [PubMed]
26. Spiering, M.J.; Moon, C.D.; Wilkinson, H.H.; Schardl, C.L. Gene Clusters for Insecticidal Loline Alkaloids in the Grass-Endophytic Fungus *Neotyphodium uncinatum*. *Genetics* **2005**, *169*, 1403–1414. [CrossRef] [PubMed]
27. Zheng, T.; Hou, Y.; Zhang, P.; Zhang, Z.; Xu, Y.; Zhang, L.; Niu, L.; Yang, Y.; Liang, D., Yi, F., et al. Profiling single-guide RNA specificity reveals a mismatch sensitive core sequence. *Sci. Rep.* **2017**, *7*, 40638. [CrossRef]
28. Guo, T.; Feng, Y.-L.; Xiao, J.-J.; Liu, Q.; Sun, X.-N.; Xiang, J.-F.; Kong, N.; Liu, S.-C.; Chen, G.-Q.; Wang, Y.; et al. Harnessing accurate non-homologous end joining for efficient precise deletion in CRISPR/Cas9-mediated genome editing. *Genome Biol.* **2018**, *19*, 1–20. [CrossRef]
29. Lemos, B.R.; Kaplan, A.C.; Bae, J.E.; Ferrazzoli, A.E.; Kuo, J.; Anand, R.P.; Waterman, D.P.; Haber, J.E. CRISPR/Cas9 cleavages in budding yeast reveal templated insertions and strand-specific insertion/deletion profiles. *Proc. Natl. Acad. Sci. USA* **2018**, *115*, E2040–E2047. [CrossRef] [PubMed]
30. Davis, K.A.; Sampson, J.K.; Panaccione, D.G. Genetic reprogramming of the ergot alkaloid pathway of metarhizium brunneum. *Appl. Environ. Microbiol.* **2020**, *86*. [CrossRef]
31. Khan, H.; McDonald, M.C.; Williams, S.J.; Solomon, P.S. Assessing the efficacy of CRISPR/Cas9 genome editing in the wheat pathogen Parastagonspora nodorum. *Fungal Biol. Biotechnol.* **2020**, *7*, 4–8. [CrossRef]
32. Tsai, H.-F.; Siegel, M.R.; Schardl, C.L. Transformation of Acremonium coenophialum, a protective fungal symbiont of the grass Festuca arundinacea. *Curr. Genet.* **1992**, *22*, 399–406. [CrossRef] [PubMed]
33. Christensen, M.; Leuchtmann, A.; Rowan, D.D.; Tapper, B. Taxonomy of Acremonium endophytes of tall fescue (*Festuca arundinacea*), meadow fescue (*F. pratensis*) and perennial ryegrass (*Lolium perenne*). *Mycol. Res.* **1993**, *97*, 1083–1092. [CrossRef]
34. Florea, S.; Schardl, C.L.; Hollin, W. Detection and isolation of epichloë species, fungal endophytes of grasses. *Curr. Protoc. Microbiol.* **2015**, *38*, 19A.1.1–19A.1.24. [CrossRef] [PubMed]
35. Al-Samarrai, T.H.; Schmid, J. A simple method for extraction of fungal genomic DNA. *Lett. Appl. Microbiol.* **2000**, *30*, 53–56. [CrossRef]
36. Zimin, A.V.; Puiu, D.; Luo, M.-C.; Zhu, T.; Koren, S.; Marçais, G.; Yorke, J.A.; Dvořák, J.; Salzberg, S.L. Hybrid assembly of the large and highly repetitive genome of Aegilops tauschii, a progenitor of bread wheat, with the MaSuRCA mega-reads algorithm. *Genome Res.* **2017**, *27*, 787–792. [CrossRef]
37. Zimin, A.V.; Marçais, G.; Puiu, D.; Roberts, M.; Salzberg, S.L.; Yorke, J.A. The MaSuRCA genome assembler. *Bioinformatics* **2013**, *29*, 2669–2677. [CrossRef]

38. Bolger, A.M.; Lohse, M.; Usadel, B. Trimmomatic: A flexible trimmer for Illumina sequence data. *Bioinformatics* **2014**, *30*, 2114–2120 [CrossRef]
39. Schmieder, R.; Edwards, R.A. Quality control and preprocessing of metagenomic datasets. *Bioinformatics* **2011**, *27*, 863–864 [CrossRef] [PubMed]
40. Wu, X.; Kriz, A.J.; Sharp, P.A. Target specificity of the CRISPR-Cas9 system. *Quant. Biol.* **2014**, *2*, 59–70. [CrossRef]
41. Panaccione, D.G.; Johnson, R.D.; Wang, J.; Young, C.A.; Damrongkool, P.; Scott, B.; Schardl, C.L. Elimination of ergovaline from a grass-Neotyphodium endophyte symbiosis by genetic modification of the endophyte. *Proc. Natl. Acad. Sci. USA* **2001**, *98*, 12820–12825. [CrossRef]
42. Schmieder, R.; Edwards, R. Fast Identification and Removal of Sequence Contamination from Genomic and Metagenomic Datasets. *PLoS ONE* **2011**, *6*, e17288. [CrossRef] [PubMed]
43. Fletcher, L.; Sutherland, B. Sheep responses to grazing ryegrass with AR37 endophyte. *Proc. N. Z. Grassl. Assoc.* **2009**, *71*, 127–132. [CrossRef]
44. Latchs, G.C.M.; Christensen, M.J. Artificial infection of grasses with endophytes. *Ann. Appl. Biol.* **1985**, *107*, 17–24. [CrossRef]
45. Chung, K.-R.; Hollin, W.; Siegel, M.R.; Schardl, C.L. Genetics of Host Specificity in Epichloë typhina. *Phytopathology* **1997**, *87*, 599–605. [CrossRef] [PubMed]
46. An, Z.Q.; Siegel, M.R.; Hollin, W.; Tsai, H.F.; Schmidt, D.; Schardl, C.L. Relationships among non-Acremonium sp. fungal endophytes in five grass species. *Appl. Environ. Microbiol.* **1993**, *59*, 1540–1548. [CrossRef] [PubMed]

toxins

MDPI

Article

Evolution of the Ergot Alkaloid Biosynthetic Gene Cluster Results in Divergent Mycotoxin Profiles in *Claviceps purpurea* Sclerotia

Carmen Hicks [1,†], Thomas E. Witte [1,†], Amanda Sproule [1], Tiah Lee [1], Parivash Shoukouhi [1], Zlatko Popovic [2], Jim G. Menzies [2], Christopher N. Boddy [3], Miao Liu [1] and David P. Overy [1,*]

1 Ottawa Research & Development Centre, Agriculture & Agri-Food Canada, 960 Carling Ave., Ottawa, ON K1A 0C6, Canada; carmen.hicks@agr.gc.ca (C.H.); thomas.witte@agr.gc.ca (T.E.W.); amanda.sproule@agr.gc.ca (A.S.); tlee076@uottawa.ca (T.L.); parivash.shoukouhi@agr.gc.ca (P.S.); miaomindy.liu@agr.gc.ca (M.L.)
2 Morden Research and Development Centre, Agriculture & Agri-Food Canada, 101 Route 100, Morden, MB R6M 1Y5, Canada; zlatko.popovic@agr.gc.ca (Z.P.); jim.menzies@agr.gc.ca (J.G.M.)
3 Department of Chemistry & Biomolecular Sciences, University of Ottawa, Ottawa, ON K1N 6N5, Canada; chrisotpher.boddy@uottawa.ca
* Correspondence: david.overy@agr.gc.ca; Tel.: +1-613-759-1857
† Co-first author (C.H. & T.E.W.).

Abstract: Research into ergot alkaloid production in major cereal cash crops is crucial for furthering our understanding of the potential toxicological impacts of *Claviceps purpurea* upon Canadian agriculture and to ensure consumer safety. An untargeted metabolomics approach profiling extracts of *C. purpurea* sclerotia from four different grain crops separated the *C. purpurea* strains into two distinct metabolomic classes based on ergot alkaloid content. Variances in *C. purpurea* alkaloid profiles were correlated to genetic differences within the *lpsA* gene of the ergot alkaloid biosynthetic gene cluster from previously published genomes and from newly sequenced, long-read genome assemblies of Canadian strains. Based on gene cluster composition and unique polymorphisms, we hypothesize that the alkaloid content of *C. purpurea* sclerotia is currently undergoing adaptation. The patterns of *lpsA* gene diversity described in this small subset of Canadian strains provides a remarkable framework for understanding accelerated evolution of ergot alkaloid production in *Claviceps purpurea*.

Keywords: *Claviceps purpurea*; fungal plant pathogen; biosynthetic gene cluster; ergot alkaloids; mycotoxins; untargeted metabolomics; mass spectrometry; sclerotia

Key Contribution: Metabolomic analysis of Canadian *Claviceps purpurea* sclerotia extracts indicates there are two distinct chemical phenotypes associated with divergent ergot alkaloid profiles within the species. Genomic comparison of select strains links specific mutations within the ergot alkaloid biosynthetic gene cluster to the emergence of ergot alkaloid diversity.

Citation: Hicks, C.; Witte, T.E.; Sproule, A.; Lee, T.; Shoukouhi, P.; Popovic, Z.; Menzies, J.G.; Boddy, C.N.; Liu, M.; Overy, D.P. Evolution of the Ergot Alkaloid Biosynthetic Gene Cluster Results in Divergent Mycotoxin Profiles in *Claviceps purpurea* Sclerotia. *Toxins* **2021**, *13*, 861. https://doi.org/10.3390/toxins13120861

Received: 27 October 2021
Accepted: 26 November 2021
Published: 2 December 2021

1. Introduction

Ergot fungi of the genus *Claviceps* are phytopathogenic ascomycetes with the ability to parasitize over 400 monocotyledonous plant species, most notably affecting a large number of grasses and common cereals [1,2]. *Claviceps* infections are characterized by the formation of recalcitrant resting structures known as sclerotia. Sclerotia form as fungal hyphae invade unfertilized ovules of host plants via stigmas and pollen tubes during anthesis, replacing developing host seeds with compact masses of hardened fungal mycelium [3,4]. While dormant in the sclerotial state, the fungus is protected, ensuring its survival though environmental extremes until appropriate weather conditions are presented [5]. Fruiting bodies (ascomata) then develop on the sclerotium that forcefully eject ascospores into the air, which are then dispersed by the wind to infect flowering host plants, thus continuing the life cycle.

Ergot infection of cereal crops directly impacts grain quality and yield and has become increasingly problematic across Canadian provinces over the past two decades [6,7]. Ergot-infested grain must have sclerotia removed prior to shipping, or grain shipments become downgraded or rejected at the point of sale, resulting in decreased returns to farmers [6]. However, yield reduction is seen as a secondary importance when compared to the toxic effects of accidental consumption by humans and other animals [8]. Ergot sclerotia contain a wide variety of ergot alkaloids that constitute 0.5–2% of the entire sclerotium mass [9]. Sharing structural similarities to neurotransmitters, ergot alkaloids interact antagonistically or agonistically to α-adrenergic, dopamine, serotonin 5-HT and other receptors, impairing both psychological and motor functions [10]. Preventing or minimizing the occurrence of ergot alkaloids in our food and animal feed is, therefore, of great importance.

The ergot alkaloid biosynthetic pathway is well characterized (Figure 1), and associated biosynthetic gene clusters were identified from the genomes of several *Claviceps* spp. [11]. The presence or absence of various genes in the gene cluster has a direct impact on the diversity of ergot alkaloids produced by a species. Ergot alkaloids are generally divided into three subclasses: simple clavines, ergoamides (simple lysergic acid amides) and the more complex peptide-containing ergopeptides; however, all share the same initial biosynthetic pathway steps involved in the tetracyclic ring formation [12,13]. Ergoamide and ergopeptide biosynthesis initiates with LpsB, a unique monomodular non-ribosomal peptide synthetase (NRPS) that recognizes D-lysergic acid as a substrate, activates it as the AMP-ester and loads it on the LpsB carrier protein. Unique among NRPS pathway, the two downstream NRPSs, LpsA or LpsC, compete for LpsB binding and the subsequent elaboration of lysergic acid [14]. LpsC is a monomodular NRPS responsible for the incorporation of a single amino acid that produce the ergoamides. LpsA is a trimodular NRPS that incorporates three amino acids units through the progressive elongation to form ergopeptams, which the mono-oxygenase EasH then converts into the ergopeptines via oxidation and spontaneous cyclization [15]. Two homologs of *lpsA* encoding LpsA1 and LpsA2 are present in the biosynthetic gene cluster. Sequence variations in the second adenylation domain (A-domain) of the two LpsA enzymes are responsible for incorporation of phenylalanine or leucine in the ergopeptams and ergopeptines [16]. Characterization of an *lpsA1* deletion mutant in *C. purpurea* P1 suggests that LpsA1 is responsible for formation of the phenylalanine containing ergotamine and ergocristine ergopeptines [17].

In Canada, guidelines are placed on grain sclerotia content allowances that range from 0.01 to 0.1% depending on the host crop [18]. Previous studies showed that ergot count and weight are not predictive of diagnostically relevant ergot alkaloid concentrations [19] as ergot alkaloid constituents and concentrations within sclerotia are highly variable and directly dependent upon the producing species. *Claviceps purpurea* is documented as producing a structurally diverse range of ergopeptides due to the adenylation site promiscuity of LpsA and to the tandem duplication and neofunctionalization of the *lpsA* gene; these were recently denoted *lpsA1* and *lpsA2* [11,17]. In addition to increased alkaloid content and variability, *C. purpurea* has the widest reported host range of any *Claviceps* species, making it a major contributor of ergot alkaloid food contamination across Canada [8,20]. Due to the wide associated host range, *C. purpurea* is a species complex that has been the focus of taxonomic evaluation. Multigene phylogenetic studies have confirmed cryptic speciation within the taxon, recorded as G1–7, wherein G1 represents *C. purpurea sensu stricto* [21,22]. The European Food Safety Agency (Panel on Contaminants in the Food Chain) recently called for increased monitoring and improved knowledge on the relation between *C. purpurea* sclerotia and ergot alkaloids of concern, in particular: ergometrine, ergotamine, ergosine, ergocristine, ergocryptine (α- and β-isomers), ergocornine and their corresponding -inine (S)-epimers [23]. Information on variability of *C. purpurea* ergot alkaloid content across crop hosts is, therefore, required if knowledgeable regulatory decisions are to be made regarding acceptable levels of ergot alkaloids for livestock and human consumption [24].

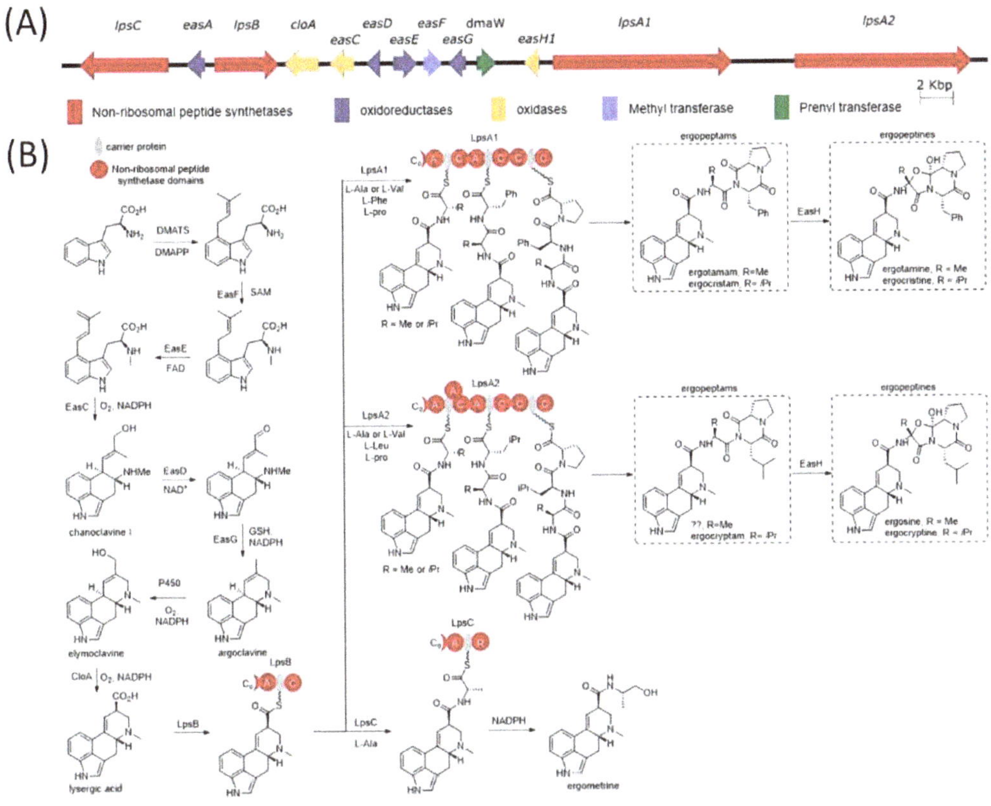

Figure 1. Lysergic acid and ergot alkaloid biosynthesis. (**A**) The organization of the ergot alkaloid biosynthetic gene cluster from *C. purpurea* 20.1. (**B**) The biosynthetic pathway for lysergic acid and the ergot alkaloids (C = condensation; A = adenylation; R = reductase domain; C_0 = C-terminal portion of a C domain).

In this study, we used UPLC-HRMS and an untargeted metabolomics approach to profile the alkaloid content of *C. purpurea* sclerotia from four different grain crops. Consistent trends in sclerotia ergot alkaloid content among grain crops separated the *C. purpurea* strains into two distinct metabolomic classes. *Claviceps purpurea* alkaloid profiles were correlated to genetic differences of the ergot alkaloid biosynthetic gene cluster from previously published isolate genomes and from newly sequenced, long-read genome assemblies of Canadian strains. Based on the gene cluster composition and unique polymorphisms, we hypothesize that the alkaloid content of *C. purpurea* sclerotia is undergoing adaptation that contributes to the diversity of ergopeptides produced.

2. Results

2.1. Untargeted Metabolomics Profiling

An initial unsupervised Euclidean Ward hierarchical clustering was performed individually for each host crop to reveal the underlying structure within the metabolomics data. Hierarchical clustering of the binary mass feature data showed a dichotomy in metabolite profiles amongst the isolates, separating the sclerotium extracts into two putative "metabolomic classes", which was consistent for all four host crops (Figure 2A, and

Figures S2A, S3A and S4A). Class 1 consisted of isolates LM04, LM207, LM232 and LM233, while Class 2 consisted of isolates LM30, LM60, LM469 and LM474. The results of the discriminant multivariate analyses supported the observed Class dichotomy in *C. purpurea* sclerotia. The top 30 ranking mass features with the greatest importance in the random forest model classification (based on both mean decrease accuracy and mean decrease Gini) were found to be highly conserved between host crops and demonstrated large fold changes between classes (Figure 2B, and Figures S2B, S3B and S4B). From the orthogonal partial least squares discriminant analysis (OPLS–DA) models generated for each host crop, the resulting S-plots separated the mass features associated with latent variable class separation (black dots in Figure 2C, and Figures S2C, S3C and S4C), where positive values on the Y-axis represent mass features with greater association to Class 1, and negative values are mass features with greater association to the Class 2 isolates. When overlaid on the OPLS-DA S-plot, the top 30 mass features derived from the random forest analysis were found to occur at opposite ends of the Y-axis (red and blue dots in Figure 2C), indicating a high correlation of the selected mass features with the formation of the observed classes in both discriminant multivariate analyses.

An untargeted consensus binary heatmap representing mass feature frequency of occurrence in sclerotia produced on the four host crops was constructed to further explore the overall mass feature diversity amongst the strains and between classes (Figure 2D). When hierarchical cluster analysis was applied, the resulting class dichotomy was consistent across individual plant crop hosts. Clear groupings of mass features were found to occur at high frequencies (on all four hosts) and were consistent between *C. purpurea* isolate classes. A larger section of mass features was found to be infrequently observed (occurring on a single host by a specific isolate) and likely resulted from inherent differences in crop host- and/or *C. purpurea* isolate-specific metabolism. Most notable, however, were the consistent production of class specific mass features, having a high frequency of occurrence on all four hosts but only occurring in sclerotia from either Class 1 or Class 2 *C. purpurea* isolates (highlighted in Figure 2D). These mass features were the same mass features designated as being responsible for class separation in the discriminant multivariate analyses derived from the non-binary mass feature data. Further investigation (see below) confirmed that these class-specific mass features represented metabolites derived from ergot alkaloid biosynthesis (Table S2).

2.2. Ergot Alkaloid Profiles Drive the Separation of Claviceps purpurea Isolate Classes

A molecular network was constructed to visualize ergot alkaloid/*C. purpurea* class associations using representative strains of both Class 1 and Class 2 sclerotium extracts from rye (Figure 3). Ergocristine/ergocristinine and ergotamine were the major ergopeptines produced and unique to LM60 (Class 2) sclerotia and ergovaline/ergovalinine, ergocornine/ergocorninine, ergocornam and ergocryptam were the major ergopeptines and ergopeptams produced and unique to LM04 (Class 1) sclerotia. Both ergosine and ergocryptine were also major metabolites produced by LM04, with minor production observed from LM60 sclerotium extracts from rye.

Clear and consistent trends in ergot alkaloid production were also evident in the frequency consensus heatmap (Figure 4). When produced, all of the ergopeptine- and ergopeptam-associated mass features were typically detected in sclerotium extracts from all four host crops surveyed. Trends observed in ergopeptine and ergopeptam production between *C. purpurea* classes were consistent with the trends observed from the molecular network. Ergotamine and ergocristine were the predominant ergopeptines having most abundant peak areas in Class 2 sclerotia and ergocryptine and ergocornine were the most abundant ergopeptines in Class 1 sclerotia. Both ergocornine (with the exception of LM30 and LM60) and ergocryptine and were detected in both Class 1 and 2 sclerotia from all four hosts; however, Class 2 sclerotia from all four hosts contained a significantly greater abundance of ergocornine and ergocyptine, as only very minor peak intensities were observed in Class 1 sclerotium extracts (LM469, LM474). Ergosine was observed in both

Class 1 and Class 2 sclerotia, with a greater abundance associated with Class 1 isolates on all four crop hosts. Finally, ergoamide mass features associated with ergometrine were also detected in all Class 2 sclerotia and only two strains of the Class 1 sclerotia, with only minor production observed from all strains on all four host crops. The two Class 1 strains that did not produce ergometrine (LM04 and LM232, Figure 4) were reported to have an internal stop codon within the *lpsC* gene, the NRPS responsible for biosynthesis of ergoamides such as ergometrine ([25]—this issue).

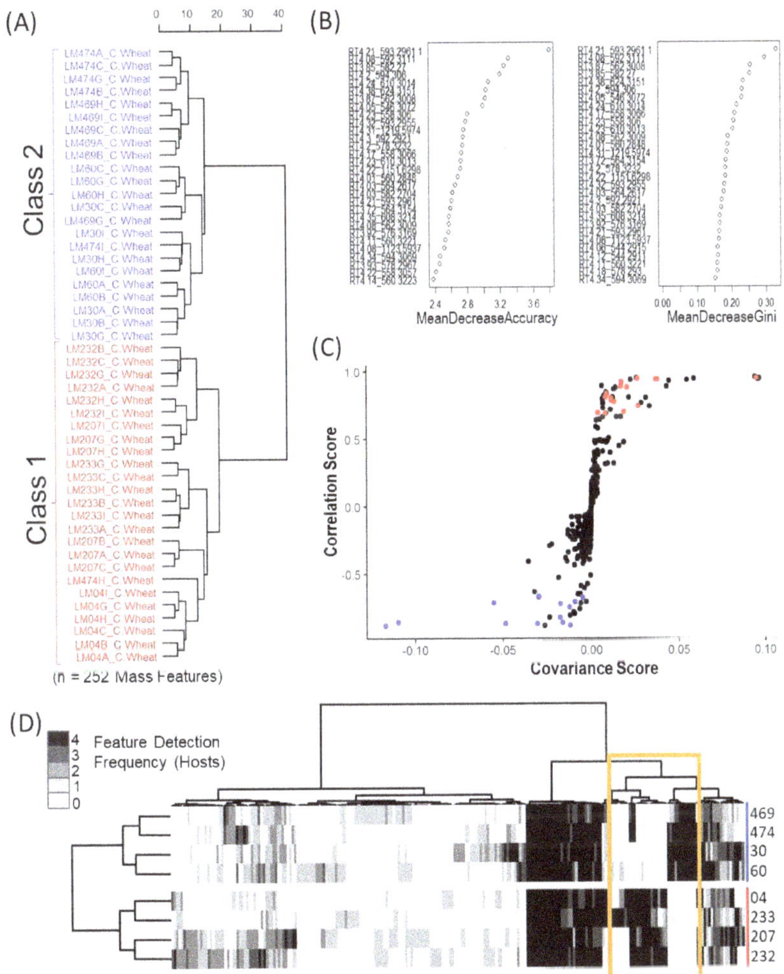

Figure 2. Untargeted analysis support two putative classes of *C. purpurea* strains grown on 'common' wheat. (**A**) Euclidean Ward hierarchical clustering of pseudo-binary (presence/absence) mass feature data of replicate sclerotia. Labels in Red and Blue represent specimens identified in Class 1 and Class 2, respectively, with (n = 6) replicates per specimen. (**B**) Top 30 mass features associated with class formation as determined via random forest analysis of raw data (left: mean decrease in accuracy; right: mean decrease in Gini of raw data of replicate sclerotia). (**C**) S-plot from OPLS-DA highlighting top 30 features identified by random forest analysis of replicate sclerotia, features in red and blue represent specimens identified in Class 1 and Class 2. (**D**) Consensus phenotype heatmap constructed using Ward D2 hierarchical clustering of Euclidean distances between all conserved mass features across the 4 hosts (n = 518 mass features). The mass features indicated in yellow are the likely drivers of this dichotomy.

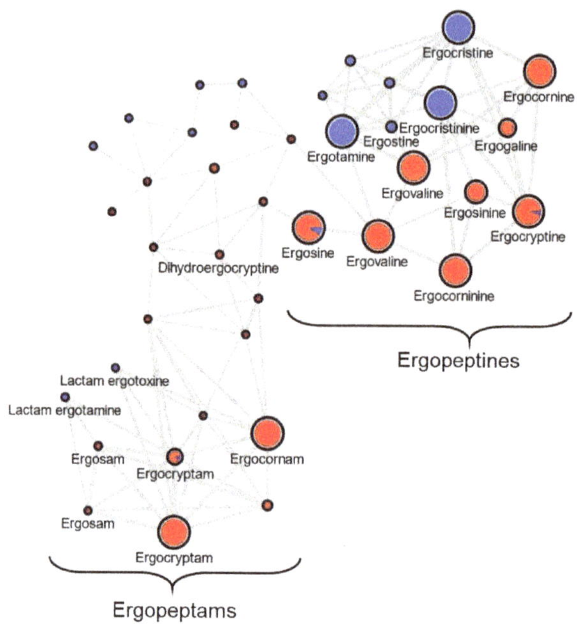

Figure 3. Molecular networking map of ergot alkaloids detected in sclerotium extracts from LM04 (Class 1—red) and LM60 (Class 2—blue) and grown on rye. The sizes of the pie charts reflect the intensity of the precursor ions (summed) and pie chart ratio is calculated from the m/z peak areas. Lines connect nodes if cosine scores comparing MS^2 scans are above 0.7, with thickness increasing as the cosine score increases. Some annotations are duplicated where the MS^2 analysis was unable to differentiate putative stereoisomers.

Major ergopeptine signals from the Class 2 group were almost universally not detected in Class 1 isolate extracts, whereas the ergopeptines that were characteristic of Class 1 isolate extracts were consistently detected in the Class 2 extracts, albeit at very low intensities. When the molecular structures of the ergopeptines and ergopeptams associated with the Class dichotomy are compared, a clear trend is observed. Ergopeptines and ergopeptams in Class 1 lack phenylalanine incorporation at the second amino acid position (aa2) of the tripeptide moiety.

2.3. *Ergot Alkaloid Gene Clusters Show Increased Rates of Mutation Associated with LpsA Gene Modules*

A comparison of the ergot alkaloid biosynthetic gene clusters from LM04 and LM60, selected as representative isolates from Classes 1 and 2, respectively, as well as *C. purpurea* isolate 20.1 [26] and LM72 ([25]—this issue), revealed a higher degree of polymorphism and repetitive DNA elements surrounding the *lpsA* genes (Figure 5A). Higher proportions of inter-strain polymorphisms were localized to specific *lpsA* modules, showing as much as 20% polymorphic sites (Figure 5A, red line graphs), well above the 1.5% 'background' polymorphic frequency calculated by comparison of two other NRPS genes, *lpsB* and *lpsC*, present in the cluster. Patterns of site divergence between *lpsA* genes correlated with the metabolomics data in two ways. First, we detected a frameshift mutation caused by an indel in the sequence encoding the second condensation domain of LM04 *lpsA2* (Figure 5A) that prematurely terminates the encoded NRPS. As such, only LpsA2 is not expected to be expressed as a functional full length NRPS in the Class 1 strain LM04. Second, the sequence encoding the second adenylation, carrier protein and condensation domains of *lpsA1* in LM04 had approximately 18% polymorphic sites compared to the Class 2 strain LM60 as well as LM72 and *C. purpurea* 20.1. These polymorphisms included missense mutations that change the A-domain's 'specificity code', which is predictive of the domain's amino acid specificity during NRPS biosynthesis [27] (Table S3).

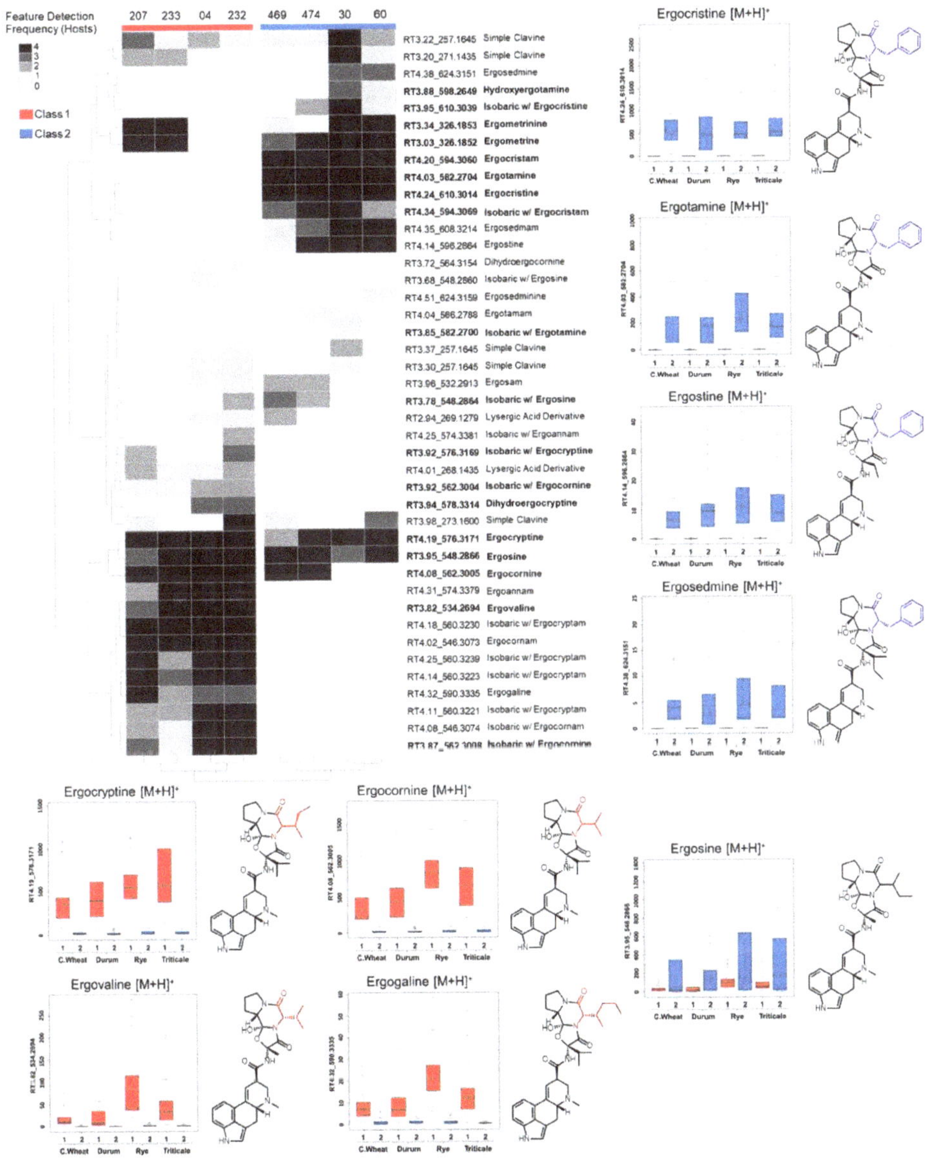

Figure 4. Consensus phenotype heatmap constructed using hierarchical clustering of representative mass features of ergot alkaloids and related derivatives demonstrating the frequency of occurrence and dichotomy in *C. purpurea* classes. Boxplots comparing frequency of occurrence by host crop of mass features accompanied with associated ergopeptine. In the boxplots, outliers are represented by black ovals; key structural differences in the ergopeptine aa2 residue are coloured by class association (blue = Class 1, red = Class 2).

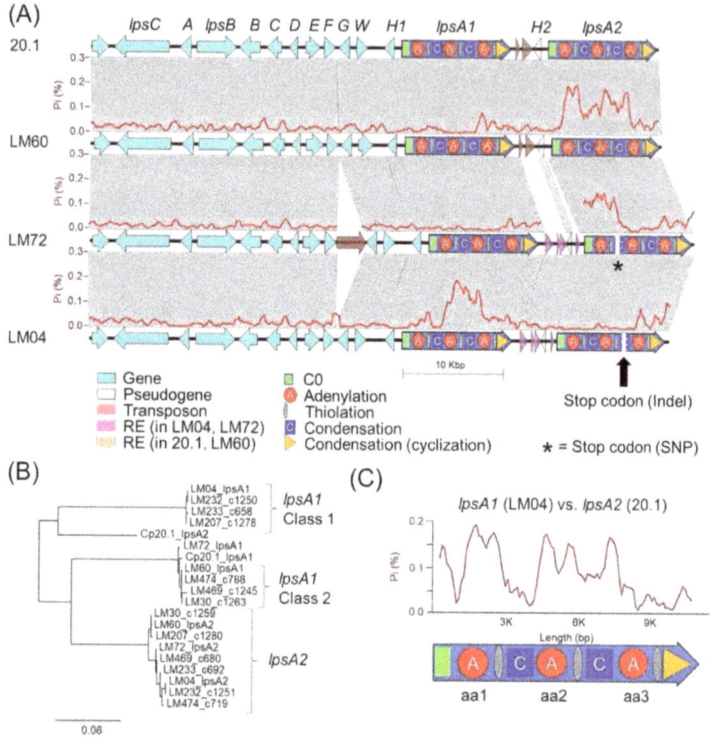

Figure 5. (**A**) Synteny analysis between ergot alkaloid biosynthetic gene clusters in *C. purpurea* strains 20.1, LM60, LM72 and LM04. Dark blue genes represent *lpsA* NRPSs, all other genes are labelled in light blue (see Figure 1 for gene classifications). Gene letters have the prefix '*eas*' removed for brevity, with the exception of *B*, which represents *cloA*, and *W*, which represents *dmaW*. Transparent arrows represent repetitive elements. Grey blocks represent syntenic DNA above 88% nucleotide identity, with the exception of the block connecting *easH2* pseudogenes located between LM60 and LM72, which have only 50% nucleotide identity. The red line graph overlaid onto the syntenic blocks represents a running average of the proportion of polymorphic DNA sites as calculated in 100 bp stepwise increments of a 500 bp sliding window, in each instance comparing the sequences above and below, and should be interpreted as aligning to the track below. NRPS domains are overlaid on *lpsA* genes that were predicted by antiSMASH 6.0. A yellow diamond is placed where a frameshift mutation was predicted to have disrupted the second condensation domain of lpsA2 in LM04 via premature stop codon insertion. (**B**) A phylogenetic comparison of the nucleotide sequence encoding the second A-domain (aa2) of all *lpsA* genes from all strains included in this study. The tree shows the clustering of all aa2 A-domains from predicted *lpsA1* genes correlates to the metabolomic clustering of strains identified as 'Class 1' vs. 'Class 2'. (**C**) Comparison of polymorphic site frequencies between *lpsA2* of *C. purpurea* strain 20.1 and *lpsA1* of LM04.

Sequence analysis of the second A-domains from LM04 *lpsA1* shows it encodes residues that most closely match the leucine 'specificity code' identified by Stachelhaus et al. [27,28]. While substrate specificity predictions of A-domains in bacterial NRPS systems are fairly robust, they are known to be less so in fungal NRPS systems. Fungal A-domain predictions are often limited to gross physicochemical properties [28] and, in some cases, they fail entirely [29]. Together, these observations provide a genetic framework for understanding the metabolomic patterns detected in strains from Class 1 and Class 2. The Class 1 strain LM04 has only one functional copy of *lpsA*, *lpsA1*, which encodes substrate specificity in the second A-domain of LpsA1 for leucine and other branched hydrophobic amino acids rather than phenylalanine, leading to the formation of the observed Class 1 ergopeptines. This contrasts with the documented phenylalanine selectivity of LpsA2 in *C. purpurea* P1 [17].

To verify whether LpsA1 second A-domain sequences, from other strains grouped in Class 1, contained a similar leucine selective A-domain, we extracted the corresponding protein sequences from *lpsA1* and *lpsA2* from the long-read genomes compared in this study and previously published Illumina short-read genomes for the *C. purpurea* strains LM30, LM60, LM207, LM461, LM232, LM233, LM469 and LM474 [30]. The short-read genomes were fragmented at the *lpsA* regions; however, the second A-domain encoding sequences appeared intact. A phylogenetic tree of the aligned sequences (Figure 5B) supports the presence of two divergent *lpsA1* lineages correlating to Class 1 and Class 2 strains. This result supports the hypothesis that all Class 1 strains inherited a divergent second A-domain sequence with substrate specificity for leucine rather than phenylalanine. Additionally, we observed that the *lpsA2* second A-domain of the *C. purpurea* strain 20.1 was highly divergent from all of the other A-domains included in the analysis. It also lacked the predicted specificity for phenylalanine. However, the high levels of sequence divergence between *C. purpurea* 20.1 *lpsA2* and LM04 *lpsA1* suggests that it is unlikely that these two altered A-domain share a common origin (Figure 5C).

Following the prediction of a frameshift mutation in LM04 *lpsA2*, we looked for evidence of *lpsA2* disruption in the genomes of other Class 1 strains to explore the possibility that this lineage has consistent pseudogenization of *lpsA2*. The same C2-domain frameshift mutation was detected in one other genome, LM232 (Figure S5). Alignment of the fragments of putative *lpsA1* and *lpsA2* genes from LM207 and LM233 to the LM04 genes showed no obvious signs of disruption; however, we note that the Illumina assemblies from LM207 and LM233 are fragmented and would benefit from long-read sequencing to thoroughly investigate this possibility. Although LM72 was not selected for metabolomic studies, our genome analysis indicated there is a premature stop codon in *lpsA2* in this strain, disrupting a condensation domain (Figure 5A).

2.4. Transposons and Transposon Fossils in the LpsA1/LpsA2 Region

The intergenic regions between *lpsA1* and *lpsA2* contained numerous small (200 1000 bp) sequences that are duplicated in other regions in the respective genomes, including one that is inserted into the remnants of the truncated and pseudogenized *easH2* in LM04. This suggests these sequences are potentially transposable element fossils. Repetitive elements in the *C. purpurea* 20.1 and LM60 intergenic regions show some similarity to repetitive elements annotated as DNA transposons of the 'MULE-MuDR' type, whereas repetitive elements in the LM72 and LM04 genomes show similarity to the 'TcMar-Tc1' type DNA transposons annotated in the LM04 genome. The *easH2* copies, which were truncated compared to *EasH1* and predicted to be pseudogenized, had low sequence similarity, approximately 50% nucleotide identity between shared sequence lengths, and between 20.1/LM60 and LM72/LM04 isoforms. A putative transposon in LM72 between *easF* and *easG* was observed. As this study did not include metabolomic analysis of LM72 sclerotia, we did not pursue the potential effects of this putative transposon on ergot alkaloid production.

3. Discussion

In this study we classified Canadian *C. purpurea sensu stricto* strains into two broad categories based on ergot alkaloid profiles. We arbitrarily designated these groups as 'Class 1' and 'Class 2', and showed that the classes are dominated by either aliphatic hydrophobic residue-incorporating or phenylalanine-incorporating ergot alkaloids, respectively. Our identification of diversified intra-species ergot alkaloid profiles is generally consistent with previously published analyses of *C. purpurea* strains. Historically, ergopeptide production in *C. purpurea* has been quite diverse and varied between geographic locations and broader population groupings. A prominent early example of this trend compared strain 'P1', which was found to produce primarily ergotamine and ergocryptine, with strain 'Ecc93', a producer primarily of ergocristine [16]; however, we note the lineage of Ecc93 may need to be further investigated in light of newer *Claviceps* nomenclature [21]. *Claviceps purpurea* strain 20.1 appears to have a similar ergot alkaloid profile as strain P1, reportedly producing ergotamine, ergocryptine and ergometrine as major products [26]. The presence of divergent ergot alkaloid profiles was also previously described from broader genetic groupings of *C. purpurea sensu lato* (G1–G4); however, the profiles of *Claviceps purpurea sensu stricto* (G1) were more difficult to define [31,32]. For example, early work by Pažoutová et al. [31] described *C. arundinis* (G2a) sclerotia as consistent producers of ergocristine, ergosine and small amounts of ergocryptine; however, the G1 strains showed highly variable ergot alkaloid presences and relative abundances. More recent work by Negård et al. [2] showed that Norwegian *C. purpurea* G1 sclerotia contained a complex mixture of ergotamine, ergosine, ergocornine, ergocryptine and ergocristine as major constituents. In light of the present study, we suggest that the profiles generated by Pažoutová and Negård indicate that G1 strains have diverse LpsA A-domain specificities, which results in a broad spectrum of chemo-phenotypes. More recently, a detailed high resolution LCMS analysis of ergot alkaloid profiles from various *Claviceps* species was published, which included six strains of *C. purpurea* isolated from Canadian cereals [33]. The metabolomic classes we defined here broadly correlate with their results: four of the six *C. purpurea* strains' profiles exhibited abundant ergotamine, ergocristine and ergostine mass feature peak areas (consistent with Class 2), and one strain showed abundant ergocornine, ergocryptine and ergovaline peak areas (consistent with Class 1). Taken together, our results contribute to a continuing understanding of the diversity detected in this species, and highlight the tension inherent in classifying strains based on chemotaxonomic vs. genetic criteria.

Our genetic analysis of the ergot alkaloid gene clusters from long-read genomes provides a framework to understand the underpinnings of chemical phenotype diversity within the *C. purpurea* G1 group. The ergot alkaloid biosynthetic gene clusters extracted from long-read assemblies of LM04 and LM60, chosen as representatives from the Class 1 and Class 2 metabolomic groupings, support a hypothesis that the class-specific ergot alkaloid profiles result from sequence diversity in the *easH*/*lpsA* tandem-duplicated region [16,17]. We predict that these mutations cause the shifting of substrate specificities of the *lpsA1* second A-domain to favor leucine and related aliphatic hydrophobic substrates such as isoleucine and valine, and additionally propose that *lpsA2* has been pseudogenized, at least in LM04, restricting ergotamine biosynthesis to interactions between LpsB and LpsA1. Given that the specific frameshift mutation described above is only detected in the genomes of LM04 and LM232, but not the other Class 1 strains LM233 or LM207, it is possible that the frameshift is not a causal feature for the metabolomic shift away from phenylalanine-incorporating ergot alkaloids across all Class 1 strains, but rather, could be a symptom of an overall relaxed selection for LpsA2 functionality due to some other mechanism. Importantly, this relaxed functionality is potentially not limited to Class 1 strains, but may also be present in Class 2 strains, for which both the *lpsA1* and *lpsA2* second A-domains are predicted to have phenylalanine specificity. The identification of a premature stop codon due to a single nucleotide polymorphism in LM72 *lpsA2* supports this hypothesis. Although LM72 has yet to associated with a given metabolomic class, we hypothesize it is in Class 2 based on A-domain selectivity predictions. The lack of detected phenylalanine-containing ergotamines in Class 1 metabolomes strongly suggests that the *lpsA2* gene is nonfunctional at least in those strains. Notably, this pattern departs from the known ergot alkaloid biosynthetic profiles of *C. purpurea* P1, in which researchers found that the *lpsA1* and *lpsA2* genes were both active, and the ergot alkaloid output of strains with *lpsA1* knocked out lost the ability to produce phenylalanine containing ergotamine but still produced leucine containing ergocryptine [17,34].

The ergot alkaloid chemical phenotype patterns observed in this study indicate that a representative portion of the Canadian populations of *C. purpurea* produce either exclusively short-chain alkyl group-bearing products (Leu, Ile, Val), or primarily phenylalanine-incorporating products with trace amounts of short-chain alkyl group-bearing products. At this time, the advantages of either phenotype on plant colonization, if any, are unknown, limiting the utility of speculation on the selective pressures shaping the evolution of *lpsA*. One important question that remains unanswered is: what are the origins of the mutations? A larger population of genomic and metabolomic data is likely needed to form a robust hypothesis; however, the presence of transposable elements with similarity to the MULE and TcMar families of DNA transposons detected in the intergenic space of the *lpsA* genes strongly suggests that the TE-mediated transposition or mutations associated with TE insertions are potential contributors to the *lpsA* diversity. Notably, the repetitive element-rich, highly polymorphic *lpsA1/lpsA2* intergenic spaces associated with either *C. purpurea* 20.1/LM04 or LM60/LM72 (Figure 5) are tightly linked with divergent *lpsA2* C0 domain sequences (Figure S5), thereby linking the presence of repetitive element insertions to nearby gene-coding sequence changes. Although the pattern of linked intergenic and *lpsA2* C0 domains is reflected in all genomes included in this study, the resulting groupings do not perfectly correlate with the assigned metabolomic classes of the strains. The domain-specific, highly variable proportions of *lpsA1/2* mutations detected when comparing strains (for example, see the comparison of *lpsA1* between LM04 and LM72, Figure 5) suggests that *lpsA* genes are likely undergoing recombinational shuffling [35,36]. Our analysis highlights the highly dynamic nature of the tandem-duplicated *easH*/*lpsA* region and the nature of ergot alkaloid NRPS module-specific evolution.

All ergot alkaloids contribute to and are responsible for a broad spectrum of biological activities as they can act as agonists, partial agonists and antagonists of multiple receptor sites, located throughout the body in various tissue types [37]. Therefore, the effects of ergot alkaloids on mammalian systems such as humans and livestock (i.e., cattle, sheep, horses and goats) are diverse, where, in the case of livestock, observed symptomatic responses to ergot alkaloid exposure can be highly variable and, therefore, historically difficult to diagnose [38]. Gangrenous ergotism occurs most frequently from acute ergot alkaloid exposure, caused by general blood vessel vasoconstriction and dysfunction resulting in tissue necrosis (dry gangrene) of the extremities such as the ear tips, tail, lower limbs and hooves [37–39]. In other instances, the consumption of subtle quantities of—and the prolonged exposure of livestock to—ergot alkaloids negatively impacts energy metabolism, feed efficiency and livestock productivity, as evidenced by decreases in food intake, live weight gain, circulating prolactin, reproductive performance, milk production and hyperthermia [37–39]. Pregnant and lactating animals are most susceptible to ergot alkaloid exposure due to increased risk of abortion and agalactia syndrome (the absence or failure to secrete milk); there is a direct correlation of ergopeptide exposure with a decrease in prolactin, growth hormone and luteinizing hormone secretion, and the inhibition of milk production [37]. Increasing our current understanding regarding the variability in production of ergot alkaloid content in *C. purpurea* sclerotia derived from different cereal crops used for food and feed production is, therefore, of continued relevance.

Ergot alkaloid profiles from *Claviceps* sclerotia consist of complex mixtures of minor stereoisomers, constitutional isomers and transient intermediate products, resulting in numerous peaks with the same m/z being detected, often at very low abundances when compared to the dominant m/z peak [31]. This was reflected in our analysis as numerous mass features were observed that shared duplicated isobaric m/z with ergotamine, ergocornine, ergocryptine and ergosine isomers, as well as isomers of other lactam and precursor molecules. However, many of these isobaric mass features were found to have random occurrences across strains and plant crop hosts, often occurring at a low abundance just above mass detection thresholds. The main ergoamides and ergopeptides of regulatory concern [23], namely ergometrine, ergotamine, ergosine, ergocristine, ergocryptine and ergocornine, when produced, were found to occur in *C. purpurea* sclerotia regardless of host crop. Of the *C. purpurea* strains examined, ergotamine and ergocristine were consistently the most abundant ergopeptides observed in Class 2 sclerotia, and ergocornine and ergocryptine were the most abundant in Class 1 sclerotia, from all four host crops tested. Based on the limited sample size, these four ergopeptides are likely to be of greatest concern in food and feed commodities derived from rye, triticale, durum wheat and 'common' wheat; however, a more extensive sampling of these crops using a larger breath of *C. purpurea s.s.* isolates will first be required to bring credence to this assumption.

4. Conclusions

The patterns of *lpsA* gene diversity described in this small subset of Canadian strains provides a remarkable framework for understanding the accelerated evolution of ergot alkaloid profiles. The results of this study provide insight into the variation in alkaloid content across Canadian *C. purpurea* isolates that could help guide future experiments in the exploration of secondary metabolite production. This research is crucial for providing developments in the understanding of the toxicological impacts of *Claviceps* species in order to improve consumer safety. *Claviceps purpurea* alkaloid content is an important topic of investigation because of its recognition as a major impactor of Canadian crops. The eight *C. purpurea* isolates that were analyzed within this study show clear differentiation of ergot alkaloid production. To impact the current regulatory mandate, the next steps would involve analyzing a larger sample size, with isolates across a broader range of hosts and host replicates as well as a broader geographic distribution to develop a better understanding of metabolomic class distribution. An additional recommendation that can be made based on the dichotomy in the observed ergot alkaloid production from *C. purpurea s.s.* isolates is that care should be observed when extrapolating taxonomic inferences based solely on ergot alkaloid profiles from sclerotia of unknown origin.

5. Materials and Methods

5.1. Strain Selection and Greenhouse Inoculation

Eight isolates of *C. purpurea* obtained from different Canadian provinces were selected from the Canadian Fungal Culture Collection (DAOMC) and the Liu lab research collection for in planta inoculation and untargeted metabolomic profiling (Table S1). From each isolate, representative sequences of the *RPB2* and *TEF1-α* protein coding genes were obtained, concatenated, and used to construct a maximum parsimony analysis with various *Claviceps* spp.; all eight selected isolates were classified as *C. purpurea s.s.* (Figure S1). In planta inoculations were completed on four different crop species: rye (*Secale cereale* L.), the 'common' wheat cultivar AC Cadillac (*Triticum aestivum* L.), durum wheat (*Triticum durum* Desf.) and triticale (*x Triticosecale* Wittmack). *Claviceps purpurea* isolates were grown on potato dextrose broth to obtain spores and spore viability was checked by plating onto malt extract agar. Viable spores were inoculated at a concentration of 10,000 conidia per mL on 20 florets per spike with 3 replicates per host species. Five healthy spikelets on each side of the spike were selected and the primary and secondary florets were filled with the conidial suspension using a syringe and hypodermic needle [40]. The sclerotia were collected at crop maturity when the infection was successful. Greenhouse inoculation trials yielded sufficient *C. purpurea s.s.* sclerotia (n = 6/host, with the exception of a single isolate LM60, which only yielded three sclerotia when inoculated on durum wheat) from the four cereal host plants. Each harvested sclerotium were cut into 5 mg fragments, with one fragment subjected to DNA extraction and *easE* gene sequencing to confirm fungal identity and the second fragment reserved for alkaloid profiling.

5.2. Isolate Identity Confirmation

Genomic DNA was extracted from sclerotium fragments using a Macherey–Nagel NucleoMag 96 Trace Kit (Macherey–Nagel, Düren, Germany) following a modified protocol as described [22]. Polymerase chain reaction (PCR) amplification was performed on all extracted gDNA samples in 10 μL reactions. Final concentrations of 1× Titanium Taq buffer (with 3.5 mM $MgCl_2$), 0.1 mM dNTPs, 0.08 μM of both forward and reverse primers, 1× Titanium Taq polymerase (Clontech, Mountain View, CA, USA) and 0.01 mg bovine serum albumin (BSA) were obtained with 1 μL of DNA template. FAD-linked oxidoreductase *easE*, a single copy ergot alkaloid synthesis gene, was amplified using designed primers: easE996f and easE1895r [22]. PCRs were run on a Mastercycler Pro S (Eppendorf, Mississauga, ON, Canada) using a touchdown protocol with initial denaturation at 95 °C for 3 min, followed by 5 cycles of 95 °C for 1 min, annealing at 63 °C (decrease of 1 °C per cycle) for 45 s and extension at 72 °C for 1 min 30 s, and then 30 cycles of 95 °C for 1 min, annealing at 58 °C for 45 s and extension at 72 °C for 1 min 30 s, with a final extension at 72 °C for 8 min. The PCR products were visualized on a 1% agarose gel with ethidium bromide treatment for 30 min at 200 V.

PCR products were amplified for Sanger sequencing using the ABI BigDye Terminator 3.1 cycling sequencing kit in a reaction volume of 10 μL, with BigDye Seq Mix diluted 1:8 with Seq buffer (Thermo Fisher Scientific, Ottawa, ON, Canada). The final volumes of each reagent were as follows: 1.75 μL of 5× sequencing buffer, 0.5 μL of BigDye Seq Mix and 0.5 μL of 3.2 μM primer, with reaction volumes increased to 9 μL using sterile high-performance liquid chromatography (HPLC) water. One microlitre of PCR product was added directly from the initial PCR amplification. The thermocycler profile used for the sequencing reactions had an initial denaturation step at 95 °C for 3 min, followed by 40 cycles at 95 °C for 30s, annealing at 58 °C for 15 s and extension at 60 °C for 3 min. An Applied Biosciences Prism 3130xl Genetic Analyzer (Life Technologies, Streetsville, ON, Canada) was used to generate DNA sequences and chromatograms. DNA sequences were edited and aligned to reference sequences in Geneious 11.1.5 to confirm identity. Once identity was confirmed, additional PCR and sequencing reactions were completed; DNA-directed RNA polymerase II subunit (*RPB2*) and translation elongation factor 1-α (*TEF1-α*) were amplified and sequenced using the above methods for phylogenetic construction.

5.3. Phylogenetic Analysis

A total of 43 reference sequences were downloaded and 2 outgroups were downloaded for each sequence, which included both the second largest subunit of the RNA polymerase II (*RPB2*) and elongation factor 1-α (*TEF1-α*) genes, and were aligned individually using MAFFT version 7 (online service https://mafft.cbrc.jp/alignment/server/ accessed on 22 January 2020). The Auto alignment strategy (FFT–NS–1, FFT–NS–2, FFT–NS–i or L–INS–i; depends on data size) was chosen. The alignments were visualized, verified and the two genes were concatenated using Geneious 11.1.5. Parsimony analysis was completed using PAUP* 4.0b10 [41]; heuristic searches with 50 replicates of random stepwise addition and tree bisection-reconnection branch swapping were conducted with a limit of 1,000,000 re-arrangements set for each replicate. Bootstrapping analyses were set with 100 replicates with full heuristic search of random stepwise addition of 50 replicates and a limit of 1,000,000 rearrangements per replicate.

5.4. HRMS Profiling of Sclerotium Extracts

Sclerotium fragments were individually frozen (−80 °C) and pulverized before being transferred into a 2 mL HPLC amber vial and extracted in 300 μL acetone:water (4:1, *v:v*) on a rotary shaker for 1 h (100 rpm). Extracts were centrifuged to pellet debris and 50 μL of the supernatant was transferred into a new HPLC vial for UPLC-HRMS profiling. Sclerotium extracts were profiled in a randomized injection order with solvent blanks interspersed between every six samples to assess for sample carryover and to aid in data curation during metabolomics processing. A reserpine standard was injected at the beginning of each sample sequence to confirm accurate calibration of the mass spectrometer and to aid in data alignment. Chemical profiling was completed using a Thermo Ultimate 3000 UPLC coupled to a Thermo LTQ Orbitrap XL HRMS and an UltiMate Corona VeoRS charged aerosol detector (Thermo Fisher Scientific Inc, Waltham, MS, USA). Chromatography was performed on a Phenomenex C_{18} Kinetex column (50 mm × 2.1 mm ID, 1.7 μm) with a flow rate of 0.35 mL/min, running a gradient of H_2O (+0.1% formic acid) and ACN (+0.1% formic acid). The gradient started at 5% ACN, increased to 95% ACN over 4.5 min, was held at 95% ACN until 8.0 min, returned to 5% ACN by 9 min and was left to equilibrate for until 10 min before the next injection. The HRMS was operated in ESI$^+$ mode (with a 100–2000 m/z range) using the following parameters: sheath gas (40), auxiliary gas (5), sweep gas (2), spray voltage (4.2 kV), capillary temperature (320 °C), capillary voltage (35 V) and tube lens (100 V). MS^n fragmentation was performed in high resolution on select ions in subsequent experiments using CID at 35 eV.

Additional experiments were performed to investigate the alkaloid diversity from LM04 and LM60 sclerotium extracts derived from rye. Samples were filtered through 0.2 μm PTFE membrane filters prior to analysis by nanoLC coupled to the Q-Exactive Plus mass spectrometer (Thermo Fisher Scientific). Chromatographic separation of metabolites was performed on a Proxeon EASY nLC II System (Thermo Fisher Scientific) equipped with a Thermo Scientific™ Acclaim™ PepMap™ RSLC C18 column (P/N ES800A), 15 cm × 75 μm ID, 3 μm, 100 Å, employing a H_2O/ACN gradient (with 0.1% formic acid). Chromatography ran for 60 min at a flow rate of 0.25 μL/min: initiated with a linear gradient from 10 to 100% of ACN for 45 min, held at 100% ACN until 50 min, then decreased from 100 to 10% of ACN by 55 min and held at 10% until 60 min to equilibrate to starting conditions. The mass spectrometer used positive electrospray ionization (ESI) at an ion source temperature of 250 °C and an ionspray (Thermo Scientific™ EASY spray) voltage of 2.1 kV. The FTMS scan type was full MS/data-dependent (dd)-MS^2. The parameters of the full mass scan were as follows: a resolution of 70,000, an auto gain control target under 3×10^6, a maximum isolation time of 100 ms and an m/z range of 100–1500. The parameters of the dd-MS^2 scan were as follows: a resolution of 17,500, an auto gain control target under 1×10^5, a maximum isolation time of 100 ms, a loop count of top 10 peaks, an isolation window of m/z 2, a normalized collision energy of 35 and a dynamic exclusion duration of 10 s.

5.5. Metabolomics: Data Pre-Processing

UPLC–HRMS profiles of sclerotium extracts were compiled into a representative data matrix by converting metabolite mass spectral data into mass features (consisting of a retention time and a mass/charge ratio; RT and m/z), where each metabolite was represented by one or more mass features associated with various pseudomolecular ions such as protonated mass, salt adducts, neutral loses and charged fragments. UPLC-HRMS profiles of *C. purpurea* sclerotium extracts from the four grain crops were preprocessed together. Preprocessing of the acquired Xcalibur raw data files was carried out using MZmine 2.53 [42]. Masses were detected with a noise threshold of 1×10^4. Chromatogram building was completed with the ADAP algorithm using a minimum group size of 6, a group intensity threshold of 1×10^6 and a minimum highest intensity of 5×10^6. RT and m/z tolerances were consistently set to 0.01 min and 0.005 m/z (or 5 ppm), respectively, throughout processing. Chromatogram deconvolution was carried out using the ADAP wavelets algorithm with a signal-to-noise threshold of 10, a minimum feature height of 2×10^6, a coefficient/area threshold of 110, a peak duration rage of 0.00–0.50 and an RT wavelet range of 0.00–0.03. Isotopic peaks were then removed followed by alignment of peaks using the JOIN aligner method and a 20:10 ratio for m/z to RT weight. Peaks missing from the data matrix were then back filled using the same RT and m/z range gap-filling algorithm.

5.6. Metabolomics: Data Reduction

The resulting data matrix was exported from MZmine as a csv file and imported into R Studio where all mass features were normalized by mean. Mass features with retention times below 1 min and after 6 min were also removed due to the amplification of background noise at these time points. Prominent mass features occurring above cut-off thresholds in solvent blanks were subtracted from all feature values for a given mass feature to reduce false positives as well as to remove "system peaks" via in-house scripts. Correlation analysis was also preformed to group adducts and fragments. The applied correlation analysis used Pearson correlations (coefficient threshold set to 0.85) with an RT window of 0.02 to group mass features. Representative mass features for each of the groupings were selected based on evaluation of the aligned chromatograms for the most consistent peak shape, relative intensity and preference for the most likely $[M + H]^+$ ion by manual determination.

5.7. Metabolomics: Binary Matrix Data Transformation

Mass feature peak area values were converted to binary presence/absence matrices (to facilitate comparison between the various host crops) using a peak area threshold value of 5 (where values lower then 5 were denoted as 0 and those greater than denoted as 1). For each individual host, mass features of host crop sclerotium replicates were individually averaged to generate a single representative value for each isolate. Mass features with averaged values greater then 0.65 were denoted as 1, while values lower were set to 0, to ensure that only mass features consistently produced by isolates were included in the resulting host sclerotia binary matrix (4/6 replicates). An untargeted consensus binary heatmap representing mass feature frequency of occurrence in sclerotia produced on the four host crops was constructed to further explore the overall mass feature diversity amongst the strains and between classes. The mass features across the four host binary matrices were then summed for each isolate to create a consensus frequency phenotype matrix consisting of frequency values ranging from 0 to 4, with 0 indicating the feature as not detected for the isolate on any of the examined hosts and 4 indicating the presence in sclerotia derived from all hosts.

5.8. Metabolomics: Multivariate and Univariate Analyses

Multivariate and univariate analyses was completed using the "MUMA" R package with applied Pareto scaling and half minimum imputation on zero values. An initial unsupervised Euclidean Ward hierarchical clustering was performed individually for each host crop to reveal the underlying structure within the data. Metabolomic phenotype heatmaps were generated using the binary data matrices in the "pheatmap" R package, with row and column dendrograms calculated using "ward.D2" clustering of the Euclidean distance matrix. A second metabolomic phenotype heatmap was also generated using only mass feature associated with ergot alkaloid biosynthesis to further demonstrate the dichotomy in ergot alkaloid production across the two classes. Discriminant multivariate analyses (random forest and OPLS DA) were performed using the original non-binary data matrix to determine which mass features contributed most to class separation for each crop host. The supervised random forest analysis was completed using the "randomForest" R packages with the identified classes used for trained data and the OPLS-DA was executed using the "MUMA" R package.

In the random forest analysis, two indices (mean decrease accuracy and mean decrease Gini) are used to assess variable importance associated with class separation. Mean decrease accuracy uses permuting "out of bag" samples to compute the importance of each variable in the predictive accuracy of the random forest [43]. Mean decrease Gini is a measure of how important a variable is in contributing to node and leaf homogeneity across all of the decision trees, where the larger the Gini index value, the greater the importance the variable has in terms of classification in the random forest model [43]. Both mean decrease accuracy and mean decrease Gini were used to rank the influence of the mass features on class separation between sclerotium extracts based on variances in relative peak intensity patterns.

5.9. Metabolomics: Mass Feature Annotation

Mass feature annotations were completed using an in-house *C. purpurea* reference database created from literature reports [33,44]. Putative ergot alkaloid annotations were assigned based on the mass spectral accuracy (±5 ppm) and relative elution order. MS^n experiments were used to confirm ergot alkaloid annotations based on common fragmentation patterns expected from reports in the literature. MassWorks^{TM} software (v5.0.0, Cerno Bioscience, Las Vegas, NV, USA) was used to improve spectral accuracy and confirm the molecular formulas of annotated ions. The sCLIPS searches were performed in dynamic analysis mode with allowances for the elements C, H, N and O set at a minimum of 1 and a maximum of 100. Charge was specified as 1, mass tolerance was set to 5 ppm and the profile mass range was −1.00 to 3.50 Da.

To support of ergot alkaloid annotations, the obtained nanoLC-HRMS/MS data of rye sclerotia extracts from LM04 and LM60 were compared to the ESI+ LCMS spectral database downloaded from the MassBank of north America website in August 2021. MS^2 spectral comparisons were made using MSDIAL [45] and again using default parameters in the GNPS web-based workflow for molecular networking [46]. Top database matches supported annotations for ergometrine, methylergometrine, ergovaline, ergosine, ergotamine, ergocornine, ergocryptine, ergocristam and ergocristine. Most ergot alkaloid peptam/peptide precursors are not represented in spectral databases at this time. Additionally, MS/MS spectral matching is unable to reliably differentiate between very similar spectra from the same parent ion m/z, such as the '-ine' and '-inine' form of ergot alkaloids.

5.10. Genomics: DNA Isolation

Isolates were cultured on PDA and harvested and ground using liquid N_2. Genomic DNA was extracted using a cetyltrimethyl ammonium bromide (CTAB) protocol [47,48]. DNA samples were assessed by running on a 1% agarose gel for the presence of DNA shearing and RNA, while DNA integrity was evaluated using TapeStation and Genomic DNA ScreenTapes (Agilent, Santa Clara, CA, USA) and impurities were evaluated using DropSense 16 (Trinean, Pleasantan, CA, USA). gDNA was quantified using a Qubit® 2.0 Fluorometer (Invitrogen by Life Technologies, Carlsbad, CA, USA) before submission for in-house sequencing (Molecular Technologies Laboratory, Ottawa Research & Development Centre, Agriculture and Agri-Food Canada) where DNA libraries were prepared and loaded onto a FLO-MIN106 flow cell and run with a MinION (Oxford Nanopore Technologies, Oxford, UK) for 48 h.

5.11. Genomics: Genome Assembly and Ergot Alkaloid BiosynthetiFc Gene Cluster Annotation

Genome assembly of LM72, LM04 and LM60 was performed with CANU v1.8 [49] using the sequenced Nanopore reads with default settings and an estimated genome size of 35 Mb. Two rounds of correction were applied to the resulting assemblies: the first round was performed using Nanopolish v. 0.13.2 [50], and the second round used Pilon v1.23 [51], which corrected the nanopolished CANU assemblies using Illumina reads that were mapped with BWA v0.7.17 [52]. Assemblies were annotated using the Funannotate v1.8.8 pipeline [53] using default settings, with predicted proteins from *C. purpurea* isolate 20.1 supplied as protein evidence to assist gene modeling. Libraries of repetitive elements were generated using RepeatModeler2 [54], and were identified, where possible, using the 2018 Repbase database of annotated transposons [55]. Repetitive elements in intergenic regions between *lpsA1* and *lpsA2* were annotated by searching for blast hits in the RepeatModeler2-generated repetitive element libraries. Illumina-based genome assemblies for LM04, LM30, LM60, LM207, LM232, LM233, LM464 and LM479 were previously published and are publicly available in the NCBI database [30].

Supplementary Materials: The following are available online at https://www.mdpi.com/article/10.3390/toxins13120861/s1, Figure S1: Concatenated *RPB2* and *TEF1-α* gene sequence MP (maximum parsimony) algorithm phylogenetic analyses of 44 *Claviceps* spp., Figure S2: Untargeted metabolomics analysis of *C. purpurea* strains grown on Durum wheat, Figure S3: Untargeted metabolomics analysis of *C. purpurea* strains grown on rye, Figure S4: Untargeted metabolomics analysis of *C. purpurea* strains grown on triticale, Figure S5: Phylogenetic tree of aligned nucleotides from C0 domains with associated intergenic 'genotypes', Table S1: Origin of *Claviceps purpurea* isolates, Table S2: Stachelhaus codes and predicted substrate specificities for all lpsA genes extracted from long-read genomes.

Author Contributions: Conceptualization, C.H., T.E.W. and D.P.O.; methodology, C.H., T.E.W., T.L. and D.P.O.; validation, T.E.W. and D.P.O.; formal analysis, C.H., T.E.W., A.S. and T.L.; investigation,

C.H., T.E.W., A.S., P.S. and Z.P.; resources, P.S., Z.P., J.G.M., M.L. and D.P.O.; data curation, C.H., T.E.W. and A.S.; writing—original draft preparation, C.H., T.E.W. and D.P.O.; writing—review and editing, C.H., T.E.W., A.S., J.G.M., C.N.B., M.L. and D.P.O.; visualization, C.H., T.E.W., C.N.B. and D.P.O.; supervision, J.G.M., C.N.B., M.L. and D.P.O.; project administration, D.P.O.; funding acquisition, J.G.M., M.L. and D.P.O. All authors have read and agreed to the published version of the manuscript.

Funding: This research was funded by Agriculture & Agri-Food Canada (Project ID#002272: Fungal and Bacterial Biosystematics-bridging taxonomy and "omics" technology in agricultural research and regulation).

Institutional Review Board Statement: Not Applicable.

Informed Consent Statement: Not Applicable.

Data Availability Statement: Annotated ergot alkaloid biosynthetic gene cluster sequences for LM04, LM60 and LM72 were uploaded to the NCBI nucleotide database under the following accession numbers respectively: OK662595, OL348384, and OL348385.

Acknowledgments: We thank the Molecular Technologies Laboratory (MTL) at the Ottawa Research & Development Centre of Agriculture & Agri-Food Canada.

Conflicts of Interest: The authors declare no conflict of interest.

References

1. Schiff, P.L. Ergot and its alkaloids. *Am. J. Pharm. Educ.* **2006**, *70*, 98. [CrossRef] [PubMed]
2. Negård, M.; Uhlig, S.; Kauserud, H.; Andersen, T.; Høiland, K.; Vrålstad, T. Links between genetic groups, indole alkaloid profiles and ecology within the grass–parasitic *Claviceps purpurea* species complex. *Toxins* **2015**, *7*, 1431–1456. [CrossRef] [PubMed]
3. Didek–brumec, M.; Gaberc–porekar, V.; Alačević, M. Relationship between the *Claviceps* life cycle and productivity of ergot alkaloids. *Crit. Rev. Biotechnol.* **1996**, *16*, 257–299. [CrossRef]
4. Miedaner, T.; Geiger, H.H. Biology, genetics, and management of ergot (*Claviceps* spp.) in rye, sorghum, and pearl millet. *Toxins* **2015**, *7*, 659–678. [CrossRef]
5. McCrea, A. The reactions of *Claviceps purpurea* to variations of environment. *Am. J. Bot.* **1931**, *18*, 50–78. [CrossRef]
6. Menzies, J.G.; Klein–Gebbnick, H.W.; Gordon, A.; O'Sullivan, D.M. Evaluation of *Claviceps purpurea* isolates on wheat reveals complex virulence and host susceptibility relationships. *Can. J. Plant Pathol.* **2017**, *39*, 307–317. [CrossRef]
7. Xue, A.G.; Chen, Y.; Ai–Rewashdy, Y. Diseases of spring wheat in central and eastern Ontario in 2017. *Can. Plant Dis. Surv.* **2017**, *98*, 148–149.
8. Menzies, J.G.; Turkington, T.K. An overview of the ergot (*Claviceps purpurea*) issue in western Canada: Challenges and solutions. *Can. J. Plant Pathol.* **2015**, *37*, 40–51. [CrossRef]
9. Komarova, E.L.; Tolkachev, O.N. The Chemistry of peptide ergot alkaloids. Part 1. Classification and chemistry of ergot peptides. *Pharm. Chem. J.* **2001**, *35*, 504–513. [CrossRef]
10. Katzung, B.G. *Basic & Clinical Pharmacology*, 14th ed.; McGraw–Hill Education: New York, NY, USA, 2018; pp. 277–299.
11. Florea, S.; Panaccione, D.G.; Schardl, C.L. Ergot alkaloids of the family Clavicipitaseae. *Phytopathology* **2017**, *107*, 504–518. [CrossRef] [PubMed]
12. Young, C.A.; Schardl, C.L.; Panaccione, D.G.; Florea, S.; Takach, J.E.; Charlton, N.D.; Jaromczyk, J. Genetics, genomics and evolution of ergot alkaloid diversity. *Toxins* **2015**, *7*, 1273–1302. [CrossRef] [PubMed]
13. Arroyo-Manzanares, N.; Gámiz-Gracia, L.; García-Campaña, A.M.; Diana di Magunvu, J.; De Saeger, S. Ergot alkaloids: Chemistry, biosynthesis, bioactivity, and methods of analysis. In *Fungal Metabolites*; Mérillon, J.M., Ramawat, K., Eds.; Reference Series in Phytochemistry; Springer International Publishing Switzerland: Berlin/Heidelberg, Germany, 2016; pp. 1–43.
14. Ortel, I.; Keller, U. Combinatorial assembly of simple and complex D–lysergic acid alkaloid peptide classes in the ergot fungus *Claviceps purpurea*. *J. Biol. Chem.* **2009**, *284*, 6650–6660. [CrossRef] [PubMed]
15. Robinson, S.L.; Panaccione, D.G. Diversification of ergot alkaloids in natural and modified fungi. *Toxins* **2015**, *7*, 201–218. [CrossRef] [PubMed]
16. Haarmann, T.; Machado, C.; Lubbe, Y.; Correia, T.; Schardl, C.L.; Panaccione, D.G.; Tudzynski, P. The ergot alkaloid gene cluster in *Claviceps purpurea*: Extension of the cluster sequence and intra species evolution. *Phytochemistry* **2005**, *66*, 1312–1320. [CrossRef] [PubMed]
17. Haarmann, T.; Lorenz, N.; Tudzynski, P. Use of a nonhomologous end joining deficient strain (Δku70) of the ergot fungus *Claviceps purpurea* for identification of a nonribosomal peptide synthetase gene involved in ergotamine biosynthesis. *Fungal Genet. Biol.* **2008**, *45*, 35–44. [CrossRef] [PubMed]
18. Official Grain Grading Guide. Available online: https://www.grainscanada.gc.ca/en/grain-quality/official-grain-grading-guide/ (accessed on 22 October 2021).
19. Gruise, T.J.; Cowan, V.; Singh, J.; McKinnon, J.; Blakley, B. Proportions of predominant ergot alkaloids (*Claviceps purpurea*) detected in Western Canadian grains from 2014 to 2016. *World Mycotoxin J.* **2018**, *11*, 259–264. [CrossRef]

20. Franzmann, C.; Wächter, J.; Natascha, D.; Humpf, H.U. Ricinoleic acid as a marker for ergot impurities in rye and rye products. *J. Agric. Food Chem.* **2010**, *58*, 4223–4229. [CrossRef]
21. Pažoutová, S.; Pešicová, K.; Chudíčková, M.; Šrůtka, P.; Kolařík, M. Delimitation of cryptic species inside *Claviceps purpurea*. *Fungal Biol.* **2015**, *119*, 7–26. [CrossRef] [PubMed]
22. Shoukouhi, P.; Hicks, C.; Menzies, J.G.; Zlatko, P.; Chen, W.; Seifert, K.A.; Assabgui, R.; Liu, M. Phylogeny of Canadian ergot fungi and a detection assay by real-time polymerase chain reaction. *Mycologia* **2019**, *111*, 493–505. [CrossRef]
23. European Food Safety Authority; Arcella, D.; Gómez Ruiz, J.Á.; Innocenti, M.L.; Roldán, R. Human and dietary exposure to ergot alkaloids. *EFSA J.* **2017**, *15*, 7. [CrossRef]
24. Proposal—Contaminant Standards for Aflatoxins, Deoxynivalenol, Fumonisins, Ergot Alkaloids and *Salmonella* in Livestock Feeds. Available online: https://inspection.canada.ca/animal-health/livestock-feeds/consultations/contaminant-standards-for-aflatoxins-deoxynivaleno/eng/1500908795245/1500908795965 (accessed on 22 October 2021).
25. Liu, M.; Findlay, W.; Dettman, J.; Wyka, S.A.; Broders, K.; Shoukouhi, P.; Dadej, K.; Kolařík, M.; Basnyat, A.; Menzies, J.G. Mining indole alkaloid synthesis gene clusters from genomes of 53 *Claviceps* strains revealed redundant gene copies and an approximate evolutionary hourglass model. *Toxins* **2021**, *13*, 799. [CrossRef]
26. Schardl, C.L.; Young, C.A.; Hesse, U.; Amyotte, S.G.; Andreeva, K.; Calie, P.J.; Fleetwood, D.J.; Haws, D.C.; Moore, N.; Oeser, B.; et al. Plant-symbiotic fungi as chemical engineers: Multi-genome analysis of the Clavicipitaceae reveals dynamics of alkaloid loci. *PLoS Genet.* **2013**, *9*, e1003323. [CrossRef] [PubMed]
27. Stachelhaus, T.; Mootz, H.D.; Marahiel, M.A. The specificity-conferring code of adenylation domains in nonribosomal peptide synthetases. *Chem. Biol.* **1999**, *6*, 493–505. [CrossRef]
28. Röttig, M.; Medema, M.H.; Blin, K.; Weber, T.; Rausch, C.; Kohlbacher, O. NRPSpredictor2—A web server for predicting NRPS adenylation domain specificity. *Nucleic Acids Res.* **2011**, *39*, W362–W367. [CrossRef] [PubMed]
29. Le Govic, Y.; Papon, N.; Le Gal, S.; Bouchara, J.-P.; Vandeputte, V. Non-ribosomal peptide synthetase gene clusters in the human pathogenic fungus *Scedosporium apiospermum*. *Front. Microbiol.* **2019**, *10*, 2062. [CrossRef]
30. Wyka, S.A.; Mondo, S.J.; Liu, M.; Dettman, J.; Nalam, V.; Broders, K.D. Whole-genome comparisons of ergot fungi reveals the divergence and evolution of species within the genus *Claviceps* are the result of varying mechanisms driving genome evolution and host range expansion. *Genome Biol. Evol.* **2021**, *13*, evaa267. [CrossRef] [PubMed]
31. Pažoutová, S.; Olšovská, J.; Linka, M.; Kolínská, R.; Flieger, M. Chemoraces and habitat specialization of *Claviceps purpurea* populations. *Appl. Environ. Microbiol.* **2000**, *66*, 5419–5425. [CrossRef] [PubMed]
32. Pažoutová, S.; Cagaš, B.; Kolínská, R.; Honzátko, A. Host specialization of different populations of ergot fungus (*Claviceps purpurea*). *Czech J. Genet. Plant Breed.* **2002**, *38*, 75–81. [CrossRef]
33. Uhlig, S.; Rangel-Huerta, O.D.; Divon, H.H.; Rolén, E.; Pauchon, K.; Sumarah, M.W.; Vrålstad, T.; Renaud, J.B. Unraveling the ergot alkaloid and indole diterpenoid metabolome in the *Claviceps purpurea* species complex using LC–HRMS/MS diagnostic fragmentation filtering. *J. Agric. Food Chem.* **2021**, *69*, 7137–7148. [CrossRef] [PubMed]
34. Lorenz, N.; Haarmann, T.; Pazoutová, S.; Jung, M.; Tudzynski, P. The ergot alkaloid gene cluster: Functional analyses and evolutionary aspects. *Phytochemistry* **2009**, *70*, 1822–1832. [CrossRef] [PubMed]
35. Berry, D.; Mace, W.; Grage, K.; Wesche, F.; Gore, S.; Schardl, C.L.; Young, C.A.; Dijkwel, P.P.; Leuchtmann, A.; Bode, H.B.; et al. Efficient nonenzymatic cyclization and domain shuffling drive pyrrolopyrazine diversity from truncated variants of a fungal NRPS. *Proc. Natl. Acad. Sci. USA* **2019**, *51*, 25614–25623. [CrossRef]
36. Baunach, M.; Chowdhurry, S.; Stallforth, P.; Dittmann, E. The landscape of recombination events that create nonribosomal peptide diversity. *Mol. Biol. Evol.* **2021**, *38*, 2116–2130. [CrossRef]
37. Klotz, J.L. Activities and effects of ergot alkaloids on livestock physiology and production. *Toxins* **2015**, *7*, 2801–2821. [CrossRef] [PubMed]
38. Coufal-Majewski, S.; Stanford, K.; McAllister, T.; Blakley, B.; McKinnon, J.; Chaves, A.V.; Wang, Y. Impacts of cereal ergot in food animal production. *Front. Vet. Sci.* **2016**, *3*, 15. [CrossRef] [PubMed]
39. Yonpiam, R.; Gobbet, J.; Jadhav, A.; Desai, K.; Blakley, B.; Al-Dissi, A. Vasoactive effects of acute ergot exposure in sheep. *Toxins* **2021**, *13*, 291. [CrossRef] [PubMed]
40. Liu, M.; Overy, D.P.; Cayouette, J.; Shoukouhi, P.; Hicks, C.; Bisson, K.; Sproule, A.; Wyka, S.A.; Broders, K.; Popovic, Z.; et al. Four phylogenetic species of ergot from Canada and their characteristics in morphology, alkaloid production, and pathogenicity. *Mycologia* **2020**, *112*, 974–988. [CrossRef]
41. Swofford, D.L. *PAUP*. Phylogenetic Analysis Using Parsimony (*and Other Methods)*, Version 4.0 b10; Sinauer Associates: Sunderland, MA, USA, 2002.
42. Pluskal, T.; Castillo, S.; Villar-Briones, A.; Orešič, M. MZmine 2: Modular framework for processing, visualizing, and analyzing mass spectrometry-based molecular profile data. *BMC Bioinform.* **2010**, *11*, 395. [CrossRef] [PubMed]
43. Breiman, L. Random forests. *Mach. Learn.* **2001**, *45*, 5–32. [CrossRef]
44. Sulyok, M.; Krska, R.; Schuhmacher, R. A liquid chromatography/tandem mass spectrometric multi-mycotoxin method for the quantification of 87 analytes and its application to semi-quantitative screening of moldy food samples. *Anal. Bioanal. Chem.* **2007**, *389*, 1505–1523. [CrossRef] [PubMed]
45. Tsugawa, H.; Ikeda, K.; Takahashi, M.; Satoh, A.; Mori, Y.; Uchino, H.; Okahashi, N.; Yamada, Y.; Tada, I.; Bonini, P.; et al. A lipidome atlas in MS-DIAL 4. *Nat. Biotechnol.* **2020**, *38*, 1159–1163. [CrossRef] [PubMed]

46. Nothias, L.F.; Petras, D.; Schmid, R.; Dührkop, K.; Rainer, J.; Sarvepalli, A.; Protsyuk, I.; Ernst, M.; Tsugawa, H.; Fleischauer, M.; et al. Feature-based molecular networking in the GNPS analysis environment. *Nat. Methods* **2020**, *17*, 905–908. [CrossRef] [PubMed]
47. Doyle, J.J.; Doyle, J.L. A rapid DNA isolation procedure for small quantities of fresh leaf tissue. *Phytochem. Bull.* **1987**, *190*, 11–15. [CrossRef]
48. Wingfield, B.D.; Liu, M.; Nguyen, H.D.T.; Lane, F.A.; Morgan, S.W.; De Vos, L.; Wilken, P.M.; Duong, T.A.; Aylward, J.; Coetzee, M.P.A.; et al. Nine draft genome sequences of *Claviceps purpurea s. lat.*, including *C. arundinis*, *C. humidiphila*, and *C. cf. spartinae*, pseudomolecules for the pitch canker pathogen *Fusarium circinatum*, draft genome of *Davidsoniella eucalypti*, *Grosmannia galeiformis*, *Quambalaria eucalypti*, and *Teratosphaeria destructans*. *IMA Fungus* **2018**, *9*, 401–418. [CrossRef] [PubMed]
49. Koren, S.; Walenz, B.P.; Berlin, K.; Miller, J.R.; Bergman, N.H.; Phillippy, A.M. Canu: Scalable and accurate long-read assembly via adaptive k-mer weighting and repeat separation. *Genome Res.* **2017**, *27*, 722–736. [CrossRef]
50. Loman, N.J.; Quick, J.; Simpson, J.T. A complete bacterial genome assembled de novo using only nanopore sequencing data. *Nat. Methods* **2015**, *12*, 733–735. [CrossRef]
51. Walker, B.J.; Abeel, T.; Shea, T.; Priest, M.; Abouelliel, A.; Sakthikumar, S.; Cuomo, C.A.; Zeng, Q.; Wortman, J.; Young, S.K.; et al. Pilon: An integrated tool for comprehensive microbial variant detection and genome assembly improvement. *PLoS ONE* **2014**, *9*, e112963. [CrossRef] [PubMed]
52. Li, H.; Durbin, R. Fast and accurate short read alignment with Burrows-Wheeler transform. *Bioinformatics* **2009**, *25*, 1754–1760. [CrossRef] [PubMed]
53. Palmer, J.M.; Stajich, J.E. Funannotate v1. 8.1: Eukaryotic genome annotation. *Zenodo* **2020**. [CrossRef]
54. Flynn, J.M.; Hubley, R.; Goubert, C.; Rosen, J.; Clark, A.G.; Feschotte, C.; Smit, A.F. RepeatModeler2 for automated genomic discovery of transposable element families. *Proc. Natl. Acad. Sci. USA* **2020**, *117*, 9451–9457. [CrossRef]
55. Bao, W.; Kojima, K.K.; Kohany, O. Repbase Update, a database of repetitive elements in eukaryotic genomes. *Mob. DNA* **2015**, *6*, 11. [CrossRef] [PubMed]

toxins

MDPI

Article

Mining Indole Alkaloid Synthesis Gene Clusters from Genomes of 53 *Claviceps* Strains Revealed Redundant Gene Copies and an Approximate Evolutionary Hourglass Model

Miao Liu [1,*], Wendy Findlay [1], Jeremy Dettman [1], Stephen A. Wyka [2], Kirk Broders [3], Parivash Shoukouhi [1], Kasia Dadej [1], Miroslav Kolařík [4], Arpeace Basnyat [1] and Jim G. Menzies [5]

1 Ottawa Research & Development Centre, Agriculture and Agri-Food Canada, Ottawa, ON K1A 0C6, Canada; wendy.findlay@agr.gc.ca (W.F.); jeremy.dettman@agr.gc.ca (J.D.); parivash.shoukouhi@agr.gc.ca (P.S.); kasia.dadej@agr.gc.ca (K.D.); abasn046@uottawa.ca (A.B.)
2 Department of Agricultural Biology, Colorado State University, Fort Collins, CO 80523, USA; stephenwyka@gmail.com
3 USDA, Agricultural Research Service, National Center for Agricultural Utilization Research, Mycotoxin Prevention and Applied Microbiology Research Unit, 1815 N. University St., Peoria, IL 61604, USA; kirk.broders@usda.gov
4 Institute of Microbiology of the Czech Academy of Sciences CAS, 14220 Prague, Czech Republic; mkolarik@biomed.cas.cz
5 Morden Research and Development Centre, Agriculture and Agri-Food Canada, Morden, MB R6M 1Y5, Canada; jim.menzies@agr.gc.ca
* Correspondence: miaomindy.liu@agr.gc.ca; Tel.: +1-613-759-1385

Citation: Liu, M.; Findlay, W.; Dettman, J.; Wyka, S.A.; Broders, K.; Shoukouhi, P.; Dadej, K.; Kolařík, M.; Basnyat, A.; Menzies, J.G. Mining Indole Alkaloid Synthesis Gene Clusters from Genomes of 53 *Claviceps* Strains Revealed Redundant Gene Copies and an Approximate Evolutionary Hourglass Model. *Toxins* **2021**, *13*, 799. https://doi.org/10.3390/toxins13110799

Received: 21 October 2021
Accepted: 10 November 2021
Published: 13 November 2021

Abstract: Ergot fungi (*Claviceps* spp.) are infamous for producing sclerotia containing a wide spectrum of ergot alkaloids (EA) toxic to humans and animals, making them nefarious villains in the agricultural and food industries, but also treasures for pharmaceuticals. In addition to three classes of EAs, several species also produce paspaline-derived indole diterpenes (IDT) that cause ataxia and staggers in livestock. Furthermore, two other types of alkaloids, i.e., loline (LOL) and peramine (PER), found in *Epichloë* spp., close relatives of *Claviceps*, have shown beneficial effects on host plants without evidence of toxicity to mammals. The gene clusters associated with the production of these alkaloids are known. We examined genomes of 53 strains of 19 *Claviceps* spp. to screen for these genes, aiming to understand the evolutionary patterns of these genes across the genus through phylogenetic and DNA polymorphism analyses. Our results showed (1) varied numbers of *eas* genes in *C.* sect. *Claviceps* and sect. *Pusillae*, none in sect. *Citrinae*, six *idt/ltm* genes in sect. *Claviceps* (except four in *C. cyperi*), zero to one partial (*idtG*) in sect. *Pusillae*, and four in sect. *Citrinae*, (2) two to three copies of *dmaW*, *easE*, *easF*, *idt/ltmB*, *itd/ltmQ* in sect. *Claviceps*, (3) frequent gene gains and losses, and (4) an evolutionary hourglass pattern in the intra-specific *eas* gene diversity and divergence in *C. purpurea*.

Keywords: ergot alkaloids; ergot fungi; gene divergence; gene diversity; indole diterpenes; phylogeny; secondary metabolites

Key Contribution: Indole alkaloid gene clusters from a wide range of *Clavicep* spp. were identified through genome screening. Six indole diterpene/lolitrem genes, *idt/ltmP, Q, B, C, S,* and *M*, were commonly present in various species in *C.* sect. *Claviceps*. Micro-evolution of *eas* genes within *Claviceps purpurea* revealed that their evolutionary rates fit an hourglass model.

1. Introduction

Fungi in the genus *Claviceps* (Clavicipitaceae, Hypocreales, Sordariomycetes) infect the florets of cereal crops, nonagricultural grasses (Poaceae), sedges (Cyperaceae), and rushes (Juncaceae) [1], followed by occupying the unfertilized ovaries and eventually replacing the seeds with fungal resting bodies, i.e., sclerotia, known as ergots [2]. In light of molecular phylogenetics, 63 named species [3,4] are classified into four sections, i.e., *Claviceps* sect. *Claviceps*, *C.* sect. *Citrinae*, *C.* sect. *Paspalorum*, and *C.* sect. *Pusillae*, on the basis of morphological, ecological, and alkaloid-producing features [3]. Ergot bodies or sclerotia contain a wide spectrum of alkaloids toxic to humans and animals, making them unwelcome pathogens in agricultural and food production [5,6], but also important resources for pharmaceuticals [7,8]. Among the alkaloids produced by *Claviceps*, ergot alkaloids (EAs) are the major culprit for the mass food/feed poisoning in human and livestock, as well as a number of tragedies in human history [9,10]. EAs are indole compounds characterized by a tricyclic or tetracyclic ring system. Over 80 different EAs found in nature fall into three structural groups: clavines, lysergic acid amides, and ergopeptides [8,11], corresponding to their structural complexity. Clavines are the intermediates or derivatives of the intermediates in the lysergic acid amide pathway, whereas ergopeptines are the most complex group [11]. Intensive investigations on biochemistry and molecular genetics have elucidated the EA biosynthetic pathways in EA producers especially *Claviceps* spp. [12,13]. A cluster of 12 functioning EA synthesis (*eas*) genes (*cloA*, *dmaW*, *easA*, *easC–H*, *lpsA–C*) in *C. purpurea* strain 20.1 were considered to encode all the enzymes needed for the end-product ergotamine and ergocryptine [14]. The four early steps, requiring *dmaW*, *easF*, *easC*, and *easE*, are responsible for the closure of the third ring resulting in chanoclavine, followed by middle steps, requiring *easD*, *easA*, *easG*, and *cloA*, for forming tetracyclic clavines, and later steps for producing the lysergic acid amides, dihydroergot alkaloids, and complex ergot peptines [13] (Figure 1). Among the 12 genes, the homologs of nine were found in *C. fusiformis* in a cluster. In *C. paspali*, two additional genes (*easP* and *easO*) were found; however, *easE* was defective. The presence or absence of *eas* genes has proven to be correlated with EA profiles in several *Claviceps* spp. and strains [13,14]. However, the investigation of *eas* gene clusters in a wide range of *Claviceps* spp. is lacking, and less is reported about the evolution of the individual gene in these clusters among and within the species.

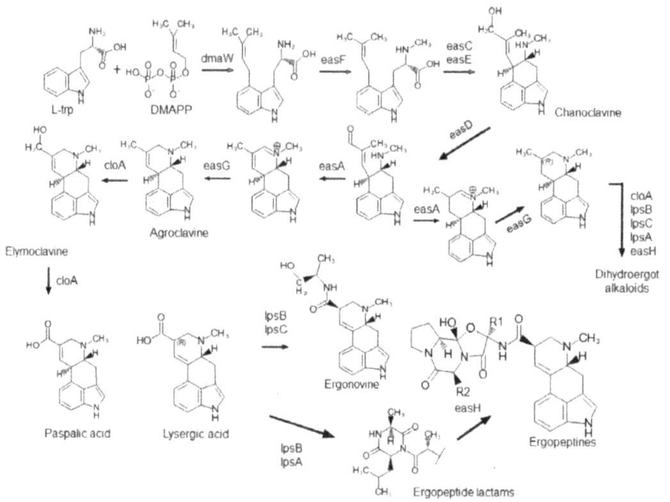

Figure 1. The ergot alkaloid biosynthetic pathway in *Claviceps* spp. Modified from Young et al. [13] and Robinson and Panaccione [15].

Indole diterpenes (IDTs) are another large group of bioactive compounds with diverse structural variations, triggering toxicity in animals and insects through interfering with ion channels [16,17]. In the literature, there are copious reports that certain species in *Claviceps* (i.e., *C. paspali* and *C. cynodontis*) and close relatives in *Epichloë* (*Neotyphodium* as the asexual name before implementation of the International Code of Nomenclature for algae, fungi, and plants (Shenzhen Code) [18]) produce the paspaline-derived IDTs, such as paspalitrem, lolitrem, and paxilline, causing ataxia and staggers in livestock that feed on the grasses infected by those fungal species [19–21]. Biosynthetic pathways and associated gene clusters of these paspaline-derived IDTs have been investigated [22–24], resulting in the discovery of at least 10 genes involved in IDT production in *Epichloë* spp. and the prediction that *ltmG, M, C,* and *B* were responsible for the synthesis of paspaline, the basic structural backbone of IDTs, whereas *ltmP* and Q were essential for the production of lolitrem and *ltmF, J, K,* and *E*, which are required for more complex structures [25,26]. The proposed scheme for the biosynthesis of paspalitrem in *C. paspali* involved seven genes including the initial formation of paspaline through *ltmG, M, C,* and *B*, followed by the sequential functioning steps of *ltmP, Q,* and *F* [22]. Recently, the pre-paspaline steps were further resolved as three sequential steps: starting from *ltmG* converting farnesyl diphosphate (FPP) to Geranylgeranyl diphosphate (GGPP), followed by *ltmC* transferring GGPP to 3-geranylgeranylindole, and finally through *ltmM* and *B* yielding paspaline [27]. In addition to *C. paspali* and *C. cynodontis*, other *Claviceps* spp., i.e., *C. arundinis*, *C. humidiphila*, and *C. purpurea*, could also produce indole diterpenes or paspaline-like compounds [28–30]. The genome investigation of *C. purpurea* 20.1 revealed the presence of *ltmM, C, B, P, Q,* and an extra gene *ltmS* [14]. It is not known whether these genes are consistently present in various strains of *C. purpurea* and other *Claviceps* species. In addition, two other classes of alkaloids, i.e., lolines (LOL) and peramines (PER), produced by *Epichloë* spp., are known to function as insecticides, but are not associated with any toxicity symptoms in grazing mammals [31,32]. Given the close relationship between *Epichloë* and *Claviceps*, it is reasonable to raise the question of whether any of the loline or peramine gene homologs are present in any of the *Claviceps* spp. even though those two classes of alkaloids have not been reported in *Claviceps*.

The 'hourglass model' borrowed from ontogeny refers to the pattern that the morphological divergence of mid-development stages of an embryo are more conserved compared with earlier and later stages, resembling an hourglass with a narrow waist, but broad ends [33,34]. Before the hourglass model (HGM) was proposed in the 1990s, the early conservation model (ECM) was widely accepted, which echoed von Baer's third law [35], i.e., embryos progressively diverge in morphology during ontogeny. The debates about these two, along with other models, i.e., adaptive penetrance model [36] and unconstrained model (random) [37], are still ongoing, although recent evidence at molecular and genomic levels has provided support for the presence of the phylotypic stage (the waist stage of development) in fungi, insects, plants, and vertebrates [38–40]. According to Haeckel's biogenetic hypothesis, ontogeny recapitulates phylogeny [41]. The evident similarities between the development of an individual and the evolution of the whole biological system have been addressed by many generations [42] to verify that these models in ontogeny are recapitulated in other evolutionary processes. For example, studies on gene evolution in *Drosophila* spp. recaptured the hourglass model in that the early maternal genes showed a higher level of diversity than zygotic genes [43]. Here, we propose the biosynthesis of complex biological compounds as an analogy of the development of an organism, and ask whether any of the models fit to the evolution of the genes involved in the biosynthesis.

The objectives of the present study were to shed light on the presence of four classes of alkaloid genes (clusters) in 53 strains of 19 *Claviceps* species, and to understand the evolutionary patterns of these genes at inter- and intraspecific levels. This information helps build the foundation for future studies on chemo- and genotype associations and for developing gene-based chemotyping and toxin detection.

2. Results

2.1. Genome Assemblies

The 37 genome sequences assembled in this study resulted in 1362 to 2581 contigs, N50 values ranged from 19,946 to 55,909 bp, and the completeness measured by Benchmarking Universal Single-Copy Orthologs (BUSCO) over the fungal database (fungi odb10) ranged from 97% to 99.1% (Table 1, available in GenBank https://www.ncbi.nlm.nih.gov/

(accessed on 9 November 2021) as accessions JAIURI000000000–JAIUSS000000000 available upon publication of the article). The quality of the assemblies was equivalent to the assemblies of 17 genomes from previous studies (Table 1) [44,45]. Overall, the 54 assemblies of 53 strains (two versions of assemblies for CCC1102 were included because certain genes were obtained from one or the other assemblies) of 19 *Claviceps* species included in this study belong to three sections: C. sect. *Citrinae*, C. sect. *Claviceps*, and C. sect. *Pusillae*, from six continents (Africa, Asia, Australia, Europe, North America, and South America) and on host plants in 26 genera (Table S1).

2.2. *Presence of Four Classes of Alkaloid Genes in 53 Genomes*

One thousand sequences of 19 loci were extracted from the 53 genome assemblies as detailed below. The DNA sequences of each genes were submitted to Genbank associated with accession numbers: *cloA* (49 sequences) MZ882098–MZ882146, *dmaW* (118 sequences) MZ871640–MZ871757, *easA* (51 sequences) MZ851397–MZ851447, *easC* (50 sequences) MZ851807–MZ851856, *easD* (51 sequences) MZ871767–MZ871817, *easE* (66 sequences) MZ877968–MZ878033, *easF* (88 sequences) MZ881959–MZ882046, *easG* (50 sequences) MZ882047–MZ882096, *easH1* (50 sequences) MZ934760–MZ934809, *easH2* (32 sequences) MZ934810–MZ934841, *lpsB* (48 sequences) MZ934842–MZ934889, *lpsC* (44 sequences) MZ934890–MZ934933, *idt/ltmB* (55 sequences) MZ935033–MZ935087, *idt/ltmC* (47 sequences) MZ935088–MZ935134, *idt/ltmG* (three sequences) MZ935227–MZ935229, *idt/ltmM* (47 sequences) MZ935135–MZ935181, *idt/ltmP* (46 sequences) MZ934987–MZ935032, *idt/ltmQ* (60 sequences) MZ934934–MZ934986, and *idt/ltmS* (45 sequences) MZ935182–MZ935226.

2.2.1. Ergot Alkaloid Genes (eas)

More consistency in terms of presence/absence of *eas* genes was observed in *C. sect. Claviceps* than C. sect. *Pusillae*. The results from BLASTn searches using in-house script (see Section 4.2 for details) and Geneious mapping (https://www.geneious.com, accessed on 9 November 2021) with reference genes showed the genomes of all isolates from C. sect. *Claviceps* contained at least 10 *eas* genes matching the *C. purpurea* 20.1 reference sequences (Table 2; *lpsA1* and *lpsA2* were excluded from analyses as they were heavily fragmented due to the significantly long length. A study on long-read sequencing of several selected strains by Hicks et al. was focused on these two genes (in this Special Issue). The 10–12 genes were assembled on two to three contigs. For most strains, nine genes (*lpsC*, *easA*, *lpsB*, *cloA*, and *easC-G*) were on the same contig. Genes after *dmaW*, i.e., *easH1*, *easH2*, and fragments of *lpsA*, were on different contigs. The *easH2* gene was either not detected or on a separate contig possibly due to the long length of *lpsA1* because it was located between *lpsA1* and *A2* in the reference genome *C. purpurea* 20.1. The exceptions were *C. humidiphila* LM576, *C. spartinae* CCC535, *C. purpurea* LM461, and *C. ripicola* LM220 and LM454, in which *lpsC* was on a different contig, or *lpsC* along with the next three to four genes were on the same contig separated from other genes (Table 2).

Both inter- and intraspecific variation was observed, regardless of the general consistency of presence of *eas* genes. Species-specific features included all three strains of *C. occidentalis* have two partial copies of *dmaW* (~658 bp, ~641 bp composed of a partial exon 1 and full-length exon 2 and 3) and a single copy of all other *eas* genes except *easH*. Of a relevant note, all partial genes detected in the present study were located at the end of contigs. Moreover, all three strains of *C. quebecensis* had a second partial nonfunctioning copy of *easE* (275, 275, and 1208 bp) and two partial copies of *easF* with good open reading frames (ORFs), and they were lacking *easH2* (Table 2).

Table 1. Statistics of genome assemblies screened.

Species	Strain	BioSample	WGS #	Contigs	Total Length (bp)	Largest Contig (bp)	N50 (bp)	L50	GC (%)	Coverage (x)	Complete BUSCO's (%)
C. arundinis	CCC1102	SAMN11159893	JAIUSP000000000	1406	29,878,863	375,533	55,909	156	51.42	61x	98.1
C. capensis	CCC1504	SAMN11159898	JAIUSL000000000	1497	27,462,555	202,637	39,758	198	51.69	66x	98.8
C. cyperi	CCC1219	SAMN11159895	JAIUSO000000000	2467	26,149,012	130,032	19,946	386	51.72	56x	98.3
C. monticola	CCC1483	SAMN11159896	JAIUSN000000000	1787	27,131,110	129,905	29,639	279	51.6	58x	98.8
C. occidentalis	LM77	SAMN11159879	JAIURJ000000000	2285	28,557,246	118,746	21,556	410	51.37	58x	97.8
C. occidentalis	LM84	SAMN11159876	JAIURI000000000	2119	28,639,296	133,797	23,641	389	51.39	164x	97.0
C. pazoutovae	CCC1485	SAMN11159897	JAIUSM000000000	1619	27,544,752	151,196	35,477	229	51.7	61x	98.3
C. purpurea s.s.	Clav04	SAMN11159846	JAIUSJ000000000	2581	30,594,081	349,533	29,378	296	51.69	46x	98.8
C. purpurea s.s.	Clav26	SAMN11159847	JAIUSI000000000	1821	30,253,558	299,368	36,369	242	51.48	59x	98.8
C. purpurea s.s.	Clav46	SAMN11159848	JAIUSG000000000	1887	30,292,940	231,314	36,582	246	51.08	58x	99.1
C. purpurea s.s.	Clav52	SAMN11159849	JAIUSE000000000	1714	29,291,845	175,165	35,956	250	51.42	60x	98.9
C. purpurea s.s.	Clav55	SAMN11159850	JAIUSD000000000	2023	30,195,775	203,523	33,461	261	51.55	59x	98.4
C. purpurea s.s.	LM14	SAMN11159853	JAIUSC000000000	1888	30,259,282	163,532	32,812	268	51.74	49x	97.9
C. purpurea s.s.	LM207	SAMN11159861	JAIUSB000000000	1910	30,165,540	260,847	31,428	273	51.74	53x	98.7
C. purpurea s.s.	LM223	SAMN11159862	JAIURY000000000	1894	30,223,423	195,661	31,693	291	51.73	74x	98.4
C. purpurea s.s.	LM232	SAMN11159863	JAIURX000000000	1911	30,304,653	216,996	33,376	265	51.73	53x	98.8
C. purpurea s.s.	LM233	SAMN11159864	JAIURW000000000	1928	30,249,987	183,378	33,023	273	51.74	49x	98.3
C. purpurea s.s.	LM30	SAMN11159855	JAIURV000000000	1816	30,203,936	160,353	35,005	265	51.75	64x	98.4
C. purpurea s.s.	LM33	SAMN11159856	JAIURU000000000	2011	30,162,301	157,176	28,954	306	51.75	45x	97.9
C. purpurea s.s.	LM39	SAMN11159857	JAIURT000000000	1797	30,183,718	168,047	34,902	258	51.75	81x	98.3
C. purpurea s.s.	LM4	SAMN11159851	JAIURS000000000	1856	30,197,808	200,831	31,054	281	51.74	64x	98.3
C. purpurea s.s.	LM46	SAMN11159858	JAIURR000000000	1842	30,109,785	205,399	32,503	270	51.76	79x	98.6
C. purpurea s.s.	LM461	SAMN11159865	JAIURQ000000000	2041	30,157,824	190,247	28,928	307	51.74	37x	97.7
C. purpurea s.s.	LM469	SAMN11159866	JAIURP000000000	1836	30,218,091	199,880	34,408	269	51.74	75x	98.3
C. purpurea s.s.	LM470	SAMN11159867	JAIURO000000000	2482	30,086,038	123,014	23,231	384	51.75	26x	97.9
C. purpurea s.s.	LM474	SAMN11159868	JAIURN000000000	1917	30,149,711	232,504	30,855	283	51.75	64x	98
C. purpurea s.s.	LM5	SAMN11159852	JAIURM000000000	1817	30,171,863	188,144	34,174	271	51.74	67x	98.4
C. purpurea s.s.	LM60	SAMN11159859	JAIURL000000000	1871	30,274,458	180,242	31,977	275	51.73	81x	98.6
C. purpurea s.s.	LM63	SAMN20436330	JAIUSS000000000	1674	30,276,205	210,630	40,954	218	51.79	68x	98.4
C. purpurea s.s.	LM65	SAMN20436331	JAIUSR000000000	1822	30,277,382	206,609	37,976	241	51.78	71x	98.4
C. purpurea s.s.	LM71	SAMN11159860	JAIURK000000000	1919	30,241,564	172,997	32,324	282	51.76	168x	98
C. purpurea s.s.	LM72	SAMN20436332	JAIUSQ000000000	1986	30,160,156	282,506	36,805	249	51.81	63x	98.4
C. quebecensis	Clav32	SAMN11159882	JAIUSH000000000	1362	28,435,427	248,888	42,252	192	51.61	64x	99.0
C. quebecensis	Clav50	SAMN11159881	JAIUSF000000000	1404	28,499,699	294,425	47,797	178	51.6	59x	98.8
C. ripicola	LM219	SAMN11159874	JAIUSA000000000	1847	30,428,256	154,690	34,898	254	51.39	55x	97.1
C. ripicola c.f.	LM220	SAMN11159873	JAIURZ000000000	1662	30,409,961	205,881	43,971	211	51.43	91x	97.7
C. spartinae	CCC535	SAMN11159888	JAIUSK000000000	2017	28,974,645	142,723	28,332	300	51.38	60x	98.1
					Assemblies from previous studies						
C. arundinis	CCC1102	SAMN11159893	SRPS01	1406	29,878,863	375,533	55,909	156	51.42	61x	97.7 *
C. africana	CCC489	SAMN11159887	SRPY01	5329	31,933,801	98,049	12,225	752	44.68	56x	95 *
C. arundinis	LM583	SAMN08798359	QEQZ01	1613	30,055,381	164,904	39,306	223	51.42	69x	96.9
C. citrina	CCC265	SAMN11159885	SRQA01	4830	25,056,896	81,802	8747	871	47.57	64x	92.2*
C. digitariae	CCC659	SAMN11159892	SRPT01	3821	31,170,596	116,859	16,077	572	45.5	57x	95.9 *
C. humidiphila	LM576	SAMN08798355	QERB01	1831	30,488,243	190,085	34,787	261	51.51	77x	97.9
C. lovelessii	CCC647	SAMN11159891	SRPU01	3201	34,575,813	65,439	5747	1781	43.61	53x	91.6 *
C. maximensis	CCC398	SAMN11159886	SRPZ01	2317	29,114,417	192,851	37,101	230	46.66	58x	98.3 *
C. occidentalis	LM78	SAMN08800200	QEQY01	2321	28,571,683	125,459	21,416	422	51.37	64x	97.3
C. perihumidiphila	LM81	SAMN08800226	QEQX01	1423	30,694,913	232,029	46,526	192	51.5	140x	96.9
C. purpurea s.s.	LM28	SAMN08797627	QERD01	1930	30,251,797	260,842	31,815	274	51.74	49x	97.9
C. purpurea s.s.	LM582	SAMN08798357	QERA01	2207	30,199,509	132,072	27,199	334	51.74	89x	98.6
C. pusilla	CCC602	SAMN11159889	SRPW01	9171	37,319,484	83,555	5659	1917	41.84	52x	90.9 *
C. quebecensis	LM458	SAMN08851611	QEQW01	1700	35,882,593	1,850,351	41,784	191	51.87	78x	98.0
C. ripicola	LM454	SAMN08798353	QERC01	2108	30,692,668	189,162	28,587	314	51.37	156x	97.9
C. ripicola	LM218	SAMN08798202	QERE01	1630	30,598,250	206,723	39,763	232	51.4	146x	97.6
C. sorghi	CCC632	SAMN11159890	SRPV01	7206	31,897,900	112,296	6643	1389	45.24	60x	89.9 *

* BUSCO completeness for these strains was based on the Dikaryon fungal database; see Wyka et al. [44] for details.

Table 2. The *eas* gene copies and their locations in 18 species in *Claviceps* sect. *Claviceps* and sect. *Pusillae*.

Section	Organism	Asbl *	Sample	*lpsC*	*easA*	*lpsB*	*cloA*	*easC*	*easD*	*easE*		*easF*			*easG*	*dmaW*			*easH*	
										E1	E2	F1	F2	F3		W1	W2	W3	H1	H2
Claviceps	*C. arundinis*	WF	CCC1102	418/411	411	411	411	411	411	411		411	273		411	411/583	273	707	583	
		SW	CCC1102		157	157	157	157	157	157		157	73		157	157	73	305	157	458
		BW	LM583		455	455	455	455	455	455		455	187		455/805	805	187		805	822
	C. capensis	WF	CCC1504	173	173	173	173	173	173	173		173			173	1354			347	
	C. cyperi	WF	CCC1219	277	277	277	277	277	277	277		277			277	277/2037	2094/696		525	
	C. humidiphila	BW	LM576	599	259	259	259	259	259	259		259	390		259	259	390		259	701
	C. monticola	WF	CCC1483	367	367	367	367	367	367	367	986	367	986		367	367/1745	986/966		494	
	C. occidentalis	WF	LM77	202	202	202	202	202	202	202		202			202	1262	702		1887	1693
		BW	LM78	192	192	192	192	192	192	192		192			192	1273	722		1871	1675
		WF	LM84	290	290	290	290	290	290	290		290/1715			1715	1340	721		1779	1618
	C. pazoutovae	WF	CCC1485	307	307	307	307	307	307	307		307	767		307	307/1479	767/622		430	
	C. perihumidiphila	BW	LM81		114	114	114	114	114	114		114			114/604	604	359		604	710
	C. purpruea	WF	Clav52	131	131	131	131	131	131	131	739	131/1395	739		1395	1395/923	879		923	835
		WF	Clav04	71	71	71	71	71	71	71	1407	71	1514/2293		71	71/993	2459/594		993	874
		WF	Clav26	105	105	105	105	105	105	105		105	816/1596		105	105/1453	1596		759	836
		WF	Clav46	201	201	201	201	201	201	201	1255	201/1358	1255/1668		1358	1358/928	1481/1294		928	1043
		WF	Clav55	419	419	419	419	419	419	419	1226	419/1416	1226/1797		1416	1416/1316	1797/1399		1316	1168
		WF	LM14	85	85	85	85	85	85	85		85			85	85/699	1391/539		699	716
		WF	LM207	374	374	374	374	374	374	374		374			374	374/1169	1625/1506		1169	1183
		WF	LM223	444	444	444	444	444	444	444		444/1556	1027/1718		1556	1556/783	1460		783	762
		WF	LM232	112	112	112	112	112	112	112		112			112	112	1843		1148	908
		WF	LM233	89	89	89	89	89	89	89		89			89/658	658	533		658	692
		BW	LM28	126	126	126	126	126	126	126		126			126	126	1874/563		126	1208
		WF	LM30	88	88	88	88	88	88	88		88			88/1022	1022	610		1022	1180
		WF	LM33	106	106	106	106	106	106	106		106			106	106/1343	1781/649		1343	1210
		WF	LM39	100	100	100	100	100	100	100	855	100/1391	855/1766	1340	1391	1391/753	1340/546		753	703
		WF	LM4	152	152	152	152	152	152	152		152			152	152/1570	1797		857	726
		WF	LM46	209	209	209	209	209	209	209		209			209	209	1386		209	1025
		WF	LM461	121	121	121	121	121	121/1443	1443	1006	1443	1006	1643	1006	1006/1505	1846/1490		1505	1402
		WF	LM469	91	91	91	91	91	91	91	980	91/1419	980/1609	1727	1419	1419/1018	1727/419		1018	680
		WF	LM470	294	294	294	294	294	294	294	2080	294/1428	2080/2165		1428	1428	2358		949	736
		WF	LM474	158	158	158	158	158	158	158	829	158/1493	829/1365		1493	1493/788	1782/41		788	719
		WF	LM5	121	121	121	121	121	121	121		121			121	121/1217	1546/679		1217	1198
		BW	LM582	98	98	98	98	98	98	98		98			98	98	1036		800	
		WF	LM60	333	333	333	333	333	333	333		333			333/1791	1790/1266	1480/676		1266	1242
		WF	LM71	159	159	159	159	159	159	159		159			655	655	931		655	727
		WF	LM63	160	160	160	160	160	160	160	1178	160	1178	985	621/160	621	985		621	918
		WF	LM65	21	21	21	21	21	21	21		21	1018	254	21	21	254		21	
		WF	LM72	190	190	190	190	190	190	190	1156	190	1156		5	5	912		5	
	C. quebecensis	WF	Clav32	308	308	308	308	308	308	308	753	308/201	753/730		201	201	730		201	
		WF	Clav50	227	227	227	227	227	227	227	731	227/231	731/679		231	231	679		231	
		BW	LM458	165	165	165	165	165	165	165	1446	165	1446/1061		498	498	1061		498	
	C. ripicola	BW	LM218	81	81	81	81	81	81	81		81			81	81	527		81	820
		WF	LM219	78	78	78	78	78	78	78		78			78	78	533		78	
		WF	LM220	77	77	77	588	588	588	588		588			588	588	95		588	285
		BW	LM454		120	120	120	623	623	623		623			623	623	107		623	
	C. spartinae	WF	CCC535	1156/1375	853	853	853	680	680	680	27	680	27		680	680	27		680	
Pusillae	*C. africana*	SW	CCC489		424			424	424	424		424			424	424			424/1076	
	C. digitariae	SW	CCC659		403											403				
	C. lovelessii	SW	CCC647		1632	1632	2885	2885/4933	556/4933	556		556			556	556			143	
	C. maximensis	WF	CCC398																	
	C. pusilla	SW	CCC602		3688	1435	3688	3968/3891	3968	3968/3180					3180	1918			1809	
	C. sorghi	SW	CCC632		2692		2475	2475	186	186		186			186	186			3370	

* The assembly versions: BW was from Wingfield et al. [45], SW was from Wyka et al. [44], and WF was generated in the present study; values in the cells denote contig numbers; the 2nd contig number was led by a / when the fragment is on two contigs; green color represents full-length genes, light orange represents partial or gapped sequences, and no fill represents no gene matches; hatches denote fragments containing frameshifts or internal stop codons. None of the genes were detected in *C. citrina* (*C.* sect. *Citrinae*, not listed).

Intraspecific variation among the 27 strains of *C. purpurea* was evident as most strains contained one copy of *lpsC, easA, lpsB, cloA, easC, D, G, H1,* and *H2,* and two copies of *dmaW*. However, three strains (LM65, LM72, and LM582) lacked *easH2*. Eleven strains had a second copy of *easE* (*easE2*), six full- or near-full-length and five partial, but these gene fragments contained indels of various sizes and internal stop codons (Table 2). This would indicate that they may not be functional genes unless those variations were caused by sequencing or assembly errors. In contrast, the second copy of full-length *easF* (*easF2*) from LM72 (MZ881984) and LM461 (MZ881981) had good ORFs. The *easF2* gene of the other six strains was split on two contigs with gaps in the middle. Most of these fragments, except the second exon at the 3′ end of Clav26, Clav55, and LM470, were free of internal stop codons. Four strains had a full length (or close to full length), and one strain (LM469 652 bp) had a partial third copy, *easF3*, yet these gene fragments had a number of indels and internal stop codons (Table 2). The intraspecific variations were also found in *C. arundinis* and *C. ripicola* (Table 2).

The six genomes from *C.* sect. *Pusillae* had more variable numbers of the *eas* genes observed, but all six genomes lacked *lpsC* and *easH2* (Table 2). The strain *C. lovelessii* CCC647 had the highest number of matches, i.e., 10 full- or near-full-length matches (*cloA* 1788 bp, *easD* had an 8 bp short gap at split region), while all but *easH1* and *lpsB* had good ORFs. In contrast, *C. digitariae* CCC659 had only two gene matches: *dmaW* and *easA*, but both were full-length with good ORFs. *C. maximensis* CCC398 and *C. citrina* CCC265 (*C.* sect. *Citrinae*) had no matches for any *eas* genes (Table 2).

Examining each *eas* genes, *easA* was present the most consistently in 51 of 53 genomes as a single copy and had good ORFs, except for the one in *C. pusilla* CCC602 which had an internal stop codon. Similarly, *lpsB, cloA, easC, easD,* and *easG* were present as a single copy in all species of sect. *Claviceps* and two to four species in sect. *Pusillae* (Table 2).

For *easE,* all species in sect. *Claviceps* contained at least one copy, six strains of *C. purpurea* (LM39, LM63, LM72, LM461, LM469, and LM474)) had a full length second copy (*easE2*), and the other five strains *C. purpurea* (Clav04, Clav46, Clav52, Clav55, and LM470), all three *C. quebecensis,* one *C. spartina,* and one *C. monticola* had a second partial copy. Compared with the *C. purpurea* 20.1 *easE1* reference sequence, all the *easE2* sequences contained a large number of deletions (gaps) of various sizes in exon and intron regions, internal stop codons, and no start codon, indicating that they are likely not functional. For species in sect. *Pusillae,* one copy of *easE* was found in four species with good ORFs (*C. africana* CCC489, *C. lovelessii* CCC647, *C. pusilla* CCC602, and *C. sorghi* CCC632).

For *easF*, all species in sect. *Claviceps* contained at least one copy; however, two strains of *C. purpurea* (Clav55 and LM470) had internal stop codons near the 3′ end. Twenty-three strains of seven species (*C. arundinis, C. humidiphila, C. monticola, C. pazoutovae, C. purpurea, C. quebecensis,* and *C. spartinae*) had a second full-length or partial copy, among which 19 strains had good ORFs. In addition, a third copy was found in some *C. purpurea* strains in full length (LM39, LM63, and LM65) or partial (LM461 and LM469). Even though with 77–93% similarity to *C. purpurea* 20.1 *easF1*, none of the third copies had a correct open reading frame (not functional) (Table 2). Three species in sect. *Pusillae* (*C. africana, C. lovelessii,* and *C. sorgji*) had one functioning copy.

For *dmaW*, most species (strains) in sect. *Claviceps* contained two full-length copies or copies split on two contigs with gaps. Six strains of *C. purpurea* (Clav26, Clav52, LM223, LM232, LM4, and LM470) had a partial second copy, but all three strains of *C. occidentalis* had partial sequences (~650 bp) for both copies. One strain of *C. arundinis* (CCC1102) had a third copy in full length, with 81% and 83% similarities with *dmaW1* and *dmaW2*, but frameshifts and internal stop codons were present. Five species in sect. *Pusillae*, except *C. maximensis*, had one copy.

Interestingly, the additional copies of *easE, easF,* and *dmaW* were more or less clustered together, such that the second copies of all three genes were present on the same contig in *C. monticola* CCC1483 and *C. spartinae* CCC535 (Figure 2A). Alternatively, the *easF2* sequence was split on two contigs, which were located with *easE2* on one contig and *dmaW2* on the other, i.e., *C. purpurea* Clav55, and *C. quebecensis* Clav32, Clav50, and LM458 (Figure 2B). More commonly, *easE2* was on the same contig as *easF2*, whereas *dmaW* was on another contig, such as in seven strains of *C. purpurea* (Clav46, Clav52, LM470, LM474, and LM72; Table 2), or *easF2* co-located with *dmaW* when *easE* was a single copy (LM583; Figure 2C). In cases when the third copy of *easF* was present, they were often on the same contig with *dmaW2*, i.e., *C. purpurea* LM39, LM63, LM65, and LM469 (Figure 2D). The arrangement in LM461 was more peculiar in that the second copies of *easE* and *easF* were on the same contig with dmaW1 and *easG* (a single-copy gene), which indicates that they may all be on the primary ergot alkaloid gene cluster (Figure 2E). The third *dmaW* from CCC1102 (from SW assembly) was not connected to other *eas* genes (Table 2).

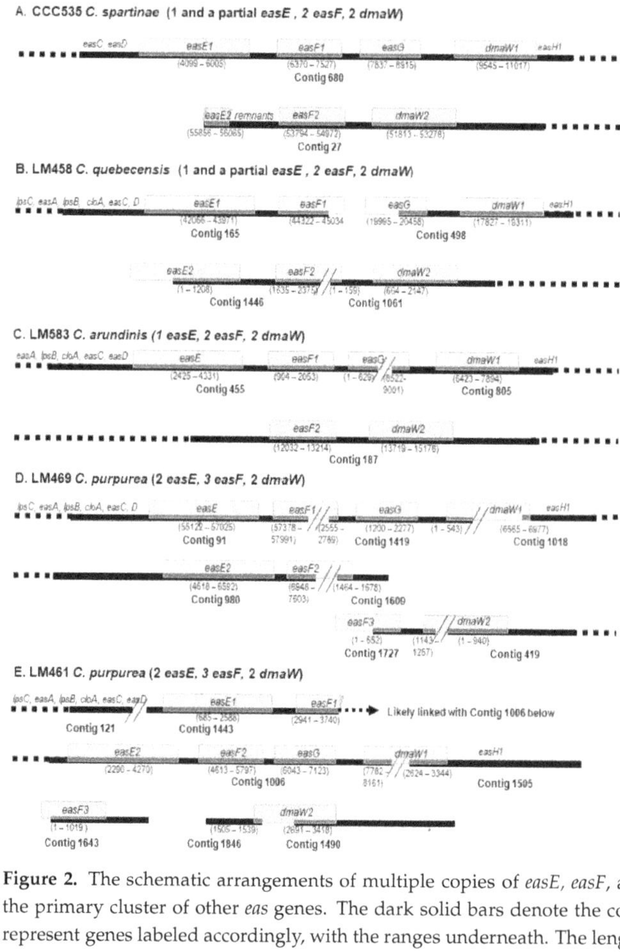

Figure 2. The schematic arrangements of multiple copies of *easE, easF*, and *dmaW* in relation to the primary cluster of other *eas* genes. The dark solid bars denote the contigs, while gray boxes represent genes labeled accordingly, with the ranges underneath. The lengths of genes and spaces are in approximate scale. The dashed bars and genes on them indicate that those genes are on the same contig; however, the details are not displayed. (**A–E**) represent different patterns of locations (see the text Section 2.2.1 for details).

For *easH, easH1* was present in 50 genomes, except *C. citrina, C. digitariae*, and *C.maximensis*; however, the genes of the four species (CCC489, CCC602, CCC632, and CCC647) in *C.* sect. *Pusillae* had numerous indels of various sizes throughout the sequence, causing frameshifts and internal stop codons. Further validation of the sequences is needed to confirm whether these are functioning. The *easH2* gene was present in 32 strains of six species (*C. arundinis*,

C. humidiphila, *C. occidentalis*, *C. perihumidiphila*, *C. purpurea*, and *C. ripicola*). The reference sequence of *easH2* from *C. purpurea* 20.1 was 840 bp, which is about 100 bp shorter than *easH1* (945 bp), and it was considered a pseudogene. Our results showed that the 32 *easH2* sequences had variable lengths and high levels of nucleotide variation (see more notes in later sections: phylogenies and gene diversity). Most of these sequences appeared not functional; however, the lengths of the sequences from two strains of *C. ripicola* (LM218 and LM220) were 954 bp and contained full-length ORFs, indicating that they are likely functioning genes.

For *lpsC*, at least one strain per species in sect. *Claviceps* (except *C. perihumidiphila*) showed one copy of *lpsC*, i.e., in total, 43 out of 46 strains contained a single copy of *lpsC*, among which three strains of *C. purpurea*, i.e., Clav26, LM4, and LM232, had a single internal stop codon; otherwise, the full range of sequences aligned very well with the reference. It is possible that the single internal stop codon could be a sequencing error. Another five strains/species, including *C. capensis* CCC1504, *C. cyperi* CCC1219, *C. humidiphila* LM576, *C. monticola* CCC1483, *C. purpurea* LM223, and *C. spartinae* CCC535 had partial sequences 1000–5000 bp long. These sequence fragments contained several indels and internal stop codons, and they are apparently not functional genes. Only one strain of *C. perihumidiphila* lacked *lpsC*.

2.2.2. Indole-Diterpene/Lolitrem (idt/ltm) Genes

Compared with *eas* genes, the presence/absence and copy numbers of *idt/ltm* genes were less variable. Through mapping genome assemblies to the reference genes, all members in sect. *Claviceps* had one copy of *ltmC*, *M*, *P*, and *S* and one or two copies of *idt/ltmB* and *Q*, except *C. cyperi* CCC1219 that lacked *ltmQ* and *S*. All members in C. sect. *Pusillae* had no matches to any *ltm* genes, whereas members of sect. *Citrinae* (*C. citrina* CCC265) had full-length matches with *ltmB*, *C*, *G*, and *M*.

Notable species-specific features were that all three strains of *C. occidentalis* (LM77, LM78, and LM84) had two partial copies *ltmQ* (1517–1518 bp); *C. arundinis* (CCC1102, LM583), *C. perihumidiphila* (LM81), *C. ripicola* (LM218, LM219, LM220, and LM454), and *C. spartinae* (CCC535) had two functional copies of *ltmB* (Table 3). The translated sequences of *ltmS* from three strains of *C. occidentalis* (LM77, LM78, and LM84) and three strains of *C. quebecensis* (Clav32, Clav50, and LM458) were 14 amino-acid residues longer than other species, and those 14 amino acids were identical among the six strains.

Intraspecific variations were observed in *C. purpurea*; four out of 27 strains showed a second copy of *ltmQ* (Table 3). In the strain Clav04, the fragment on the primary cluster (contig130) *ltmQ1* was a partial copy, whereas another copy on contig 637 was a full-length copy (*ltmQ2*) with a good ORF. Clav46 had two partial copies; ironically, the copy on contig 43 (where all other *ltm* genes co-located) had a number of short deletions causing frameshifts and internal stop codons, whereas the copy on contig 229 had good ORFs, except that the first 243 bp (including 53 residuals in exon 1 and partial exon 2) were missing. On the other hand, some of the single-copy *ltmQ* sequences, such as in *C arundinis* CCC1102, *C. pazoutovae* CCC1485, *C. perhumidiphila* LM81, *C. purpurea* LM72, *C. quebecensis* Clav32 and LM458, *C. ripicoloa* LM218 and LM219, and *C. spartinae* CCC535, had varied number of indels causing frameshifts and internal stop codons; however, phylogenetically, they still belonged to copy 1 (more details in in Section 2.3).

All six genes were clustered on the same contig in 29 strains of the 12 species in sect. *Claviceps*; otherwise, at least three genes were on the same contig. The clustered six *ltm* genes were arranged in the same order as in *C. purpurea* 20.1 [14] (Table 3; gene coordinates are not shown). In *C. citrina*, *ltmB* and *C* were on the same contig (1947), whereas *ltmM* and *G* were on separate contigs. It is not assessable whether they were in one cluster. In general, the inter-gene sequences ranged from 500–1200 bp; however, several strains had very long spaces between *ltmP* and *B*, such as 4 kb in *C. ripicola* LM220 and over 2 kb in LM218 and *C. arundinis* CCC1102 and LM583 (results not shown).

Table 3. The *idt/ltm* gene copies and their locations in *C.* sect. *Claviceps* and sect. *Citrinae.*

Section	Organism	Assembly *		*ltmQ1*	*ltmQ2*	*ltmP*	*ltmB1*	*ltmB2*	*ltmC*	*ltmS*	*ltmM*	*ltmG*
Citrinae	*C. citrina*	WF	CCC265				1947		1947		582	2211
Claviceps	*C. arundinis*	WF	CCC1102	50		50	50	332	50	50	50	
		BW	LM583	158		158	158	124	158	158	158	
	C. capensis	WF	CCC1504	29		29	29		29	29	29	
	C. cyperi	WF	CCC1219			25	25		25		25	
	C. humidiphila	BW	LM576	945		945	478		745	745	745	
	C. monticola	WF	CCC1483	568		568	591		591	591	591	
	C. occidentalis	WF	LM77	1456	1898	1456	1538		1538	1538	657	
		BW	LM78	985	1877	985	985		985	985	691	
		WF	LM84	376	1789	376	376		376	376	376	
	C. pazoutovae	WF	CCC1485	225		225	185		185	185	185	
	C. perihumidiphila	BW	LM81	27		27	27	7	27	27	27	
	C. purpruea	WF	Clav52	174		174	174		174	174	1230	
		WF	Clav04	130	637	130	130		130	130	130	
		WF	Clav26	116		116	116		116	116	116	
		WF	Clav46	43	229	43	43		43	43	43	
		WF	Clav55	1358/1838/1286	1444	1286	557		557	557	557	
		WF	LM14	243		243	243		243	243	243	
		WF	LM207	255		255	255		255	255	255	
		WF	LM223	327		327	327		327	327	327	
		WF	LM232	315		315	315		315	315	315	
		WF	LM233	7		7	7		7	7	7	
		BW	LM28	258		258	258		258	258	258	
		WF	LM30	87		87	87		87	87	87	
		WF	LM33	51		51	51		51	51	51	
		WF	LM39	192		192	192		192	192	192	
		WF	LM4	361		361	361		361	1220	1220	
		WF	LM46	29		29	29		29	29	29	
		WF	LM461	529/65	1592	965	965		965	965	1500	
		WF	LM469	37		37	37		37	37	37	
		WF	LM470	646		787	787		787	787	787	
		WF	LM474	243		243	243		243	243	243	
		WF	LM5	17		17	17		17	17	17	
		BW	LM582	112		112	112		112	112	112	
		WF	LM60	765		765	765		765	765	440	
		WF	LM71	393		1283	977		977	977	977	
		WF	LM63	433		433	433		433	433	433	
		WF	LM65	406		406	406		406	406	406	
		WF	LM72	549/1151		1151	361		361	361	361	
	C. quebecensis	WF	Clav32	56		56	56		56	56	56	
		WF	Clav50	91		91	91		91	91	91	
		BW	LM458	536		536/1563	475		475	475	475	
	C. ripicola	BW	LM218	191		191	191	136	191	191	191	
		WF	LM219	395		395	638	589	638	638	638	
		WF	LM220	368		368	368	591	368	368	368	
		BW	LM454	138		138	764	949	764	764	764	
	C. spartinae	WF	CCC535	225		225	225	47	225	1212	1212	

* The assembly versions: BW was from Wingfield et al. [45], SW was from Wyka et al. [44], and WF was generated in the present study; values in the cells denote contig numbers, two values connected by/indicate the fragment was on two contigs; green color represents full-length genes, light orange represents partial or gapped sequences, and no fill represents no gene matches; hatches denote fragments containing frameshifts or internal stop codons. None of the *idt/ltm* genes were detected in *C.* sect. *Pusillae* except for two short fragments of *ltmG* from *C. maximensis* CCC398 and *C. digitariae* CCC659 by low stringency search, which are not listed (see also Section 2.2.2).

Through the additional BLAST searches with lower stringency (E-value < E^{-50}), fragments of 483 and 501 bp of *ltmG* from *C. maximensis* CCC398 and *C. digitariae* CCC659, respectively, were pulled out by using *ltmG* from *C. paspali* RRC-1481. They were 76% and 78% similar, respectively, to the reference sequence in the coverage (comparable to the 74% similarity between *C. citrina* CCC265 and *C. paspali* RRC-1481). Running BLAST searches of these two fragments to the NCBI database indicated that 60 bp of the 483 bp from *C. maximensis* matched with *Beauveria bassiana* ARSEF 2860 geranylgeranyl pyrophosphate synthetase; 279 of 501 bp from *C. digitariae* matched with *idtG* (geranylgeranyl diphosphate synthase) from *Periglandula ipomoeae* strain IasaF13.

2.2.3. Loline Alkaloid (lol) and Peramine (per) Genes

All the searches with *lol* and *per* reference genes resulted in no hits, except for the low-stringency BLAST with *lolC* that resulted in small fractions of sequences (~150–180 bp) matched with the start of the fifth exon for seven species (strains): *C. africana* (CCC485), *C. citrina* (CCC265), *C. digitariae* (CCC659), *C. lovelessii* (CCC647), *C. maximensis* (CCC398), *C. pusilla* (CCC602), and *C. sorghi* (CCC632). These fragments matched with 80% to 92% identity to *O*-acetylhomoserine from *Purpureocillium lilacinum* (XM 018324292), *Drechmeria coniospora* (XM 040800194), and *Verticillium dahliae* (XM 009654023) in the NCBI database https://blast.ncbi.nlm.nih.gov/Blast.cgi accessed in August 2021. These sequences were not submitted to GenBank because of their short length.

2.3. Phylogenies of *eas* and *idt/ltm* Genes

The individual phylogenetic trees of 11 *eas* genes all agreed on the long-branched separation between C. sect. *Pusillae* and sect. *Claviceps*, which was congruent with the pattern inferred by the previous multigene analyses combined with morphological, ecological, and metabolic features [3] and supported by the phylogenomic analyses [44] (Figure 3a). In C. sect. *Pusillae*, all genes agreed on the close proximity of *C. fusiformis*, *C. lovelessii*, and *C. pusilla*, as well as of *C. africana* and *C. sorghi*. The main incongruence among the gene trees appeared in the uncertain placements of *C. digitariae* and *C. paspali*, as well as the variant relationships among *C. fusiformis*, *C. lovelessii*, and *C. pusillae*, which could be a result of insufficient sampling (see further explanation in Section 3; Figure 3b–d and S1).

In terms of the species relationships in the sect. *Claviceps*, considering single-copy genes, a majority of gene trees agreed on the grouping of the four major clades inferred by the previous phylogenomic study [44]. For communication convenience, we named them as four Batches to avoid confusion with species level and general use of clades: Batch humidiphila including *C. arundinis*, *C. humidiphila*, and *C. perihumidiphila*, Batch purpurea including *C. capensis*, *C. monticola*, *C. pazoutovae*, and *C. purpurea* (previously designated as Clade purpurea by Píchová et al. [3]), Batch occidentalis including *C. occidentalis* and *C. quebecensis*, and Batch spartinae including *C. ripicola* and *C. spartinae* (Figures 3a and S1). The exceptions were *C. perihumidiphila* and *C. cyperi* that had uncertain placement on different gene trees (Figure S1b,d,f,g). The more notable disparities among the gene trees appeared in the order of divergence of the four Batches from C. sect. *Pusillae* or sect. *Paspalorum* (Figures 4, S1 and S2). Previous phylogenomic analyses resulted in the topology of a twice bifurcate pattern, ((Batch humidiphila)(Batch spartinae); (Batch occidentalis)(Batch purpurea)) [44], and this pattern was only supported by *easG* (Figure 4a). A slight variation of the *easA* tree appeared in that Batch humidiphila was an earlier diverged lineage than Batch spartinae, and these two formed a paraphyletic group instead of a monophyletic group (Figure 4b). All other genes supported the derived position of Batch humidiphila and Batch spartinae (Figure 4c–e). Furthermore, eight genes (*cloA*, *dmaW1*, *easC*, *easE1*, *easH1*, *lpsC*, and *ltmB1*) placed Batch purpurea at a more ancestral position than Batch occidentalis, whereas six genes (*easF1*, *lpsB*, *ltmM*, *ltmP*, *ltmS*, and *ltmQ1*) reversed the divergence order of these two Batches (Figure 4c,d). The other three genes (*easD*, *lpsC*, and *ltmC*) showed an unresolved order of divergence (Figure 4e).

As for genes with multiple copies, the most complex was *dmaW*. The *dmaW2* sequences were separated into two groups. Group I included 16 strains of eight species (all non-*C. purpurea dmaW2* except *C. monticola* CCC1483 and *C. pazoutovae* CCC1485), forming a parallel lineage with their *dmaW1* counterpart and representing one gene duplication at node ① (Figures 5a and S2a). Group II included *C. purpurea*, *C. monticola* CCC1483, and *C. pazoutovae* CCC1485, as well as one strain of each *dmaW1* (LM60) and *dmaW3* (CCC1102). This group diverged from *C. purpurea dmaW1*, representing the second duplication at node ②. Within group II, the otherwise consistent close relationship between *C. monticola* and *C. pazoutovae* was broken by seven strains of *C. purpurea*. This can be explained by a third duplication at node ③. The presence of *dmaW3* of *C. arundinis* CCC1102 and *dmaW1 C. purpurea* LM60 in group II indicated extra duplication events at nodes ④ and ⑤ (Figure 5a).

The second and third copies of *easF* (*easF2*, *easF3)* grouped in one clade diverged from *C. cyperi easF1*. Within this clade, *C. purpurea easF2* (14 strains) appeared as a paraphyletic group, from which diverged a clade composed of *C. purpurea easF3* (five strains) and a subclade *easF2* of *C. quebecensis*, *C. humidiphila*, *C. arundinis*, *C. spartinae*, *C. pazoutovae*, and *C. moticola*. From this tree topology, at least two gene duplication events were inferred (Figures 5b and S2b).

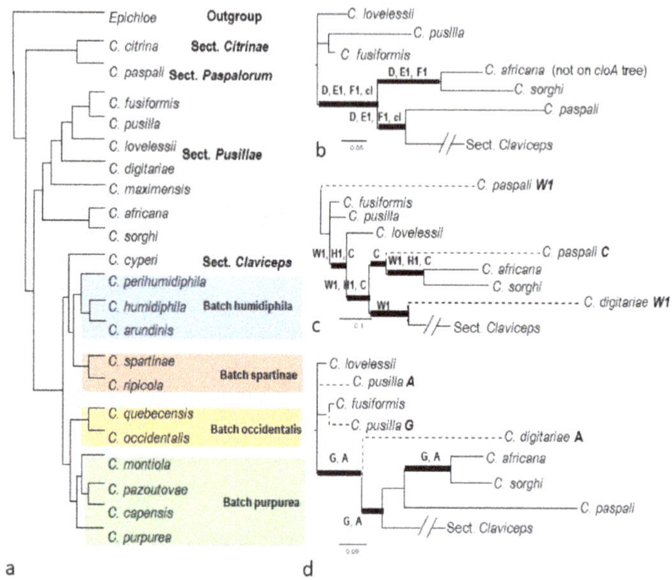

Figure 3. (**a**) The hypothetical species relationships of *Claviceps* spp. inferred by orthologous genes from Wyka et al. [44]. (**b–d**) Variant species relationships in Sect. *Pusillae* summarized from phylogenies inferred by each *eas* gene trees (Supplementary Figures S1 and S2). The thickened branches denote bootstrapping values >80%. The letters next to thick branches denote the genes supporting the grouping, abbreviated as *A*, *C—H1* = *easA*, *easC—H1*; *cl* = *cloA*; *W1* = *dmaW1*. Dashed branches indicate that taxon was present on the gene trees listed after the species name. *lpsC* and *lpsB* are not listed here because only one or three sequences were available on the trees. DNA sequences of *C. fusiformis* and *C. paspali* were from GenBank EU006773 and JN613321.

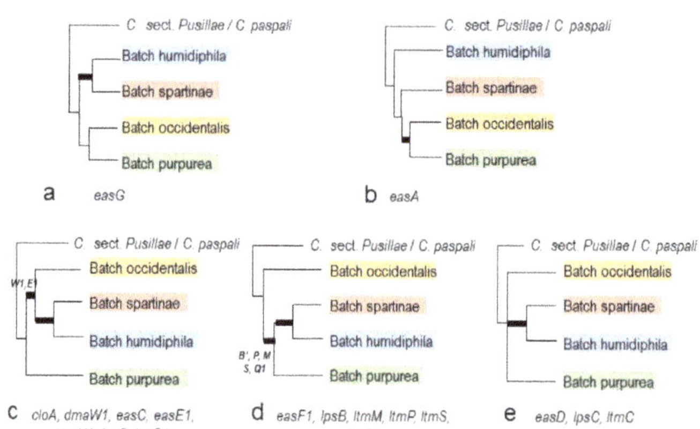

Figure 4. (**a–e**) Varied species relationships in sect. *Claviceps* summarized from phylogenetic trees of *eas* and *ltm* genes by PhyML analyses (the full trees are provided in Supplementary Figures S1 and S2). The thick branches denote bootstrapping values >80%. The letters beside the thick branches indicate that those genes had strong support for those branches; otherwise, all genes listed below the figure had strong support.

The second copy of *easE* (*easE2*) from 16 samples grouped into one clade, which diverged from *easE1* of *C. occidentalis*. However, within the *easE2* clade, *C. purpurea* samples were separated into two subclades. The sample Clav 04 appeared as an orphan clade located close to *C. quebecensis easE2*, and another 10 samples grouped together and had affinity with *C. monticola easE2*, indicating that the historical gene duplications possibly occurred twice at nodes ① and ② (Figures 5c and S2c).

The second copies of *easH* (*easH2*) were grouped into three groups that diverged three times independently. Group I includes two strains of *C. ripicola* (LM218 and LM220) that diverged from *easH1* of the clade composed of *C. capensis*, *C. moticola*, and *C. pazoutovae*. As noted earlier, the sequence lengths of *easH2* from these two strains are similar to *easH1* and contained good ORFs, indicating that they were likely from a very recent gene duplication. Group II, including three strains of *C. occidentalis*, one strain each of *C. arundinis*, *C. humidiphila*, and *C. perihumidiphila*, and 15 strains of *C. purpurea*, diverged from the *easH1* clade composed of eight species in sect. *Claviceps* (*C. occidentalis*, *C. cyperi*, *C. quebecensis*, *C. perihumidiphila*, *C. ripicola*, *C. spartinae*, *C. arundinis*, *C. humidiphila*, and *C. purpurea*). Group III, including nine strains of *C. purpurea* and the reference sequence of *C. purpurea 20.1 easH2*, diverged within the clade of *C. purpurea easH1* (Figures 5d and S2d).

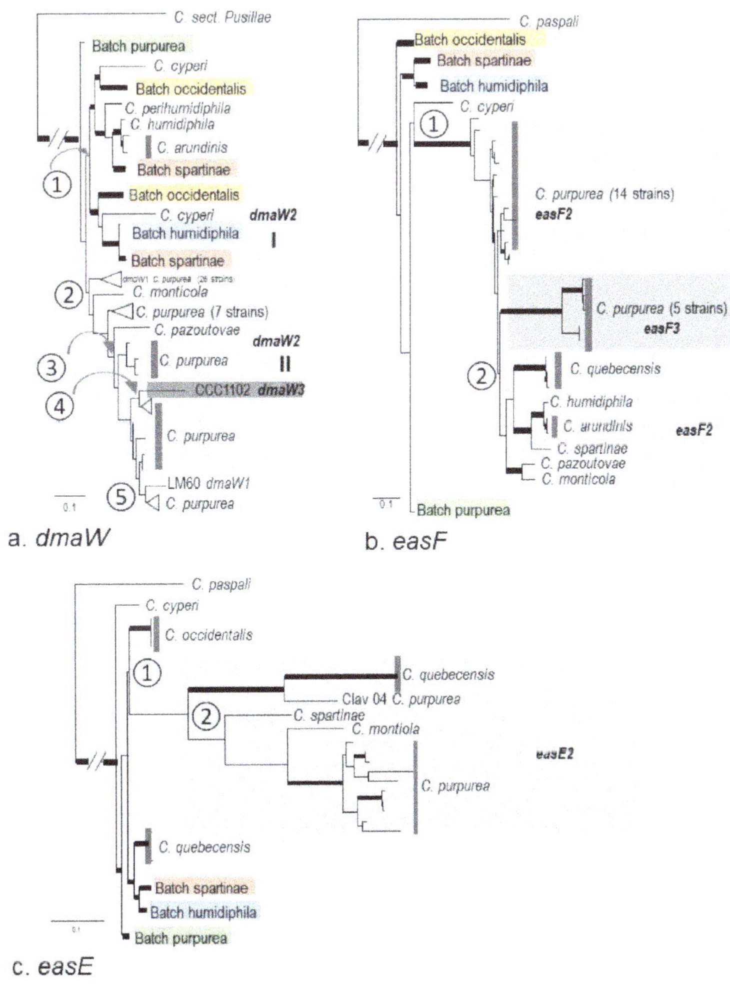

Figure 5. *Cont.*

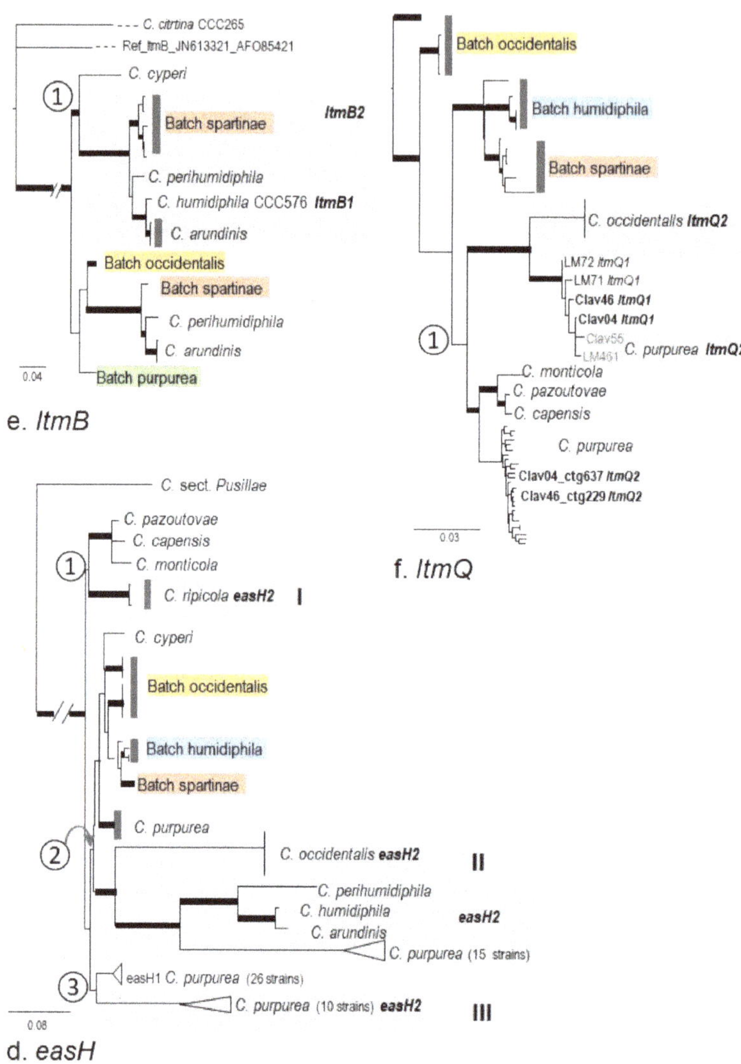

Figure 5. The simplified phylogenies of individual multicopy genes showing potential duplication events. The unedited trees generated by PhyML are presented in the Supplementary Figure S2. (**a**) *dmaW*, (**b**) *easF*, (**c**) *easE*, (**d**) *easH*, (**e**) *ltmB*, and (**f**) *ltmQ*. The thickened branches indicate bootstrapping values ≥80%; dashed and hatches branches are shorter than their real length. The lineages that are not shaded gray are the first copies of each gene.

For *idt/ltm* genes, the second copies of *ltmB* can be considered as one group arising from one gene duplication, except that *ltmB1* of *C. humidiphila* LM576 was placed in this group. This sequence was the only copy detected in LM576 and, therefore, labeled as copy one. However, it was on a separate contig (contig 478), clustered with neither *ltmP* and *ltmQ* (contig 945, Table 3) nor *ltmC*, *ltmS*, and *ltmM* (contig 745). It is very likely that this represents the second copy of this gene, and copy one was either lost or not detected (Figures 5e and S2e).

The three partial *ltmQ2* genes from three strains of *C. occidentalis* grouped closely with a clade composed of four strains *C. purpurea ltmQ1* (Clav04, Clav46, LM71, and LM72) and two *ltmQ2* (Clav55, and LM461) (Figures 5f and S2f). As noted earlier in Section 2.2.2, *ltmQ1* of Clav04 and Clav46 was either a partial gene or a nonfunctional gene, respectively, whereas the second copies were functioning genes. Here, *ltmQ2* of Clav04 and Clav46 grouped in *C. purpurea ltmQ1* clade 1. This situation can be explained by a scenario in which these two copies might have switched locations due to errors in assembling. For another two sequences, *ltmQ1* of LM71 was on a different contig with other *ltm* genes, and in LM72, the gene was split into two contigs, where one half was connected with *ltmP*, while the other half was independent. Overall, these four sequences appeared as the same copy in *C. purpurea ltmQ2* (Clav55 and LM461). If that is the case, one gene duplication event possibly happened at node ①. Alternatively, the *ltmQ2* of Clav04 and Clav26, as well as the two *ltmQ2* groups, could have resulted from independent gene duplications (Figures 5f and S2f). Long-read sequencing, i.e., Nanopore or PacBio, could bring more insight by ruling out the possible assembly errors.

2.4. Intraspecific Genetic Variation within C. purpurea

Overall, the haplotype diversities (Hd) of *eas* genes ranged from 0.936 to 1 (close to saturation), except for *easH2* that had a lower value, 0.858. Nucleotide diversity (Pi) of *eas* genes ranged from 0.08 (*easD*) to 0.168 (*easH2*), the average number of nucleotide difference (K) ranged from 7.1510 (*easD*) to 212.238 (*easE2*), tree-based divergence from COT ranged from 0.06 (*easA* and *easD*) to 0.150 (*easH2*), and tree-based diversity ranged from 0.01 (*easD*) to 0.219 (*easE2*). In general, *easD* and *easA* had lower values for divergence and/or diversity. The second copies of *dmaW*, *easE*, *easF*, and *easH* had much higher values of the four parameters. Some of those genes may not function and, therefore, had fewer functional constraints. If only the first copy of the genes was considered, the genes with the highest diversity and divergence values were Pi 0.03 (*dmaW1*), K 92.379 (*lpsC*), tree-based divergence from COT 0.0025 (*dmaW1*), and tree-based diversity 0.038 (*dmaW1*). The two genes functioning in the middle of the pathway, i.e., *easA* and *easD*, were observed to be the most conserved genes compared with the other genes in the earlier or later steps (Table 4, Figure 6a).

Compared with the first copy of *eas* genes, *idt*/*ltm* genes had a similar level of the highest diversity and divergence. Pi ranged from 0.007 (*ltmM* and *ltmS*) to 0.02 (*ltmQ1*), average number of nucleotide difference (K) ranged from 6.839 (*ltmS*) to 41.486 (*ltmQ1*), tree-based divergence from COT ranged from 0.005 (*ltmM*) to 0.066 (*ltmB1*), and tree-based diversity ranged from 0.009 (*ltmM*) to 0.04 (*ltmQ*) (Table 4, Figure 6b).

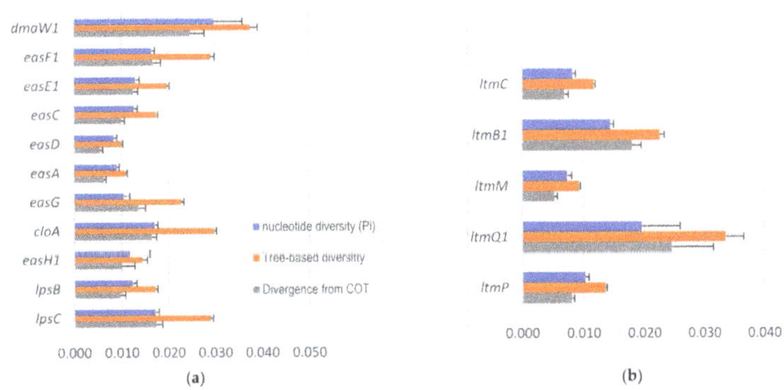

Figure 6. Nucleotide diversity and tree-based diversity and divergence for individual *eas* genes (**a**) and *idt/ltm* genes (**b**). Error bars denote the standard deviation for Pi and standard error for the other two parameters. The genes are arranged from top to bottom according to their order in the biosynthetic pathway. *ltmS* is not included in the chart as its function is unknown.

Table 4. Nucleotide polymorphism, tree-based divergence, and diversity of ergot alkaloid (*eas*) and indole-diterpene/lolitrm (*idt*/*ltm*) synthesis genes in *C. purpurea*.

Biosynthesis Genes	# of Sequences [1]	Total # of Sites	# of Sites (Excluding Indel)	Segregating Sites	Ratio	# of Haplotypes	Haplotype (Gene) Diversity	Nucleotide Diversity		Average Number of Nucleotide Differences	Tree-Based Divergence from COT [2]		Tree-Based Diversity	
	N	n°	n	s	s/n	h	Hd	Pi	Std Dev	K	Mean	Std Error	Mean	Std. Error
Ergot Alkaloid (*eas*) Genes														
dmaW	35	1516	921	196	0.213	32	0.995	0.055	0.004	50.523	0.062	0.004	0.086	0.002
dmaW1	21	1480	938	154	0.164	19	0.99	0.030	0.006	27.886	0.025	0.003	0.038	0.002
dmaW2	14	1516	923	102	0.111	13	0.989	0.042	0.004	39.11	0.042	0.008	0.065	0.004
easF	32	1232	630	121	0.192	24	0.972	0.058	0.015	36.548	0.049	0.011	0.081	0.003
easF1	25	1232	642	40	0.062	17	0.953	0.016	0.001	10.54	0.017	0.002	0.029	0.001
easF2	8	1232	634	101	0.159	8	1	0.048	0.020	30.464	0.031	0.015	0.055	0.010
easE	34	2283	1607	577	0.359	28	0.979	0.085	0.021	135.938	0.085	0.028	0.147	0.008
easE1	28	1921	1891	112	0.059	22	0.968	0.013	0.001	24.526	0.013	0.001	0.020	0.000
easE2	7	2283	1614	536	0.332	7	1	0.132	0.034	212.238	0.115	0.042	0.218	0.032
easC	28	1508	1503	89	0.059	23	0.979	0.013	0.001	10.034	0.010	0.001	0.017	0.000
easD	28	851	846	37	0.044	22	0.976	0.008	0.001	7.1510	0.006	0.001	0.010	0.000
easA	28	1143	1143	51	0.045	21	0.966	0.009	0.001	10.286	0.006	0.000	0.011	0.000
easG	28	1151	923	57	0.062	20	0.974	0.010	0.001	9.6720	0.014	0.001	0.023	0.001
cloA	28	2754	2110	173	0.082	23	0.979	0.017	0.001	35.796	0.016	0.001	0.030	0.000
easH	51	1195	569	247	0.434	23	0.936	0.132	0.014	75.151	0.102	0.013	0.154	0.004
easH1	28	947	943	64	0.068	15	0.944	0.012	0.004	11.053	0.010	0.003	0.014	0.001
easH2	23	1195	571	232	0.406	15	0.858	0.168	0.017	95.85	0.150	0.033	0.188	0.010
lpsB	28	4006	3961	255	0.064	23	0.979	0.012	0.001	48.484	0.010	0.001	0.017	0.000
lpsC	27	5431	5416	421	0.078	25	0.994	0.017	0.001	92.379	0.017	0.001	0.029	0.001
Indole-Diterpene/Lolitrem (*idt*/*ltm*) Genes														
ltmC	28	1249	1246	60	0.048	25	0.992	0.008	0.000	10.299	0.007	0.001	0.012	0.000
ltmB1	28	871	868	38	0.044	21	0.971	0.014	0.001	12.516	0.066	0.010	0.024	0.001
ltmM	28	1766	1731	74	0.043	25	0.992	0.007	0.001	12.68	0.005	0.000	0.009	0.000
ltmQ	25	2180	2119	293	0.138	24	0.997	0.019	0.006	41	0.026	0.010	0.040	0.004
ltmQ1	24	2180	2119	292	0.138	23	0.996	0.020	0.006	41.486	0.025	0.007	0.034	0.003
ltmP	28	1949	1895	111	0.059	24	0.981	0.010	0.001	19.479	0.008	0.000	0.014	0.000
ltmS	28	955	924	34	0.037	24	0.989	0.007	0.001	6.839	0.008	0.001	0.015	0.000

[1] Sequences with large gaps causing a significant reduction in the number of sites were excluded from the analyses. [2] Tree-based divergence from the center of tree (COT) and diversity were estimated by DIVIEN; other parameters were estimated by DnaSP.

3. Discussion

3.1. Correlations between the Presence/Absence of Alkaloid Genes and Alkaloid Production

It has been shown while attempting to induce EA production for pharmaceutical purposes (see review by Flieger [46]) that different ergot species produce varied types of ergot alkaloids. Simultaneously, mycologists explored the use of alkaloid chemistry for characterizing *Claviceps* species [47,48]. Pažoutová and colleagues [49] differentiated chemoraces using the qualitative and quantitative features of EA production. A systematic study on EA production in 43 *Claviceps* species confirmed that ergopeptides were produced only by the members in C. sect. *Claviceps*, whereas dihydroergot alkaloids (DH-ergot alkaloids) were produced only by certain members of C. sect. *Pusillae*, i.e., *C. africana*, *C. gigantea*, and *C. eriochloe*. Sixteen out of 28 species in C. sect. *Pusillae* were shown not to produce any EAs, including *C. maximesis*, *C. pusillae*, and *C. sorghi*. Species only producing clavines included *C. fusiformis*, *C. lovelessii*, and three other species [3]. More recent studies demonstrated that the indole alkaloid profiles supported the recognition of new species based on molecular and ecological data [29,30].

The EA genes detected in the present study were consistent with the known EA production of the included species, for the most part. For example, *C. africana* CCC489 had eight genes detected (lacking *cloA*, *easH2*, *lpsB*, and *lpsC*), and all appeared to be functional, consistent with its production of DH-ergot alkaloids. Similarly, in *C. lovelessii* CCC647, ten EA genes were detected (lacking *lpsC* and *easH2*); however, *easH1* and *lpsB* had mutations resulting in a number of internal stop codons, which is consistent with the production of *clavines*, a product of the early pathway [3]. A lack of EA production corresponded to no matches for any EA genes in *C. maximensis* CCC398 and *C. citrina* CCC265 (*C.* sect. *Citrinae*). However, for *C. pusillae* and *C. sorghi*, several functional genes were detected even though no EA production was reported [3]. In *C. pusillae* CCC602, eight genes had full-length matches (*dmaW1*, *easA*, *C*, *D*, *E*, *G*, and *H1*, and *lpsB*) and one partial match (*cloA* 332 bp), but only *dmaW1*, *easC*, and *easE* had ORFs. The lack of *easF*, the second step in the pathway encoding dimethylallyltryptophan *N*-methyltransferase, might explain the lack of production of EAs. *C. sorghi* CCC632 had seven full-length matches (*dmaW1* and *easA*, *C*, *E*, *F*, *G*, and *H1*) and two partial (*cloA* 435 bp and *easD* 653 bp). Except for *cloA* and *easH1*, all other genes had good ORFs. Theoretically, at least chanoclavine should be produced unless those genes were not expressed possibly due to a lack of triggers from physical or environmental conditions [50].

Only the members in *C.* sect *Claviceps* had *lpsC* and *easH2*, although *C. perihumidiphila* LM81, one strain of *C. ripicola* (LM454), and *C. arundinis* (LM583) lacked *lpsC*, and *C. capensis*, *C. cyperi*, *C. humidiphila*, and *C. monticola* had a partial *lpsC*. Moreover, three *C. purpurea* strains (LM65, LM72, and LM582) and three *C. quebecensis* strains (Clav32, Clav50, and LM458) lacked *easH2*. Whether the absence of these genes causes variations in their EA profiles requires a systematic investigation on the associations between *eas* genes and products in those species. It is worth noting, however, that the possibility of false negatives in genome screening cannot be ruled out. For instance, for *C. arundinis* CCC1102, *lpsC* was detected in the WF version of the genome assembly (created in the present study), but not in the previous version (SW [44], Table 2). The opposite also occurred in that a full length of *dmW3* was detected in SW assemblies, but only partially (360 bp) in WF assemblies (this study).

The production of indole diterpenoid compounds in ergot fungi was reported in a small number of species, i.e., *C. arundinis*, *C. cynodontis*, *C. humidiphila*, *C. paspali*, and *C. purpurea* [21,28–30]. Our genome mining showed that *ltmQ*, *P*, *B*, *C*, *M*, and *S* were present in all species in *C.* sect. *Claviceps* except *C. cyperi*. Furthermore, *ltmB*, *C*, and *M* and a nonfunctioning *ltmG* were detected in *C. citrina*, while a partial *ltmG* was detected in *C. maximensis* CCC389 and *C. digitariae* CCC659. According to the proposed pathway, to produce paspaline, the first step requires *ltmG*, followed by *ltmC*, *ltmM*, and *ltmB* [27]. The absence of *ltmG* could stop production unless GGPP is present through other resources. This might be the case in the producers of indole diterpenoid compounds listed above. In the same way, it is very likely that most of the species in C. sect. *Claviceps* and the three species in sect. *Citrinae* and sect. *Pusillae* could also produce some forms of indole diterpenoid compounds.

3.2. Macro-Evolution of the Gene Clusters—Frequent Gene Duplications and Losses

Ergot alkaloid diversity among diverse producers, i.e., species in Hypocreales, Eurotiales, and Xylariales, was formed by three major processes: gene gains, gene losses, and gene sequence changes [13,14]. This is true within the genus *Claviceps*. A recent genus-level genome comparison hypothesized that unconstrained tandem gene duplications were caused by putative loss of repeat-induced point mutations in C. sect. *Claviceps* [44]. This pattern of duplication was confirmed here by the presence of a cluster of second or third copies of *easE*, *easF*, and *dmaW*, as well as second copies of *ltmQ* and *B* (Tables 2 and 3). Moreover, *easE2* and *F2* of *C. purpurea* LM461 were on the same contig as *easG* and partial *dmaW1*, suggesting that the second copies of *easE* and *F* were arranged on the primary cluster possibly as a result of tandem gene duplication. None of the extra gene copies were found in *C.* sect. *Pusillae* or sect. *Citrinae*, consistent with a previous observation that the genomes of sect. *Pusillae* and sect. *Citrinae* had much fewer gene duplication events predicted [44]. According to the phylogenies of multicopy genes, one to five gene duplications can be inferred for individual genes. The *dmaW* gene, encoding the enzyme for the first and determinant step of *EA* production, had the highest number of potential gene duplications. Even though the presence of *dmaW* was conserved across various EA producers and proven to be a monophyletic group [51], its evolutionary rate was faster than genes in the middle steps of the EA pathway.

Gene losses can be inferred through the discrepant placement of certain gene copies on the phylogenies. For instance, one copy of *ltmB* in *C. humidiphila* LM576 was detected; however, this copy grouped with *ltmB2*. It is very likely that this was the second copy of *ltmB* gene, and the first copy was either lost or not detected (Figure 5e, see also Sections 2.2.2 and 2.2.3). The *ltmQ1* from four strains of *C. purpurea* (LM71, LM72, Clav04, and Clav46) was placed in the *ltmQ2* clade. For LM71 and LM72, there was only one copy detected (*ltmQ1*); the scenario is likely similar to *ltmB* of LM576, where this single copy was the second copy, and the original gene was either lost or not detected (Figure 5f). On a related note, *ltmQ2* of Clav04 and Clav46 was located in the *ltmQ1* clade. An intuitive explanation would be that the identities of the two copies switched due to assembly artefacts (Figure 5f). Lastly, the incongruent order of divergence of the four Batches of species in *C.* sect. *Claviceps* inferred by single-copy genes could be explained as lineages sorted during the frequent gains and losses of the ancestral genotypes (Figure 4). Unlike *C.* sect. *Claviceps*, the phylogeny incongruence in *C.* sect *Pusillae* was mainly caused by the uncertain placement of *C. digitariae* and *C. paspali*. In light of the genome structure, this was likely caused by insufficient sampling instead of gene lineage sorting.

3.3. Micro-Evolution of eas Genes within C. purpurea—An Approximate Hourglass Model

The inter- and intraspecific variations of the second metabolite gene clusters in fungi are typically reported as variations in structures, gene contents, copy numbers, null alleles, and nonhomologous clusters (see review by Rokas [52]). Fewer studies have focused on the DNA sequence variations in each of the gene members. Lorenz et al. [53] identified the sequence differences in *lpsA* between two *C. purpurea* strains (P1 and ECC93) that were associated with the different alkaloid types; however, they could not find differences in *cloA* between *C. fusiformiis* and *C. hirtella* that could explain why this gene was functional in the former but not in the latter. Phylogenetic analyses of DNA sequences of four core genes (*dmaW*, *easF*, *easC*, and *easE*) from selected samples across Clavicipitaceae (with emphasis on *Epichloë*) uncovered extensive gene losses, and the origin of EA clusters on Clavicipitaceous fungi was determined to be direct descent rather than horizontal transfer [13].

The present study is the first, to our knowledge, to examine the variations of each gene on a fine scale, i.e., among 28 strains of *C. purpurea*. Both DNA polymorphism analyses of the DNA sequence alignments through DnaSP and tree-based diversity and divergence analyses using the DEVIEN software indicated that the evolutionary rate of early step genes, i.e., *dmaW* and *easF* is much higher than the middle step genes, i.e., *easA*, *C*, *D*, and *E* (Figure 6, Table 4). The pattern matches with the hourglass model in ontogeny, which was also evidenced in genomic studies [39]. The hourglass model (HGM) and early conservation model (ECM) in ontogeny are explained by developmental constraints. HGM considers that, at the middle stage, the *meta*- and *cis*-interactions reach the highest complexity, posing constraints for development [54,55], whereas ECM considers the constraints at early stage to be critical because any alterations at early stage would cause cascading effects [56]. The EA pathway was reported as an unusually inefficient one such that a high volume of certain intermediates were accumulated more than needed for producing the end-products [57]. This may impose less selective pressure on the middle steps. The sclerotia of *C. purpurea* from tall fescue contained chanoclavine (4 ± 3 μg/g) and agroclavine (2 ± 1 μg/g) in addition to the end-products, i.e., ergopeptines and ergnovine [57]. The extra amount of chanoclavine coincides with the lowest evolutionary rates of *easD* and *easA* inferred in the present study (Figure 6). The role of *easD* is to oxidize chanoclavine to chanoclavine aldehyde, followed by the reactions of *easA* and *easG* to yield agroclavine. It is likely that *easD* is under less selective pressure because plenty of supplies are available. Alternatively, it might be under a high level of functional constraints because of its pivotal position in the pathway (first step of closure of the D-ring). A different isoform of *easA* in *C. africana* and *C. gigantean* reacts differently, creating a shunt yielding dihydroergot alkaloids (Figure 1). This diversification may result from the change in ecological niches. Nevertheless, the rates of diversity and divergence of *easA* were the second lowest after *easD*, even though it is physically located in between *lpsB* and *lpsC*. Both of these later step genes had much higher rates than *easA*, possibly due to fewer constraints or more direct positive selection, as they are involved in the final steps. The *cloA* gene represents another point of the pathway where shunts may take place. Presumably depending on the different isoforms of *cloA*, varied levels of oxidation occur, resulting in different end-products [13,15]. The high rates of diversity and divergence of *cloA* may reflect a high level of positive selection.

The signatures of selective pressure in DNA sequences could be detected through neutrality tests. For instance, if the value of Tajima's D significantly deviates from zero, it indicates the presence of selective pressures, i.e., negative values suggest a positive selection, whereas positive values indicate balancing selection [58]. We conducted neutrality tests and found that none of the genes departed significantly from neutrality (results not shown). These results are contradictory to Liu et al. [59], in that *easE* and *easA* were under positive selection in Canadian and western USA *C. purpurea* populations. We speculate here that the small sample sizes in present study (28 sequences versus 200–300 in the previous study) might be the factor limiting the ability of the Tajima's D test to detect selective pressures.

Compared with *eas* gene pathways, it is difficult to evaluate whether or not the evolutionary pattern of *ltm* genes conformed with the hourglass model because the sequential order of steps was uncertain. Even if we assume that paspaline-derived compounds are the main products, in the absence of *ltmG*, there are only two to three sequential steps to paspaline. Nevertheless, *ltmM* had the lowest rate of divergence and diversity compared with earlier (*ltmC*) and later steps (*ltmP* and *Q*).

Our results provide evidence for the first time that *eas* gene evolution follows the hourglass model. Whether this pattern exists in other metabolic gene pathways and the mechanisms that underpin this or other patterns are questions to be answered in future work.

4. Materials and Methods

4.1. Genome Aquisition

Fifty-four genomes of 19 *Claviceps* spp. were studied. The assemblies of 17 genomes and the raw reads of another 34 genomes were from previous studies (Table 1) [44,45], which outlined the protocols for the DNA extraction, library preparation, and sequencing platforms. In the present study, three additional genomes were sequenced (LM63, LM65, and LM72) using a protocol similar to that described in [44]. Briefly, the gDNA samples were normalized to 300 ng and sheared to 350 bp fragments using an M220 Covaris Focused-Ultrasonicator instrument (Covaris, Woburn, MA, USA). The obtained inserts were used as a template to construct PCR-free libraries using the NxSeq AmpFREE Low DNA Library kit (LGC, Biosearch Technologies, Middleton, WI, USA)) following LGC's library protocol. Balanced libraries in equimolar ratios were pooled, and paired-end sequencing was carried on a NextSeq500/550 (Illumina, San Diego, CA, USA) using 2 × 150 bp NextSeq Mid Output Reagent Kit (Illumina, San Diego, CA, USA) according to the manufacturer's recommendations.

The new assemblies of 37 genomes were achieved using the following protocols: raw reads were trimmed using BBDuk, a component of BBTools downloaded from the Joint Genome Institute website (https://jgi.doe.gov/data-and-tools/bbtools/ accessed on

9 November 2021). Both quality-trim and kmer-trim were applied using the parameters qtrim = rl, trimq = 20, forcetrimleft = 10, minlength = 36, ftm = 5, ref = adapters/adapters.fa, ktrim = r, k = 22, mink = 11, hdist = 1, tbo tpe. The qualities of initial reads and post-trimming reads were assessed using FastQC version 0.11.9, setting parameters as quiet, noextract. Pairs of trimmed reads for each strain were assembled using the SPAdes version 3.14.0 genome assembly toolkit with the default parameters [60]. QUAST version 5.0.2 was used to evaluate the resulting assemblies and to obtain statistics about the assembled contigs [61]. To assess the completeness of the genome assemblies, BUSCO 4.1.4 was run on the contigs using the fungal database (fungi odb10) (Creation date: 10 September 2020, number of species: 549, number of BUSCOs: 758) [62].

4.2. *Alkaloid Gene Screening and Extraction*

To investigate the presence/absence of the four classes of alkaloid synthesis genes in 54 genomes, BLAST searches were conducted to interrogate the genomes with the reference genes of interest using an in-house perl script (running blastn with an E-value of E^{-99} as the cutoff). Alternatively, each individual genome assembly was mapped onto the reference genes using the 'Map to Reference' function in Geneious prime 2020.1.2 (https://www.geneious.com, accessed on 9 November 2021). The reference gene clusters were downloaded from GenBank and applied as follows: the clusters of 14 ergot alkaloid synthesis (*eas*) genes and six indole-diterpene/lolitrem genes (*IDT/ltm*) from *C. purpurea* strain 20.1 (JN186799 containing *cloA, dmaW, easA, C–G, easH1, easH2, lpsA1, lpsA2, lpsB,* and *lpsC*; JX402756 containing *idt/ltmB, C, M, P, Q,* and *S*) and *C. paspali* RRC-1481 JN186800 (*easO*) were first applied as a query to interrogate each genome. In addition, the cluster from *C. fusiformis* PRL1980 EU006773 (10 genes: *cloA, dmaW, easA, C–H,* and *lpsB*) were applied to further interrogate genomes in *C.* sect. *Pusillae* and *C. citrina*. For the *IDT/ltm* genes that were not previously reported in *Claviceps purpurea* 20.1, the reference sequences from *C. paspali* JN613321 (*ltmF* and *ltmG*) and *Epichloë* (*ltmE* and *J* on JN613318, and *K* on JN613320) were used to conduct lower stringency megablast searches (https://www.geneious.com, accessed in 9 November 2021) with E-values E^{-50} and E^{-20}. Megablast searches were also conducted for loline alkaloid genes (*lolA, D, E, M–P, T,* and *U* on JF830816, *lolC* FJ464781, and *lolF* FJ594413) and peramine (*perA* JN640287) in all 54 genomes. Genes that were present in genomes were extracted manually. Split fragments of a single gene on different contigs were concatenated on the basis of reference sequences. DNA sequences of genes extracted from the new genomes were submitted to GenBank.

When multiple copies of certain genes were present (such as *dmaW, easE, easF, ltmB,* and *ltmQ*), the copy on the main cluster was designated as copy 1, as determined by examining the contig numbers. The exception was *easH*, which was determined on the basis of the similarity to the two copies determined by previous studies [14]. Disconnected fragments shorter than 300 bps were not considered.

4.3. *Phylogenetic Analyses*

The extracted sequences for each gene were aligned individually through the Geneious Prime (https://www.geneious.com, accessed on 9 November 2021) Align/Assemble function using Global alignment with free end gaps, 93% similarity (5.0/−9.026168) as the cost matrix, a gap open penalty of 12, a gap extension penalty of 3, and two refinement iterations. This protocol is particularly suitable for aligning sequences with large gaps or shorter fragments to full-length sequences. Maximum likelihood phylogenetic trees were developed using the PhyML 3.3.20180621 [63] plugin of Geneious Prime (https://www.geneious.com, accessed on 9 November 2021). Both GTR and HKY substitution models were attempted; branch supports were evaluated through bootstrapping analyses of 100 replicates. Reference sequences of *lpsB* of *C. paspali* has only 52% similarity with *C. purpurea*, causing spurious alignment and a significantly long branch; therefore, they were not included in the analyses.

4.4. Intraspecific Gene Diversity and Divergence Analyses

Population demographic parameters are suitable for investigating genetic differentiation and gene evolution at an intraspecific level. We investigated the DNA polymorphisms, nucleotide diversity (Pi), and average number of nucleotide differences (K) among 27 strains of *C. purpurea* using DnaSP [64]. Another reason for choosing this sub-set of data, instead of all 53 samples, is that all but three strains (LM65, LM2, and LM582 lacked *easH2*) contained all 12 genes, making the results more comparable. Nonetheless, the sequences with long gaps causing a significant reduction in alignment length in *dmaW* and *easF* were excluded from the DnaSP analyses. In addition, the tree-based diversity and divergence from the center of the tree (COT) were calculated through the web-based DIVEIN software (https://indra.mullins.microbiol.washington.edu/DIVEIN/diver.html, accessed on 9 November 2021) [65]. The following parameters were applied: GTR substitution model, optimized equilibrium frequencies, the best of NNI and SPR tree improvement, and topology + branch length tree optimization algorithm. For multicopy genes (*dmaW*, *easE*, *easF*, and *easH*), we calculated the parameters for each individual copy and combined them as one gene (Table 4).

Supplementary Materials: The following are available online at https://www.mdpi.com/article/10.3390/toxins13110799/s1, Figure S1: The phylogenetic trees developed by PhyML for each individual single-copy *eas* and *idt/ltm* genes, thickened branches indicate bootstrapping values >80%. Figure S2: The phylogenetic trees developed by PhyML for each individual multi-copy *eas* and *idt/ltm* genes, Table S1: The collection information for 53 strains of *Clavicep* spp.

Author Contributions: Conceptualization, M.L.; methodology, K.D., P.S. and S.A.W.; software (pipeline), W.F.; formal analysis, M.L., W.F., P.S. and A.B.; resources, J.D., M.K., J.G.M., S.A.W. and K.B.; data curation, W.F., P.S., S.A.W. and A.B.; writing—original draft preparation, M.L.; writing—review and editing, W.F., J.D., S.A.W., K.B., P.S., K.D., M.K. and J.G.M.; funding acquisition, M.L., J.D. and K.B. All authors have read and agreed to the published version of the manuscript.

Funding: This research was funded by Agriculture and Agri-Food Canada's Growing Forward 2 for a research network on Emerging Mycotoxins (EmTox, project # J-000048), STB fungal and bacterial biosystematics J-002272, the Agriculture and Food Research Initiative (AFRI) National Institute of Food and Agriculture (NIFA) Fellowships Grant Program: Predoctoral Fellowships grant no. 2019-67011-29502/project accession no. 1019134 from the United States Department of Agriculture (USDA), and the American Malting Barley Association grant no. 17037621. Additional funding was provided by Agriculture and Agri-Food Canada grant J-001564, Biological Collections Data Mobilization Initiative (BioMob, Work Package 2). This research was supported in part by the U.S. Department of Agriculture, Agricultural Research Service.

Institutional Review Board Statement: Not applicable.

Informed Consent Statement: Not applicable.

Data Availability Statement: The genome and gene data presented in this study are openly available in NCBI upon publication of this article https://www.ncbi.nlm.nih.gov/ (accessed on 9 November 2021). Accession numbers are detailed in the text Section 2.2 and Table 1.

Acknowledgments: We thank the Molecular Technologies Laboratory (MTL) at the Ottawa Research & Development Centre of Agriculture and Agri-Food Canada, and Kassandra R. Bisson for technical assistance, Chunfang Zheng, and Frank You for bioinformatics assistance, Christopher Schardl for advice during the early stages of the study, two anonymous reviewers for reviewing the manuscript.

Conflicts of Interest: The authors declare no conflict of interest. The funders had no role in the design of the study; in the collection, analyses, or interpretation of data; in the writing of the manuscript, or in the decision to publish the results. The USDA is an equal opportunity provider and employer.

References

1. Tenberge, K.B. Biology and life strategy of the ergot fungi. In *Ergot: The Genus Claviceps*; Kren, V., Cvak, L., Eds.; Hardwood Academic Publishers (republished 2006 by Taylor and Francise e-Library): Amsterdam, The Netherlands, 1999; pp. 25–56.
2. Luttrell, E.S. Host-parasite relationships and development of the ergot sclerotium in *Claviceps purpurea*. *Can. J. Bot.* **1980**, *58*, 942–958. [CrossRef]
3. Píchová, K.; Pažoutová, S.; Kostovčík, M.; Chudíčková, M.; Stodůlková, E.; Novák, P.; Flieger, M.; van der Linde, E.; Kolařík, M. Evolutionary history of ergot with a new infrageneric classification (Hypocreales: Clavicipitaceae: *Claviceps*). *Mol. Phylogenetics Evol.* **2018**, *123*, 73–87. [CrossRef]
4. Liu, M.; Overy, D.P.; Cayouette, J.; Shoukouhi, P.; Hicks, C.; Bisson, K.; Sproule, A.; Wyka, S.A.; Broders, K.; Popovic, Z.; et al. Four phylogenetic species of ergot from Canada and their characteristics in morphology, alkaloid production, and pathogenicity. *Mycologia* **2020**, *112*, 974–988. [CrossRef]
5. Campbell, W.P.; Freisen, H.A. The control of ergot in cereal crops. *Plant Dis. Rep.* **1959**, *43*, 1266–1267.
6. European Food Safety Authority. Scientific opinion on ergot allkaloids in food and feed. *EFSA J.* **2012**, *10*, 2798.
7. Tfelt-Hansen, P.; Saxena, P.R.; Dahlöf, C.; Pascual, J.; Láinez, M.; Henry, P.; Diener, H.-C.; Schoenen, J.; Ferrari, M.D.; Goadsby, P.J. Ergotamine in the acute treatment of migraine: A review and European consensus. *Brain* **2000**, *123*, 9–18. [CrossRef] [PubMed]
8. Schiff, P.L., Jr. Ergot and its alkaloids. *Am. J. Pharm. Educ.* **2006**, *70*, 98. [CrossRef]
9. Barger, G. *Ergot and Ergotism: A Monograph Based on the Dohme Lecture Delivered in Johns Hopkins University, Baltimore*; Gurney and Jackson: London, UK, 1931.
10. Belser-Ehrlich, S.; Harper, A.; Hussey, J.; Hallock, R. Human and cattle ergotism since 1900: Symptoms, outbreaks, and regulations. *Toxicol. Ind. Health* **2013**, *29*, 307–316. [CrossRef] [PubMed]
11. Schardl, C.L.; Panaccione, D.G.; Tudzynski, P.; Geoffrey, A.C. Ergot alkaloids—Biology and molecular biology. In *The Alkaloids: Chemistry and Biology*; Academic Press: Cambridge, MA, USA, 2006; Volume 63, pp. 45–86.
12. Tudzynski, P.; Holter, K.; Correia, T.; Arntz, C.; Grammel, N.; Keller, U. Evidence for an ergot alkaloid gene cluster in *Claviceps purpurea*. *Mol. Gen. Genet.* **1999**, *261*, 133–141. [CrossRef]
13. Young, C.A.; Schardl, C.L.; Panaccione, D.G.; Florea, S.; Takach, J.E.; Charlton, N.D.; Moore, N.; Webb, J.S.; Jaromczyk, J. Genetics, genomics and evolution of ergot alkaloid diversity. *Toxins* **2015**, *7*, 1273–1302. [CrossRef] [PubMed]
14. Schardl, C.L.; Young, C.A.; Hesse, U.; Amyotte, S.G.; Andreeva, K.; Calie, P.J.; Fleetwood, D.J.; Haws, D.C.; Moore, N.; Oeser, B.; et al. Plant-symbiotic fungi as chemical engineers: Multi-genome analysis of the clavicipitaceae reveals dynamics of alkaloid loci. *PLoS Genet.* **2013**, *9*, e1003323. [CrossRef]
15. Robinson, S.L.; Panaccione, D.G. Diversification of ergot alkaloids in natural and modified fungi. *Toxins* **2015**, *7*, 201–218. [CrossRef]
16. Knaus, H.G.; McManus, O.B.; Lee, S.H.; Schmalhofer, W.A.; Garcia-Calvo, M.; Helms, L.M.; Sanchez, M.; Giangiacomo, K.; Reuben, J.P.; Smith, A.B., 3rd; et al. Tremorgenic indole alkaloids potently inhibit smooth muscle high-conductance calcium-activated potassium channels. *Biochemistry* **1994**, *33*, 5819–5828. [CrossRef]
17. Smith, M.M.; Warren, V.A.; Thomas, B.S.; Brochu, R.M.; Ertel, E.A.; Rohrer, S.; Schaeffer, J.; Schmatz, D.; Petuch, B.R.; Tang, Y.S.; et al. Nodulisporic acid opens insect glutamate-gated chloride channels: Identification of a new high affinity modulator. *Biochemistry* **2000**, *39*, 5543–5554. [CrossRef] [PubMed]
18. Turland, N.J.; Wiersema, J.H.; Barrie, F.R.; Greuter, W.; Hawksworth, D.L.; Herendeen, P.S.; Knapp, S.; Kusber, W.-H.; Li, D.-Z.; Marhold, K. *International Code of Nomenclature for Algae, Fungi, and Plants (Shenzhen Code) Adopted by the Nineteenth International Botanical Congress Shenzhen, China, July 2017. Regnum Vegetabile 159*; Koeltz Botanical Books: Glashütten, Germany, 2018; Volume 159.
19. Botha, C.J.; Kellerman, T.S.; Fourie, N. A tremorgenic mycotoxicosis in cattle caused by *Paspalum distichum* (L.) infected by *Claviceps paspali*. *J. S. Afr. Vet. Assoc.* **1996**, *67*, 36–37. [PubMed]
20. Prestidge, R.A. Causes and control of perennial ryegrass staggers in New Zealand. *Agric. Ecosyst. Environ.* **1993**, *44*, 283–300. [CrossRef]
21. Uhlig, S.; Botha, C.J.; Vrålstad, T.; Rolén, E.; Miles, C.O. Indole−diterpenes and ergot alkaloids in *Cynodon dactylon* (bermuda grass) infected with *Claviceps cynodontis* from an outbreak of tremors in cattle. *J. Agric. Food Chem.* **2009**, *57*, 11112–11119. [CrossRef] [PubMed]
22. Kozák, L.; Szilágyi, Z.; Tóth, L.; Pócsi, I.; Molnár, I. Functional characterization of the *idtF* and *idtP* genes in the *Claviceps paspali* indole diterpene biosynthetic gene cluster. *Folia Microbiol.* **2020**, *65*, 605–613. [CrossRef]
23. Saikia, S.; Takemoto, D.; Tapper, B.A.; Lane, G.A.; Fraser, K.; Scott, B. Functional analysis of an indole-diterpene gene cluster for lolitrem B biosynthesis in the grass endosymbiont *Epichloë festucae*. *FEBS Lett.* **2012**, *586*, 2563–2569. [CrossRef]
24. Young, C.A.; Bryant, M.K.; Christensen, M.J.; Tapper, B.A.; Bryan, G.T.; Scott, B. Molecular cloning and genetic analysis of a symbiosis-expressed gene cluster for lolitrem biosynthesis from a mutualistic endophyte of perennial ryegrass. *Mol. Genet. Genom. MGG* **2005**, *274*, 13–29. [CrossRef]
25. Young, C.A.; Felitti, S.; Shields, K.; Spangenberg, G.; Johnson, R.D.; Bryan, G.T.; Saikia, S.; Scott, B. A complex gene cluster for indole-diterpene biosynthesis in the grass endophyte *Neotyphodium lolii*. *Fungal Genet. Biol.* **2006**, *43*, 679–693. [CrossRef]
26. Young, C.A.; Tapper, B.A.; May, K.; Moon, C.D.; Schardl, C.L.; Scott, B. Indole-diterpene biosynthetic capability of *Epichloë* endophytes as predicted by *ltm* gene analysis. *Appl. Environ. Microbiol.* **2009**, *75*, 2200–2211. [CrossRef]

27. Jiang, Y.; Ozaki, T.; Harada, M.; Miyasaka, T.; Sato, H.; Miyamoto, K.; Kanazawa, J.; Liu, C.; Maruyama, J.-I.; Adachi, M.; et al. Biosynthesis of indole diterpene lolitrems: Radical-induced cyclization of an epoxyalcohol affording a characteristic lolitremane skeleton. *Angew. Chem. Int. Ed.* **2020**, *59*, 17996–18002. [CrossRef]
28. Uhlig, S.; Egge-Jacobsen, W.; Vrålstad, T.; Miles, C.O. Indole-diterpenoid profiles of *Claviceps paspali* and *Claviceps purpurea* from high-resolution Fourier transform Orbitrap mass spectrometry. *Rapid Commun. Mass Spectrom. RCM* **2014**, *28*, 1621–1634. [CrossRef]
29. Negård, M.; Uhlig, S.; Kauserud, H.; Andersen, T.; Høiland, K.; Vrålstad, T. Links between genetic groups, indole alkaloid profiles and ecology within the grass-parasitic *Claviceps purpurea* species complex. *Toxins* **2015**, *7*, 1431–1456. [CrossRef]
30. Uhlig, S.; Rangel-Huerta, O.D.; Divon, H.H.; Rolén, E.; Pauchon, K.; Sumarah, M.W.; Vrålstad, T.; Renaud, J.B. Unraveling the ergot alkaloid and indole diterpenoid metabolome in the *Claviceps purpurea* species complex using lc–hrms/ms diagnostic fragmentation filtering. *J. Agric. Food Chem.* **2021**, *69*, 7137–7148. [CrossRef] [PubMed]
31. Schardl, C.L.; Grossman, R.B.; Nagabhyru, P.; Faulkner, J.R.; Mallik, U.P. Loline alkaloids: Currencies of mutualism. *Phytochemistry* **2007**, *68*, 980–996. [CrossRef]
32. Tanaka, A.; Tapper, B.A.; Popay, A.; Parker, E.J.; Scott, B. A symbiosis expressed non-ribosomal peptide synthetase from a mutualistic fungal endophyte of perennial ryegrass confers protection to the symbiotum from insect herbivory. *Mol. Microbiol.* **2005**, *57*, 1036–1050. [CrossRef] [PubMed]
33. Duboule, D. Temporal colinearity and the phylotypic progression: A basis for the stability of a vertebrate Bauplan and the evolution of morphologies through heterochrony. *Development* **1994**, *1994*, 135–142. [CrossRef]
34. Slack, J.M.W.; Holland, P.W.H.; Graham, C.F. The zootype and the phylotypic stage. *Nature* **1993**, *361*, 490–492. [CrossRef] [PubMed]
35. Von Baer, K.E. *Über Entwickelungsgeschichte der Thiere: Beobachtung und Reflexion*; Bei den gebrüdern Bornträger: Königsberg, Russia, 1828; Volume 1.
36. Richardson, M.K. Vertebrate evolution: The developmental origins of adult variation. *Bioessays* **1999**, *21*, 604–613. [CrossRef]
37. Poe, S.; Wake, M.H. Quantitative tests of general models for the evolution of development. *Am. Nat.* **2004**, *164*, 415–422. [CrossRef]
38. Cheng, X.; Hui, J.H.; Lee, Y.Y.; Wan Law, P.T.; Kwan, H.S. A "developmental hourglass in fungi". *Mol. Biol. Evol.* **2015**, *32*, 1556–1566. [CrossRef] [PubMed]
39. Prud'homme, B.; Gompel, N. Genomic hourglass. *Nature* **2010**, *468*, 768–769. [CrossRef]
40. Quint, M.; Drost, H.-G.; Gabel, A.; Ullrich, K.K.; Bönn, M.; Grosse, I. A transcriptomic hourglass in plant embryogenesis. *Nature* **2012**, *490*, 98–101. [CrossRef]
41. Haeckel, E. *Generelle Morphologie der Organismen. Allgemeine Grundzüge der Organischen Formen-Wissenschaft, Mechanisch Begründet Durch Die von Charles Darwin Reformirte Descendenztheorie*; G. Reimer: Berlin, Germany, 1866; Volume 1.
42. Gould, S.J. *Ontogeny and Phylogeny*; Harvard University Press: Cambridge, MA, USA, 1985.
43. Cruickshank, T.; Wade, M.J. Microevolutionary support for a developmental hourglass: Gene expression patterns shape sequence variation and divergence in *Drosophila*. *Evol. Dev.* **2008**, *10*, 583–590. [CrossRef]
44. Wyka, S.A.; Mondo, S.J.; Liu, M.; Dettman, J.; Nalam, V.; Broders, K.D. Whole-genome comparisons of ergot fungi reveals the divergence and evolution of species within the genus *Claviceps* are the result of varying mechanisms driving genome evolution and host range expansion. *Genome Biol. Evol.* **2021**, *13*, evaa267. [CrossRef] [PubMed]
45. Wingfield, B.D.; Liu, M.; Nguyen, H.D.T.; Lane, F.A.; Morgan, S.W.; De Vos, L.; Wilken, P.M.; Duong, T.A.; Aylward, J.; Coetzee, M.P.A.; et al. Nine draft genome sequences of *Claviceps purpurea* s.lat., including *C. arundinis*, *C. humidiphila*, and C. cf. *spartinae, pseudomolecules for the pitch canker pathogen Fusarium circinatum, draft genome of Davidsoniella eucalypti, Grosmannia galeiformis, Quambalaria eucalypti,* and *Teratosphaeria destructans*. *IMA Fungus* **2018**, *9*, 401–418. [CrossRef] [PubMed]
46. Flieger, M.; Wurst, M.; Shelby, R. Ergot alkaloids–sources, structures and analytical methods. *Folia Microbiol.* **1997**, *42*, 3–29. [CrossRef]
47. Tanda, S. Mycological studies on ergot in Japan (Part 9). Distinct variety of *Claviceps purpurea* Tul. on *Phalaris arundinacea* L. and P. *arundinacea* var. *picta* L. *J. Agric. Sci. Tokyo Nogyo Daigaku* **1979**, *24*, 67–95.
48. Pažoutová, S.; Parbery, D.P. The taxonomy and phylogeny of *Claviceps*. In *Ergot: The Genus Claviceps*; Kren, V., Cvak, L., Eds.; Hardwood Academic Publishers (republished 2006 by Taylor and Francise e-Library): Amsterdam, The Netherlands, 1999; pp. 57–77.
49. Pažoutová, S.; Olsovská, J.; Linka, M.; Kolínská, R.; Flieger, M. Chemoraces and habitat specialization of *Claviceps purpurea* populations. *Appl. Envron. Microbiol.* **2000**, *66*, 5419–5425. [CrossRef] [PubMed]
50. Tudzynski, P.; Correia, T.; Keller, U. Biotechnology and genetics of ergot alkaloids. *Appl. Microbiol. Biotechnol.* **2001**, *57*, 593–605. [CrossRef] [PubMed]
51. Liu, M.; Panaccione, D.G.; Schardl, C.L. Phylogenetic analyses reveal monophyletic origin of the ergot alkaloid gene *dmaW* in fungi. *Evol. Bioinform.* **2009**, *5*, EBO–S2633. [CrossRef]
52. Rokas, A.; Wisecaver, J.H.; Lind, A.L. The birth, evolution and death of metabolic gene clusters in fungi. *Nat. Rev. Microbiol.* **2018**, *16*, 731–744. [CrossRef]
53. Lorenz, N.; Haarmann, T.; Pazoutová, S.; Jung, M.; Tudzynski, P. The ergot alkaloid gene cluster: Functional analyses and evolutionary aspects. *Phytochemistry* **2009**, *70*, 1822–1832. [CrossRef]
54. Raff, R.A. *The Shape of Life: Genes, Development, and the Evolution of Animal Form*; University of Chicago Press: Chicago, IL, USA, 2012.

55. Galis, F.; van Dooren, T.J.; Metz, J.A. Conservation of the segmented germband stage: Robustness or pleiotropy? *Trends Genet. TIG* **2002**, *18*, 504–509. [CrossRef]
56. Schlosser, G.; Wagner, G.P. *Modularity in Development and Evolution*; University of Chicago Press: Chicago, IL, USA, 2004.
57. Panaccione, D.G. Origins and significance of ergot alkaloid diversity in fungi. *FEMS Microbiol. Lett.* **2005**, *251*, 9–17. [CrossRef]
58. Tajima, F. Statistical method for testing the neutral mutation hypothesis by DNA polymorphism. *Genetics* **1989**, *123*, 585–595. [CrossRef] [PubMed]
59. Liu, M.; Shoukouhi, P.; Bisson, K.R.; Wyka, S.A.; Broders, K.D.; Menzies, J.G. Sympatric divergence of the ergot fungus, Claviceps purpurea, populations infecting agricultural and nonagricultural grasses in North America. *Ecol. Evol.* **2021**, *11*, 273–293. [CrossRef]
60. Bankevich, A.; Nurk, S.; Antipov, D.; Gurevich, A.A.; Dvorkin, M.; Kulikov, A.S.; Lesin, V.M.; Nikolenko, S.I.; Pham, S.; Prjibelski, A.D.; et al. SPAdes: A new genome assembly algorithm and its applications to single-cell sequencing. *J. Comput. Biol.* **2012**, *19*, 455–477. [CrossRef]
61. Gurevich, A.; Saveliev, V.; Vyahhi, N.; Tesler, G. QUAST: Quality assessment tool for genome assemblies. *Bioinformatics* **2013**, *29*, 1072–1075. [CrossRef]
62. Simão, F.A.; Waterhouse, R.M.; Ioannidis, P.; Kriventseva, E.V.; Zdobnov, E.M. BUSCO: Assessing genome assembly and annotation completeness with single-copy orthologs. *Bioinformatics* **2015**, *31*, 3210–3212. [CrossRef] [PubMed]
63. Guindon, S.; Dufayard, J.F.; Lefort, V.; Anisimova, M.; Hordijk, W.; Gascuel, O. New algorithms and methods to estimate maximum-likelihood phylogenies: Assessing the performance of PhyML 3.0. *Syst. Biol.* **2010**, *59*, 307–321. [CrossRef]
64. Rozas, J.; Ferrer-Mata, A.; Sanchez-DelBarrio, J.C.; Guirao-Rico, S.; Librado, P.; Ramos-Onsins, S.E.; Sanchez-Gracia, A. DnaSP 6: DNA Sequence Polymorphism Analysis of Large Data Sets. *Mol. Biol. Evol.* **2017**, *34*, 3299–3302. [CrossRef] [PubMed]
65. Deng, W.; Maust, B.S.; Nickle, D.C.; Learn, G.H.; Liu, Y.; Heath, L.; Kosakovsky Pond, S.L.; Mullins, J.I. DIVEIN: A web server to analyze phylogenies, sequence divergence, diversity, and informative sites. *Biotechniques* **2010**, *48*, 405–408. [CrossRef] [PubMed]

toxins

MDPI

Review

The Usage of Ergot (*Claviceps purpurea* (fr.) Tul.) in Obstetrics and Gynecology: A Historical Perspective

Aleksander Smakosz [1], Wiktoria Kurzyna [2], Michał Rudko [3] and Mateusz Dąsal [2,*]

1 Department of Pharmaceutical Biology and Biotechnology, Faculty of Pharmacy, Wroclaw Medical University, 50-367 Wroclaw, Poland; aleksander.smakosz@gmail.com
2 Department of Humanities and Social Science, Faculty of Pharmacy, Wroclaw Medical University, 50-367 Wroclaw, Poland; wiktoria.kurzyna@student.umed.wroc.pl
3 Department of Physical Chemistry, Faculty of Pharmacy, Wroclaw Medical University, 50-367 Wroclaw, Poland; michalzdzislawrudko@gmail.com
* Correspondence: mateusz.dasal@umed.wroc.pl

Abstract: In the past centuries consumption of bread made of ergot-infected flour resulted in mass poisonings and miscarriages. The reason was the sclerotia of *Claviceps purpurea* (Fr.) Tul.—a source of noxious ergot alkaloids (ergotamine and ergovaline). The authors have searched the 19th century medical literature in order to find information on the following topics: dosage forms of drugs based on ergot and their application in official gynecology and obstetrics. The authors also briefly address the relevant data from the previous periods as well as the 20th century research on ergot. The research resulted in a conclusion that applications of ergot in gynecology and obstetrics in the 19th century were limited to controlling excessive uterine bleeding and irregular spasms, treatment of fibrous tumors of the uterus, and prevention of miscarriage, abortion, and amenorrhoea. The most common dosage forms mentioned in the works included in our review were the following: tinctures, water extracts (Wernich's and Squibb's watery extract of ergot), pills, and powders. The information documented in this paper will be helpful for further research and helpful in broadening the understanding of the historical application of the described controversial crude drugs. Ergot alkaloids were widely used in obstetrics, but in modern times they are not used in developed countries anymore. They may, however, play a significant role in developing countries where, in some cases, they can be used as an anti-hemorrhage agent during labor.

Keywords: ergot; ethnopharmacology; abortion; childbirth

Key Contribution: This review presents a historical view on the application of ergot in 19th century official pharmacy and medicine in Europe and the USA. In this paper we highlight an initial framework for in-depth analysis of historical ergot properties and usage issues.

Citation: Smakosz, A.; Kurzyna, W.; Rudko, M.; Dąsal, M. The Usage of Ergot (*Claviceps purpurea* (fr.) Tul.) in Obstetrics and Gynecology: A Historical Perspective. *Toxins* **2021**, *13*, 492. https://doi.org/10.3390/toxins13070492

Received: 21 May 2021
Accepted: 13 July 2021
Published: 15 July 2021

1. Introduction

In the history of mankind, *materia medica* representatives have experienced their ups and downs. Plant and animal raw materials have been commonly used in medicine, pharmacy, and other economic sectors. However, fungal raw materials were usually out of the scope of interest. On the other hand, it is difficult not to appreciate the significance of *Saccharomyces cerevisiae* (Desm.) Meyen ex E.C. Hansen (brewer's yeast and baking yeast; with biotechnology, it is possible to produce various drugs with cultures of the above species).

The situation is a bit different with sclerotia of *Claviceps purpurea* (Fr.) Tul. (ergot)—a source of unknown origins and unknown ancient history. Through the ages, this medicinal fungus has been closely connected with gynecology and obstetrics.

We do not know if the ancients could identify and cultivate rye. Possibly this species was used in Thrace and Macedonia to bake bread. Dated around 600 BC, an Assyrian tablet alluded to a "noxious pustule in the ear of grain" (Could it be ergot?). In classical Latin, rye was called secale [1]. A little later, during the medieval period, it was called siligo [1].

A large number of *Poaceae* representatives may become infected by fungi belonging to the *Claviceps* genus [1,2]. However, from both the historical and economical point of view, the most important one is *Claviceps purpurea* (ergot), which causes damage to rye, wheat, and barley [2].

In the past centuries, consumption of bread made of ergot-infected flour resulted in mass poisonings. With regard to symptoms, two forms of ergot poisoning can be distinguished—*ergotismus gangrenosus* and *ergotismus convulsivus* [3]. The first stage of the poisoning is similar in both cases—gastrointestinal and abnormal crawling sensation in the limbs which later develops into pain. With time, the poisoning develops into one of the types mentioned above. In the case of gangrenous ergotism, ischemia affecting the limbs results in distal changes of skin color and then gangrene, which in turn causes the loss of limbs or even death [3]. Convulsive ergotism manifests itself in nervous system disorders. This form of poisoning causes painful involuntary muscle twitching while the body of the person affected by the illness takes abnormal postures. In some cases, mania and hallucinations occurs simultaneously [2,3].

Ergotismus gangrenosus is most probably an equivalent of the so called "St. Anthony's Fire", while *ergotismus convulsivus* is sometimes compared to "St. Vitus Dance" [1–3]. However, the majority of sources associate it with Huntington's disease [3]. The medieval epidemics of *ergotismus gangrenosus* were common in the regions of Europe west of the Rhine (France), while *ergotismus convulsivus* occurred mostly in the part of Europe east of the Rhine (Germany) and in Scandinavia [3,4]. Ergotism epidemics frequently occurred in communities where the diet was rich in rye and took place after cold and wet winters followed by wet springs, as high air moisture and wind facilitate the spreading of the *Claviceps purpurea* fungus. It was mentioned that breast-fed infants did not show any poisoning symptoms [3].

The last time ergot poisoning happened in Europe on a larger scale was in Westphalia, Hanover, and Lauenburg in 1771 where in some villages only 5 out of 120 people survived [3]. It is also implied that ergot may be responsible for the well-known "choreomania" (dancing plague) during which European villagers of the Middle Ages were falling into an involuntary dance trance [3].

2. Ethnomycology of *C. purpurea*

2.1. Nomenclature of Ergot

For centuries, many scientists had difficulties with classifying ergot. In 1816, Augustin Pyramus de Candolle (Swiss botanist who specialized in economic botany and agronomy) described *Secale Cornutum* as a fungus and named this species *Sclerotium Clavis* [5]. Another Swiss researcher—Elias Magnus Fries (professor of botany and applied economics at Uppsala University, the father of modern fungal taxonomy)—denominated it *Spermoëdia Clavus* [5]. What is crucial is that Edwin John Queckett (botanist, surgeon, and microscopist) called it *Ergotoetia abortifaciens* (*abortifaciens*—Latin term for a substance that induces abortion). In 1839 he wrote [6] the following: "I adopted the term abortans [...] is it probable that I ever should have proposed [...] name as being fitted to form the specific one of the newly discovered genus."

Miles Joseph Berkley (Anglican clergyman interested in plant pathology) was also conscious of the specific effect of the aforementioned fungus and therefore he called it *Oidium abortifaciens* [5]. In 1853, Edmond Tulasne (French mycologist and botanist) introduced his views on the development cycle of ergot. Since then, this fungus has been called *Claviceps purpurea* [5].

This name, however, was not widely accepted and editors of contemporaneous London Pharmacopoeia referred to ergot as *Acinula Clavus*—a species never described before [2]. In pharmaceutical sources, ergot was called *Clavus siliginis, Calcar, Secalis mater, Secale luxurians, Secale cornutum*, and *Grana secalis degenerate* [2,4].

2.2. Ethnopharmacology of C. purpurea

We have no clear evidence as to when ergot was first introduced into medicinal use in the West. However, its history in the East is clearer. Alexander Tschirch (pharmacist and pharmacognosist who lived at the turn of the 20th century) stated that Chou Kung (Chinese philosopher and physician) wrote about ergot circa 1100 BC [1]. According to this man of science, this source was used as an obstetrical remedy—the application was unknown in Europe until the 17th or 18th century. On the other hand, the further works associated with botany, e.g., "Thousand Golden Remedies" of Sun Simiao written in the 7th century and "Pen Ts'ao Kang Mu" written in the 16th century by Li Shih-chen (he complied 11,091 prescriptions used throughout centuries; these are listed in 52 volumes describing 443 animal substances and 1,074 plant materials) do not mention ergot [1].

Rye was primarily cultivated by the Teutons who were a Germanic tribe [1], so it should not be surprising that the first medicinal reference to ergot as a drug comes from German sources. Adam Lonicer was a German botanist and physician in the city of Frankfurt [7]. He was an author of the *Kreuterbuch*—one of the most notable herbals in the history of herbalism and pharmacy. In 1582, the fourth edition of this work was published. In the chapter concerning agricultural crops and grains (the subchapter devoted to rye), he wrote that sometimes long black grains stick out of the spike similar to long nails [7]. This description clearly indicates that Lonicer knew what ergot was. Furthermore, he also wrote that women used three sclerotia of the *Claviceps* to induce a strong uterine contraction [7]. This is the oldest evidence of *Secale cornutum* application in gynecology and obstetrics.

The oldest known illustration (woodcut) of ergot can be found in a book entitled the following: "Botanical theatre or the history of plants" (lat. *Theatrum Botanicum Sive Historia Plantarum*) edited by the son of Caspar Bauhin in 1658 [8].

3. Ups and Downs of Ergot Application in Gynecology and Obstetrics

3.1. Excessive Uterine Bleeding and Irregular Spasms

According to John Stearns the official usage of ergot in gynecology started in 1747 in Holland. It was subsequently used in France until it was interdicted in 1774 by a legislative act [9]. In the first decade of the 19th century in the state of Washington, there was a high-profile case of a Scottish woman who applied ergot medications in obstetrics practice with fatal results. These issues were associated with the overdosage of C. purpurea. In addition, this fungal resource is very variable in its active constituent content. Due to this quality, it is highly unlikely to determine what a safe dose should be. Dr. J. Stearns, a physician from the State of New York, began applying ergot-based drugs in his gynecology practice in 1807 [9]. Due to the of the rejection of this drug by previous generations, he had to collect it by hand. He started with ergot powder and then he tried the decoction, which he considered superior. The maximum dose was 10 grains (an equivalent of about 32.4 g) but the regular dose was much lower [9]. In the following years, in the USA, a clinical trial net of obstetricians who used ergot in their practice in form of powder and decoction was established [9]. Stearns called this powder "Childbirth powder" (lat. *pulvis parturiens*) [4]. In 1813, the trial results were published in *The New England Journal of Medicine and Surgery*. They were met with general approval. What is important is that some of the physicians described the cases of children that were stillborn after ergot had been applied [9]. This is not surprising—*C. purpurea* is a very potent medication. Due to prior French objections, some American doctors considered the "new" drug to be harmful and worthless [9].

The British doctors were more reserved—they valued the action of ergot, but they observed that application of ergot during pregnancy and delivery was associated with a greater death rate of infants. Apart from applying ergot in midwifery they also tried to use *C. purpurea* in cases of amenorrhoea [4].

The approach to this raw material in France changed in 1872 when the Academy of Medicine in Paris enacted Article 32 of the law of the 19th Ventose and allowed midwives to prescribe ergot [10]. The Academy stated that ergot-based medicines had a lot of advantages in obstetric practice. That regulation was in opposition to the previous laws and decrees, which restricted the prescription of poisons to physicians and veterinarians [10]. Therefore, pharmacies were able to provide gynecological drugs based on sclerotia of *C. purpurea* to a new group of medical professionals.

It is important to point out that usage of ergot was limited to cases that posed a threat to the life of both mother and child. In such cases, the risk was legitimate. Taking this into consideration, Dr. John Stearns created a list of observations regarding the use of ergot [9]. He stated that it should never be administered during labor and in quantities larger than thirty grains (an equivalent of about 48.6 g) [9]. Combined with opium and water, the sclerotia of *C. purpurea* were given when interrupted pains of regular labor occurred (dosage: teaspoonful was administered every ten min). The described drug was contraindicated when regular labor or the contractions were uninterrupted [9].

On the other hand, application of ergot was indicated in the case of extending labor when the contractions were irregular and/or too weak to advance the labor or when the contractions and pains were transferred from the uterus to the other parts of the body, giving puerperal convulsions [9].

In 1814, Dr. Henry S. Waterhouse witnessed difficulties during childbirth [9]. The patient reported vaginal bleeding and lower abdomen constriction. Then, she began to lose consciousness and started bite her tongue. In the hours that followed, the doctor began to observe alarming contractions of the muscles of her limbs, back, abdomen, neck, and lower jaw [11]. Conventional drugs of the time such as tincture of opium and tincture of asafoetida (*Ferula assa-foetida* L.) had not stopped the symptoms from progressing [9]. Eventually, he decided to give her a mixture of 30 grains of ergot with water. The administration of this medication stopped the worsening of her condition and she was soon able to deliver the baby successfully [9].

Another important remark on ergot was made by Dr. John Paterson who specialized in midwifery. In 1840, he wrote the following [11]: "I consider the ergot more to be depended on, as to its particular effects on the uterus than almost any other specific in the Pharmacopoeia."

This scientist also mentioned that he had never seen any side effects of ergot application. It is rather implausible since just a single dose can cause symptoms such as vomiting, colic, pains, headache, and hallucinations [9,11].

Considering the side effects mentioned above, many pharmacists, physicians, and herbalists tried to compound an ergot-based drug with as few side effects as possible. Dr. Rees's ethereal solution seemed to be the best choice. According to *The Retrospect of Practical Medicine and Surgery* from 1840 [11]: "The ethereal solution, the properties of which you have so well tested, was prepared by digesting 4 ounces. of the powdered ergot in 4 fluid ounces of ether for seven days. The result was a solution of the fatty matters contained in the drug: this was poured off, evaporated to dryness, and the residue again dissolved in 2 fluid ounces of ether."

One part of the preparation was the equivalent of two parts of ergot. Both ether and Claviceps possess a narcotic potential and so usage of the ethereal solution was highly hazardous [11].

Other forms of ergot-based drugs used in obstetrics practice were emulsions, mucilages, syrups, and water extracts mixed with aromatic water [2].

3.2. *Fibrous Tumor of the Uterus*

In the 19th century, treatment of a fibrous tumor of the uterus was regarded as beyond the reach of medicine. One of the most problematic symptoms associated with this illness is excessive uterine bleeding and hypertrophy. Ergot-based medicines worked via elimination (excretion) of polyp or intramural tumor [12]. Dr. Byford used a fluid extract of ergot in this treatment (half spoonful for three weeks; both by mouth and sometimes hypodermically). As a result of the therapy, the tumor was expelled and the uterus inverted. The surgeon had to remove the remaining fibroids. In some cases, the pain during therapy was intolerable and therefore the treatment had to be terminated before obtaining expected therapeutic effect. At that time, there were two main compounds used in the therapy of fibrous tumors of the uterus—Wernich's watery extract of ergot and Squibb's watery extract of ergot [12,13].

The first one was yielded by the extraction of the drug: First with ether, then with ethanol, and lastly with water; then the solvent evaporated. Squibb's compound was a simple water extract evaporated on an evaporating dish [13,14]. Both drugs were mixed with glycerin and sometimes with belladonna (*Atropa belladonna* L.) or opium [12,15]. The therapy described above was forgotten until 1953, when further studies on the effects of pure ergometrine on this type of cancer (especially associated with uterine bleeding) were conducted [14]. In this case, the dose of 0.2–0.4 mg of ergometrine was used [14].

3.3. Abortion and Poisoning

In 1844, there was a trial of a physician charged with the intention to procure an abortion [16]. The accuser held that Dr. James Calder administered noxious pills made of powdered savin juniper and essential oil (*Juniperus sabina* L.) and a decoction of ergot to an unmarried woman. In addition, he gave her powder compounded of iron (II) carbonate and cantharides (*Lytta vesicatoria* L.) [16]. In the further proceedings, it turned out that they were in a relationship and she was likely impregnated by him [16]. Therefore, he tried to induce an abortion by instructing her to take two pills a day. Due to the terrible smell and taste, she threw them into the fireplace [16].

Another example of a similar case took place in 1889 [17]. The prosecutrix—a nurse at the Selby Oak Workhouse—accused a surgeon named Cuthbert of an intentional abortion procured with the fluid extract of ergot. Due to unreliable evidence, the trial ended without a sentence. This result may have been influenced by the support of the entire medical community [17].

Poisoning (both criminal and suicidal) with ergot was (and is) uncommon. Cases in which the autopsy proved death by poisoning with this crude drug were usually connected with an attempt to procure abortion. One such case was reported in 1864 [18]. The victim—a young unmarried woman—had all the clinical signs of ergot poisoning before her death—yellow skin, vomiting, headache, dryness, and irritation of the throat, and intestine hyperemia [18]. Death occurred due to the aforementioned *ergotismus gangrenosus*. Further investigation demonstrated that she had been consuming the tincture of ergot and pennyroyal (*Mentha pulegium* L.) essential oil for 11 weeks [18]. The latter substance was used as an abortifacient from the times of Ancient Greece [19].

On the other hand, the administration of ergot in the case of profuse hemorrhage could prevent miscarriage. This kind of treatment was suggested by Dr. A. Freer who declared that he used this medicine in such cases over 200 times [20]. At the same time, he wrote, with a high level of uncertainty, about the application of ergot in abortion (10th–12th week) [20]. Some practicing physicians had the opposite experience with this remedy. Dr. John Basset published his remarks on the application of ergot as a medical abortion drug in 1872 [21]. He expressed the opinion that contractions of the muscular fiber (caused by a pulverized sclerotia) are too weak to expel the fetus. Moreover, the ergot alkaloids stopped uterine bleeding—which is a significant occurrence during the abortions performed at that time [21].

According to D. Allen and G. Hatfield (ethnobotanists), only a solitary and contemporary record of folk usage of ergot in procuring abortions (in Norfolk) has been traced, but it may have been as well a widespread practice throughout the centuries [22].

It is necessary to notice that ergot-induced poisoning is also an issue in animal husbandry. Cattle (*Bos taurus* L.) are very sensitive to ergot alkaloids (especially ergotamine, ergovaline, ergonovine, ergocristine, and ergocornine). Clinical signs of ergotism in cattle include convulsions, ataxia, gangrenous extremities, vasoconstriction, and abortion [23].

4. Contemporary Pharmacology of Ergot

As stated above, ergot was used as a crude drug until the 19th century. In the 20th century, individual alkaloids were extracted and described. Ergotoxine was discovered and described in 1906 as a single chemical compound. This state of affairs held until 1943, when individual alkaloids were isolated from it. Another alkaloid, ergotamine, was isolated in 1918 by Stoll. Lysergic and isolysergic acids were described in 1934 and 1936, respectively. In 1935, another alkaloid was discovered—ergometrine [24,25].

Ergoline is a core chemical structure of ergot alkaloids. The molecular mechanism of drug action is associated with the impact of ergot alkaloids on dopamine, noradrenaline, and serotonin receptors via structural similarities between aforementioned molecules and ergoline core [26]. Similarities of molecular structure are shown in Figure 1.

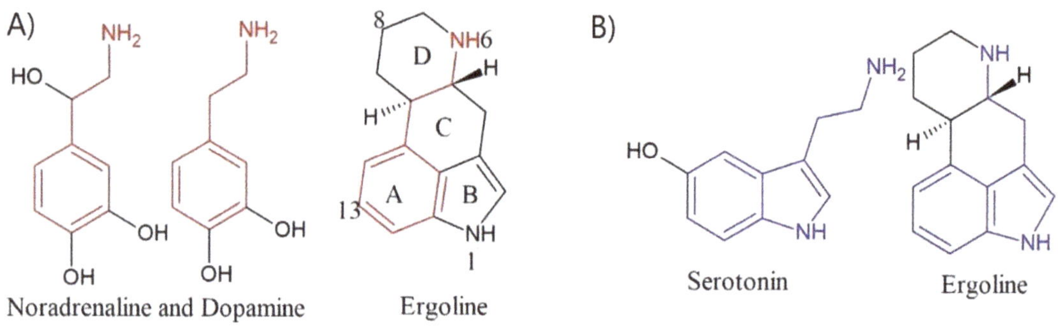

Figure 1. Comparison of dopamine, noradrenaline, and serotonin structure with ergoline. Similarities are exposed by utilizing color (**A**) dopamine, noradrenaline and ergoline; (**B**) Serotonin and ergoline.

By modifying the lysergic acid, a range of derivatives with different receptor activity could be obtained. Amide derivatives of lysergic acid with a small side chain have less adremolytic and higher 5HT antagonist activity. Hydrogenation of ergotamine class alkaloids results in a higher adrenolitic effect [27]. Nicergoline is one of the examples of hydrogenation impact on ergot alkaloids derivatives as it is administered in hypertension strong α_1-receptor blocker [26]. The introduction of elaborate moieties in C-13 or C-14 has a strong impact on weakening the interaction with dopamine receptors and results in increased selectivity towards 5-HT_2 receptors. Selectivity towards 5-HT_1 receptors can be obtained via modification of the C and D ring of ergoline core [27].

Ergotamine and dihydroergotamine are α-adrenergic agonists/antagonists, display dopamine 2 receptors agonistic activity, and show partial agonist action on 5-HT receptors. Ergometrine is an α-adrenergic partial agonist with no impact on dopamine 2 receptors and has partial agonist action on 5HT receptors [27,28].

Ergot alkaloids in obstetrics are administered at the third stage of labor to prevent postpartum hemorrhage. They must be administered with great caution because of side effects, e.g., blood pressure elevation and pain. Thus, the dose must be chosen with great care because of the side effects. Employment of ergot alkaloids is more important in developing countries where postpartum hemorrhage is the cause of many deaths and where access to modern medicine methods is limited [28]. On the other hand, the administration of oxytocin in postpartum hemorrhage prevention (in contrast to ergot alkaloids) can result in a prolonged third stage of labor [29]. Ergot alkaloids used to be widely applied in obstetrics and yet, nowadays, only a few are still used. Ethylergonovine is used as a highly effective second-line uterotonic medication (unfortunately it is associated with severe vasoconstriction) [30]. In the developing countries in some cases, ergot alkaloids can be the only anti-hemorrhage agents available during labor. The summary of the ergot alkaloids properties is shown in Table 1.

Table 1. Active compounds found in ergot, application, year of discovery, and impact on receptors for chosen compounds.

Compound	Action/Application	Year of Discovery	Refs.
Ergotamine	Uterotonic, acceleration of labor, restriction of postpartum hemorrhage	1918	[24,26]
Dihydroergotamine	Sympatholytic, shortening labor.	1943 ***	[26,31,32]
Lysergic acid	Basic compound for further synthesis.	1934	[24]
Isolysergic acid	Basic compound for further synthesis.	1936	[24]
Ergometrine (also named ergonovine)	Uterotonic, restriction of postpartum hemorrhage	1935 **	[25,26]
Ergotoxine	Restriction of postpartum hemorrhage	1906, 1943 *	[24]
Ergocristine	No significant use in obstetrics	1943 *	[24]
Ergokryptine	Lactation suppresion, shortening of gestation period in animal model.	1943 *	[24,28,33]
Ergocornine	Suppression of prolactine secretion.	1943 *	[24,29,34]

* Discovery that ergotoxine is a mixture of alkaloids, ** extraction and characterization, *** year of the first synthesis, Ref. = Reference.

5. Conclusions

In conclusion, the application of ergot in gynecology and obstetrics in the 19th century was limited to controlling excessive uterine bleeding and irregular contractions, treatment of fibrous tumors of the uterus, and prevention of miscarriage. There is little evidence that sclerotia of the *Claviceps purpurea* were used in abortion or amenorrhoea. In the cases described above, the aforementioned abortifacient was either ineffective or caused deaths of the patients. On the other hand, abortion in in the 19th century was in most cases illegal and so information about the usage of abortive medications was a type of taboo.

The most common dosage forms mentioned in the works and that are included in our review were the following: tinctures, water extracts (Wernich's watery extract of ergot, Squibb's watery extract of ergot), pills, and powders. The information documented in this paper will be helpful for further research and for broadening the understanding of the historical application of the described controversial crude drug. Twentieth and twenty-first century applications of ergot in medicine will be covered by the authors in future papers.

6. Methodology

The authors conducted a literature search within the JSTO [35] and archive.org [36] databases using keywords such as "Ergot obstetrics", "Ergot abortion", and "Ergot gynecology" between 1822 and 1896 (published in the United Kingdom and the United States of America). After reading the works, we selected the most relevant papers and books for this article. Regarding the inclusion criteria, the works were selected depending on the inclusion based on the following topics: dosage forms of drugs based on ergot and their application in official gynecology and obstetrics. Exclusion criteria include the following: All other articles that did not cover one of these topics as their primary endpoint. The binominal Latin names of plants were synchronized with The Plantlist [37] database and binominal names of fungi were synchronized with the Index Fungorum [38] database.

Author Contributions: Conceptualization, A.S. and M.D.; methodology, A.S.; formal analysis, A.S. and M.D.; investigation, A.S., W.K., M.R. and M.D.; resources, A.S., W.K., M.R. and M.D.; writing—original draft preparation, A.S., W.K., M.R. and M.D; writing—review and editing, A.S. and M.D.; visualization, M.R and M.D.; supervision, A.S. and M.D.; project administration, A.S.; funding acquisition, W.K. and M.D. All authors have read and agreed to the published version of the manuscript.

Funding: This review was funded by the Wroclaw Medical University.

Institutional Review Board Statement: Not applicable.

Informed Consent Statement: Not applicable.

Acknowledgments: The authors wish to thank Piotr Czerwik for the review.

Conflicts of Interest: The authors declare no conflict of interest.

References

1. Bove, F.J. *The Story of Ergot*; A. Karger: Basel, Germany, 1970; pp. 134–139.
2. Christison, R. *A Dispensatory or Commentary on the Pharmacopoeias of Great Britain*; Adam and Charles Black: Edinburgh, UK, 1848; pp. 409–415.
3. Eadie, M.J. Convulsive ergotism: Epidemics of the serotonin syndrome? *Lancet Neurol.* **2003**, *2*, 431–432. [CrossRef]
4. Presscot, O. *A Dissertation on the Natural History and Medicinal Effects of Secale Cornutum, or Ergot*; Cummings and Hilliard no. 1, Cornhill: Boston, MA, USA; Flagg & Gould: Andover, UK, 1813; pp. 1–10.
5. Felter, H.W.; Lloyd, J.U. *King's American Dispensatory*; Ohio Valley Co.: Cincinnati, OH, USA, 1905; Volume 1, pp. 715–726.
6. Queckett, E.J. Origin of the ergot of rye. *Lancet* **1839**, *827*, 542–544. [CrossRef]
7. Lonicer, A. *Kreuterbuch*; Heirs of Christian Egenolph: Frankfurt am Main, Germany, 1840; p. 285.
8. Bauhin, C. *Theatrum Botanicum Sive Historia Plantarum*; Apud Ioannem Konnig: Basel, Germany, 1658; p. 434.
9. Stearns, J. Observations on the Secale Cornutum, or Ergot; with Directions for its use in Parturition. *Am. Med. Rec.* **1822**, *5*, 666–673.
10. Prescription of ergot by midwives. *Br. Med. J.* **1872**, *626*, 719–720. Available online: https://www.jstor.org/stable/25233705 (accessed on 1 April 2021).
11. Braithwaite, W. *The Retrospect of Practical Medicine and Surgery, No. 1.*, 3rd ed.; Simpkin, Marshall, and Co.: London, UK, 1840; pp. 179–183.
12. The Treatment of Fibrous Tumours of the Uterus by Ergot. *BMJ* **1876**, *807*, 759–761. Available online: https://www.jstor.org/stable/25237355 (accessed on 1 April 2021).
13. Martin, F.H. Treatment of Uterine Fibroids. *JAMA* **1896**, *26*, 549–551. Available online: https://www.jstor.org/stable/20214765 (accessed on 1 April 2021). [CrossRef]
14. Heaney, N.S.; Hyman, E. Conservative Therapy of Benign Uterine Bleeding—With Special Reference to the Use of Ergot. *Calif. Med.* **1953**, *79*, 103–104. Available online: https://www.ncbi.nlm.nih.gov/pmc/articles/PMC1521989/ (accessed on 1 April 2021).
15. Smart, A.R. *Ergot, Its Physiological and Therapeutical Action: Read Before the Southern Michigan Medical Society*; Unknown Publisher: Detroit, MI, USA, 1876; pp. 76–86. Available online: https://collections.nlm.nih.gov/catalog/nlm:nlmuid-101222112-bk (accessed on 1 April 2021).
16. Shapter, T. Report of the Trial of a Medical Practitioner, on a Charge of Intent to Procure Abortion. *Prov. Med. Surg. J.* **1844**, *8*, 18–21. Available online: https://www.jstor.org/stable/25497822 (accessed on 1 April 2021).
17. The Charge of Intent to Procure Abortion at Selby Oak. *Lancet* **1889**, *3458*, 1195. [CrossRef]
18. Stephens, J. The Case of Fatal Attempt to Procure Abortion. *BMJ* **1864**, *200*, 503–505. Available online: https://www.jstor.org/stable/25204125 (accessed on 1 April 2021). [CrossRef]
19. Nelson, S.E. Persephone's Seeds: Abortifacients and Contraceptives in Ancient Greek Medicine and Their Recent Scientific Appraisal. *Pharm. Hist.* **2009**, *51*, 57–69. Available online: https://www.jstor.org/stable/41112420 (accessed on 15 May 2021).
20. Freer, A. Ergot in Abortion. *BMJ* **1872**, *612*, 327. Available online: https://www.jstor.org/stable/25232967 (accessed on 1 April 2021).
21. Basset, J. Ergot in Abortion. *BMJ* **1872**, *614*, 381. Available online: https://www.jstor.org/stable/25233062 (accessed on 1 April 2021).
22. Allen, D.E.; Hatfield, G. *Medicinal Plants in Folk Tradition*; Timber Press: Cambridge, UK, 2004; p. 51.
23. Craig, A.M.; Klotz, J.L.; Duringer, J.M. Cases of ergotism in livestock and associated ergot alkaloid concentrations in feed. *Front. Chem.* **2015**, *3*, 8. [CrossRef] [PubMed]
24. Hofmann, A. Historical view on ergot alkaloids. *Pharmacology* **1978**, *16* (Suppl. 1), 1–11. [CrossRef] [PubMed]
25. MacLean, A.B. Ergometrine. *J. Obs. Gynaecol.* **2005**, *25*, 1–2. [CrossRef]
26. Sharma, N.; Sharma, V.; Manikyam, H.; Krishna, A. Ergot Alkaloids: A Review on Therapeutic Applications. *Eur. J. Med. Plants* **2016**, *14*, 1–17. [CrossRef]
27. Tudzynski, P.; Correia, T.; Keller, U. Biotechnology and genetics of ergot alkaloids. *Appl. Microbiol. Biotechnol.* **2001**, *57*, 593–605. Available online: https://link.springer.com/10.1007/s002530100801 (accessed on 1 April 2021). [CrossRef]
28. Liabsuetrakul, T.; Choobun, T.; Peeyananjarassri, K.; Islam, Q.M. Prophylactic use of ergot alkaloids in the third stage of labor. *Cochrane Database Syst. Rev.* **2018**, *2018*. [CrossRef]
29. Salati, J.A.; Leathersich, S.J.; Williams, M.J.; Cuthbert, A.; Tolosa, J.E. Prophylactic oxytocin for the third stage of labor to prevent postpartum haemorrhage. *Cochrane Database Syst. Rev.* **2019**, *4*, 1465–1858. [CrossRef]
30. Vallera, C.; Choi, L.O.; Cha, C.M.; Hong, R.W. Uterotonic Medications. *Anesthesiol. Clin.* **2017**, *35*, 207–219. [CrossRef]
31. Silberstein, S.D.; McCrory, D.C. Ergotamine and dihydroergotamine: History, pharmacology, and efficacy. *Headache* **2003**, *43*, 144–166. [CrossRef] [PubMed]
32. Altman, S.G.; Waltman, R.; Lubin, S.; Reynolds, S.R.M. Oxytocic and toxic actions of dihydroergotamine-45. *Am. J. Obs. Gynecol.* **1952**, *64*, 101–109. [CrossRef]
33. Whitacre, M.D.; Threlfall, W.R. Effects of ergocryptine on plasma prolactin, luteinizing hormone, and progesterone in the periparturient sow. *Am. J. Vet. Res.* **1981**, *42*, 1538–1541.

34. Welsch, C.W.; Iturri, G.; Meites, J. Comparative effects of hypophysectomy, ergocornine and ergocornine-reserpine treatments on rat mammary carcinoma. *Int. J. Cancer* **1973**, *12*, 206–212. [CrossRef]
35. JSTOR. Available online: https://www.jstor.org/ (accessed on 1 April 2021).
36. Internet Archive. Available online: https://www.archive.org/ (accessed on 1 April 2021).
37. The Plant List Version 1.1. Available online: https://www.theplantlist.org/ (accessed on 1 April 2021).
38. Index Fungorum. Available online: http://www.indexfungorum.org/ (accessed on 1 April 2021).

toxins

MDPI

Article

Novel Ergot Alkaloids Production from *Penicillium citrinum* Employing Response Surface Methodology Technique

Memuna Ghafoor Shahid [1,*], Muhammad Nadeem [2], Ahmed Gulzar [3], Muhammad Saleem [4], Hafeez ur Rehman [3], Gul Zareen Ghafoor [5], Muhammad Umar Hayyat [5], Laila Shahzad [5], Rabia Arif [4] and Rubina Nelofer [2]

1 Department of Botany, GC University, Lahore 54000, Pakistan
2 Food and Biotechnology Research Center, PCSIR Laboratories Complex, Lahore 54000, Pakistan; mnadeempk@yahoo.com (M.N.); rubinanelofer@gmail.com (R.N.)
3 Department of Economics, University of Management & Technology, Lahore 54000, Pakistan; ahmedgulzar2011@gmail.com (A.G.); hafeez.rehman@umt.edu.pk (H.u.R.)
4 Department of Botany, University of the Punjab, New Campus, Lahore 54000, Pakistan; saleem.botany@pu.edu.pk (M.S.); phdgenetics@gmail.com (R.A.)
5 Sustainable Development Study Center, GC University, Lahore 54000, Pakistan; zareen.sdsc@gmail.com (G.Z.G.); umerenv@yahoo.com (M.U.H.); lailashahzad@gcu.edu.pk (L.S.)
* Correspondence: memunaghafoorshahid@gmail.com

Received: 7 April 2020; Accepted: 30 May 2020; Published: 29 June 2020

Abstract: Ergot alkaloids are novel pharmaceutical and therapeutic agents synthesized in this study using fungal species *Penicillium citrinum*. To get the maximum yield of ergot alkaloids a statistical process of response surface methodology was employed using surface culture fermentation technique. Initially, the strain of *Penicillium* was improved using physical (ultraviolet (UV) and chemical (ethyl methane sulfonate (EMS) treatments to get the maximum yield of ergot alkaloids through surface culture fermentation technique. After improving the strain, survival rate of colonies of *Penicillium citrinum* treated with UV and EMS was observed. Only 2.04% living colonies were observed after 150 min of exposure of *Penicillium citrinum* in UV light and 3.2% living colonies were observed after 20 min of the exposure in EMS. The mutated strains of *Penicillium citrinum* were screened for their production of ergot alkaloids and after fermentation experiments, maximum yield was obtained from PCUV-4 and PCEMS-1 strains. After strain improvement, Plackett–Burman design (PBD) and Box–Behnken design (BBD) of RSM were employed and 10-fold yield enhancement (35.60 mg/100 mL) of ergot alkaloids was achieved. This enhancement in yield of ergot alkaloids proved the positive impacts of RSM and UV on the yield of ergot alkaloids. The study provides a cost effective, economical and sustainable process to produce medically important ergot alkaloids which can be used in various pharmaceutical formulations to treat human diseases.

Keywords: ergot alkaloids; strain improvement; UV; EMS; *Penicillium citrinum*; response surface methodology; PBD; BBD

Key Contribution: The production of pharmaceutically and therapeutically important alkaloids using cost effective and sustainable technique can uplift the economy of a country.

1. Introduction

Secondary metabolites are organic compounds produced by different species of bacteria, fungi and plants that are helpful in their growth, development and reproduction. They play a supportive role in the long-term impairment of an organism's health, survival and formation of the offspring. The secondary metabolites are sometimes restricted to a specific set of species within a group of organisms. These are also helpful in the plant defense mechanism against insects and other organisms and can be used by human beings in various industrial, pharmacological and commercial applications [1]. Natural products or secondary metabolites are also known as bioactive compounds which are helping in the discovery and manufacturing of various drugs for the treatment of human ailments [2]. Many plant, animal and fungal species are known for producing naturally occurring bioactive compounds called alkaloids which are considered as a major group of naturally synthesized secondary metabolites [3].

Ergot alkaloids are being described as an assemblage of bioactive compounds of various species including plants and fungi. These were initially documented in *Claviceps purpurea* (a fungal species), the agent known for causing the disease of ergot of rye. These ergot alkaloids were the first time isolated and identified in the sclerotia produced on the kernels of rye plant. Many types of commercially and industrially significant ergot alkaloids were reported from the sclerotia of genus *Claviceps*. However, other species of fungi such as *Blansia, Epichole, Penicillium* and *Aspergillus* and several higher plants can also produce some quantity of ergot alkaloids [4]. In fungi, ascomycetes can efficiently produce alkaloids naturally and in laboratories. *Penicillium*, a well-known genus of Ascomycetes is significant for producing commercially valued secondary metabolites [5,6]. *Penicillium* species can produce significant amounts of alkaloids, antibiotics, hormones and mycotoxins [7]. Ergot alkaloids have been divided into three types on the basis of their structures such as clavines, lysergic acid amides and ergopeptines [8]. Ergot alkaloids are produced commercially for manufacturing of drugs employing various fermentation techniques. Alkaloids synthesis was regulated and enhanced industrially by adding different organic and inorganic ingredients in fermentation medium [3].

The yield of any product can also be enhanced by employing some statistical optimization parameters which help in the quick screening and selection of a number of fermentation factors/ingredients at one time to get the maximum yield of the product in fermentation medium. These statistical procedures also reflect the role and interaction of every individual factor in a specific fermentation method. response surface methodology (RSM) is composed of mathematical and statistical techniques for developing the empirical modeling of a fermentation process for optimizing conditions for producing industrially and commercially important secondary metabolites [9]. Methods and techniques in the synthesis of ergot alkaloids were improved over time with advancement of techniques and still efforts are being conceded, using the optimization techniques of fermentation technology, genetic improvement of strain and the use of protoplasts of the cultures [10].

Statistical designs such as Plackett–Burman design (PBD) [11] and Box–Behnken design (BBD) [12] are very effective and significant techniques for the investigation of targeted factors of fermentation medium. PBD is an efficient screening approach that reduces the number of experiments and gives information for the evaluation of target factors. Only the significant factors with positive responses are selected for further optimization [11,13]. In contrast, BBD is used to generate higher order responses using fewer experimental runs than a normal one factor at a time technique. The Box–Behnken design uses the twelve middle-edge nodes and three center-nodes to fit a 2nd order equation. Box–Behnken designs place points on the mid points of the edges of the cubical design region, as well as points at the center [14]. Another significance of the BBD is that it does not contain combinations for which all factors are simultaneously at their highest or lowest levels. Hence, these designs are useful in avoiding experiments performed under extreme conditions, for which unsatisfactory results may occur. Therefore, through these designs one can enhance the production of any product using fewer experiments at a low cost. Hence, the increasing demand of ergot alkaloids, because of their pharmaceutical and therapeutic nature, is compelling to establish a significant and cheaper process of the production of ergot alkaloids for commercial purposes [15].

The production of microbial ergot alkaloids cannot be achieved without the movement and permanence of the producers and this can be enhanced by inducing mutation in them. The effect of mutations on the biosynthesis of alkaloids is more reliable and applicable because of its speculative importance. The genetics of alkaloids formation has not been widely studied and the effect of mutations using physical (UV-light) and chemical (EMS) mutagens on the biosynthesis of alkaloids can achieve positive results using various strains of fungi such as *Penicillium roquefortii, Aspergillus niger, Trichoderma viride* [8].

In the light of the above scenario, the present study was designed after keeping in mind the significance of ergot alkaloids. The ergot alkaloids are ergotamine derivatives used to treat many ailments. They are significant in increasing the strength of uterine contractions during child birth and are used to limit postpartum bleeding. Many of the pharmaceutical companies use these derivatives and their extracts in various drugs such as for the treatment of migraine headaches as well. The main objective of the present study was the production and extraction of commercially important ergot alkaloids from fungal species using fermentation technique to contribute to the field of pharmaceutical industry. Hence, this research was designed to synthesize ergot alkaloids from fungi because the life cycle of fungi is very small than plants and it can produce large amounts of ergot alkaloids in a very short period of time. Therefore, *Penicillium citrinum* was used in the present research for the production of ergot alkaloids within a very short period of time using a sophisticated statistical technique of response surface methodology during fermentation studies. The ergot alkaloids which are produced during this study were identified as ergocryptine and ergoclavine which are very useful as pharmaceutical and therapeutic agents and can be used in various drug formulations.

2. Results

2.1. Strain Improvement

2.1.1. Impact of Physical and Chemical Mutagen on *Penicillium citrinum*

The wild strain of *Penicillium citrinum* was subjected to mutation by UV irradiation and ethyl methane sulfonate (EMS) reagent.

Impact of UV Irradiations

The survival percentage of mutated colonies of *Penicillium citrinum* was decreased with the increase in exposure time under UV light. The minimum survival rate was found as 2.04% after 150 min of UV exposure (Table 1). The colonies survived at 150 min of exposure were grown on malt extract and agar medium as shown in Figure 1.

Table 1. Survival rate of colonies of *Penicillium citrinum* after exposure in ultraviolet (UV) light.

UV Exposure Time (min)	*Penicillium citrinum*	
	No. of Colonies	Survival Rate (%)
0	49	100
15	44	89.7
30	41	83.6
45	37	75.5
60	31	63.2
75	28	57.1
90	20	40.8
105	14	28.5
120	7	14.2
135	3	6.12
150	1	2.04

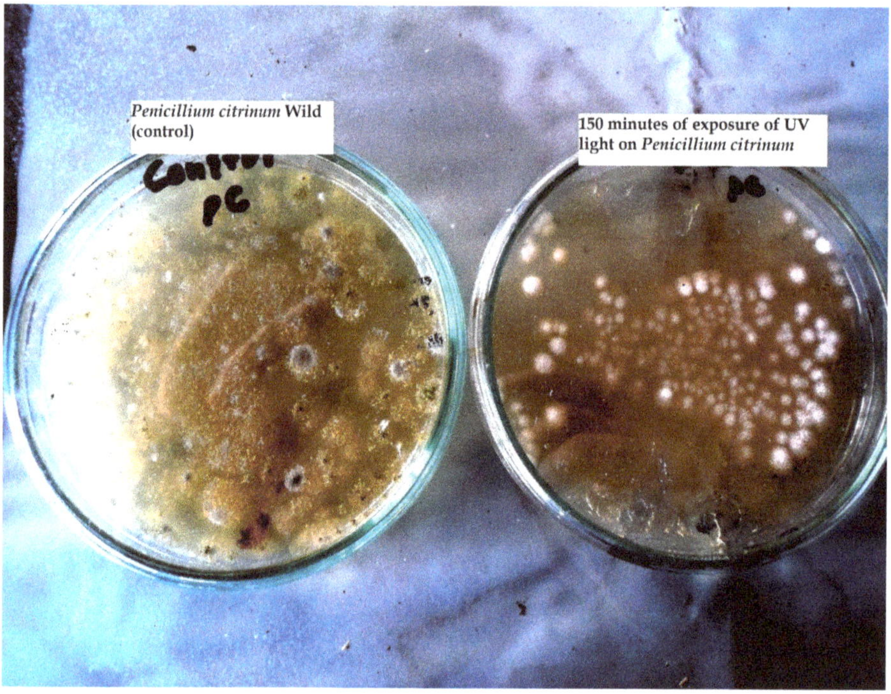

Figure 1. Colonies of *Penicillium citrinum* before and after exposure to UV.

Impact of Ethyl Methane Sulfonate (EMS)

After exposure in EMS for variable times, the survival rate was found to be decreased rapidly and after 30 min of exposure no colony of *Penicillium citrinum* were found alive (Table 2). Minimum survival rate (3.2%) was recorded after 25 min of exposure in EMS and the presence of only a few colonies were observed (Figures 2 and 3).

Table 2. Survival rate of colonies *Penicillium citrinum* in Ethyl Methane Sulfonate (EMS).

EMS Exposure Time (min)	*Penicillium citrinum*	
	No. of Colonies	Survival Rate (%)
0	31	100
10	25	80.6
15	14	45.1
20	7	22.5
25	1	3.2
30	0	0

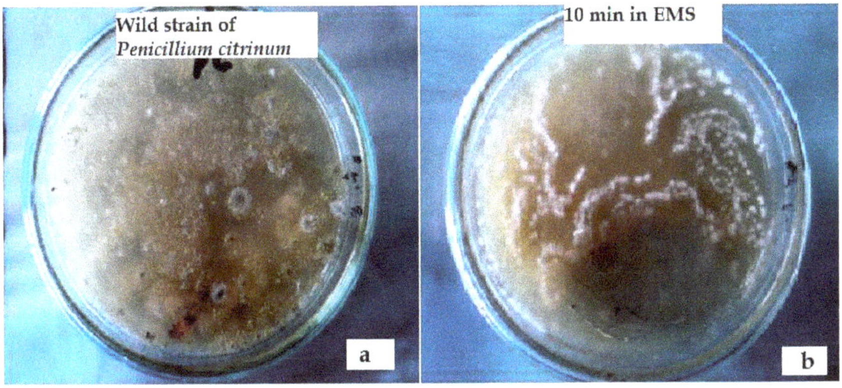

Figure 2. (**a**) Colonies of *Penicillium citrinum* without ethyl methane sulfonate (EMS) treatment. (**b**) *P. citrinum* colonies after 10 min of exposure in EMS.

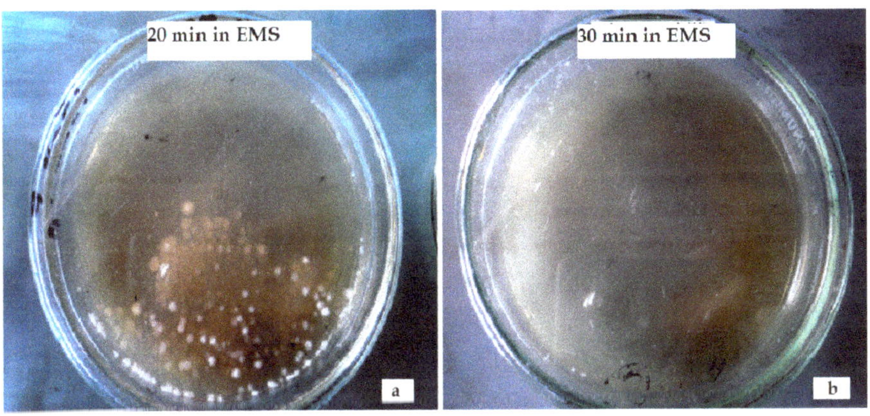

Figure 3. (**a**) Colonies of *Penicillium citrinum* after 20 min of exposure in EMS; (**b**) appearance of no fungal colonies after 30 min of exposure in EMS.

UV and EMS Mutated Strains for Ergot Alkaloids Synthesis

All of the mutated strains were screened for their ability to produce ergot alkaloids and PCUV-4 and PCEMS-3 were found as the useful mutants producing the maximum ergot alkaloids as 4.56 ± 0.01 mg/mL and 2.99 ± 0.005 mg/mL, respectively (Table 3).

Table 3. UV and EMS mutant screening.

UV treated Strains of *Penicillium citrinum*	Extracellular Extract (mg/mL)	Intracellular Extract (mg/mL)	EMS treated Strains of *Penicillium citrinum*	Extracellular Extract (mg/mL)	Intracellular Extract (mg/mL)
PCUV-1	1.49 ± 0.01	1.05 ± 0.02	PCEMS-1	1.84 ± 0.02	1.58 ± 0.01
PCUV-2	1.68 ± 0.02	1.65 ± 0.03	PCEMS-2	2.5 ±0.01	2.10 ± 0.03
PCUV-3	2.56 ± 0.05	1.65 ± 0.01	PCEMS-3	2.99 ± 0.005 *	2.78 ± 0.04 *
PCUV-4	4.56 ± 0.01 *	1.89 ± 0.03 *	Wild	2.27 ± 0.02	2.18 ± 0.02
PCUV-5	3.86 ± 0.02	1.34 ± 0.01			
Wild	2.45 ± 0.03	1.66 ± 0.01			

Each value is an average of three replicates and ± indicates the standard deviation of these replicates. And * $p < 0.05$.

2.2. *Response Surface Methodology*

The Plackett–Burman and Box–Behnken designs were used for the screening and identification of components of the fermentation medium using PCUV-4 as experimental organism.

2.2.1. Screening Step Using PBD

Plackett–Burman design (PBD) was employed for screening of fermentation factors such as yeast extract, sucrose, asparagines, succinic acid, tryptophan, KH_2PO_4, $MgSO_4$, $FeSO_4$ and pH. The highest yield (14.74 ± 0.01 mg/mL) was achieved from run No. 2 and the lowest was observed from run No. 3 (0.36 ± 0.02 mg/mL) as given in Table 4. The Pareto chart was used to show the effect of all fermentation ingredients on ergot alkaloids production (Figure 4).

Table 4. Screening of variables using Plackett–Burman design (PBD).

Run.	Yield of Ergot Alkaloids (mg/mL)
1	11.84 ± 0.1
2	14.76 ± 0.01 *
3	0.36 ± 0.03
4	6.53 ± 0.01
5	10.96 ± 0.01
6	11.95 ± 0.02
7	0.74 ± 0.04
8	5.38 ± 0.1
9	11.79 ± 0.03
10	7.76 ± 0.05
11	0.24 ± 0.02
12	13.02 ± 0.03

Each value is an average of three replicates and "±" indicates the standard deviation among three replicates. * $p < 0.05$.

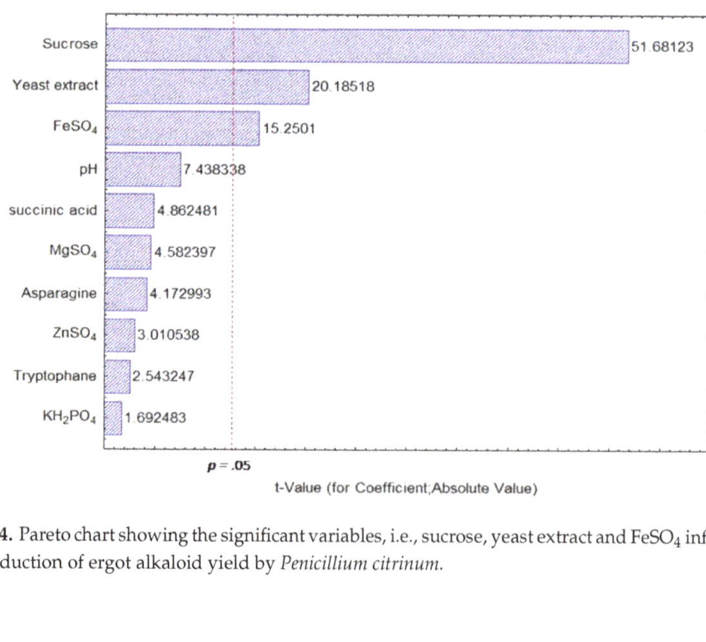

Figure 4. Pareto chart showing the significant variables, i.e., sucrose, yeast extract and $FeSO_4$ influencing the production of ergot alkaloid yield by *Penicillium citrinum*.

ANOVA for PBD Model

The ANOVA (analysis of Variance) of PBD is given in Table 5 that described the factors with $p < 0.05$ were considered significant for the production of ergot alkaloids. From Table 5, it was clearly indicated that sucrose, yeast extract and $FeSO_4$ were significantly influencing the yield of ergot alkaloids. In addition, the coefficient of determination (R^2) of the model was 0.9996 which explained 99.96% variance in data. Sucrose had a confidence level of above 95% in comparison to other variables and thus considered highly significant for ergot alkaloids production.

Table 5. Analysis of variance for using PBD.

Source	Sum of Squares	Degree of Freedom	Mean Square	F-Value	*p*-Value
Intercept	0.46	1	0.46	8.26	0.21
Sucrose	147.78	1	147.78	2670.95	0.012
yeast extract	22.54	1	22.54	4.7.44	0.032
Succinic acid	1.30	1	1.30	23.64	0.13
$MgSO_4$	1.16	1	1.16	20.96	0.14
KH_2PO_4	0.16	1	0.16	2.86	0.34
$FeSO_4$	12.87	1	12.87	232.57	0.042
$ZnSO_4$	0.50	1	0.50	9.06	0.20
Asparagine	0.96	1	0.96	17.41	0.15
Tryptophan	0.36	1	0.36	6.47	0.24
pH	3.06	1	3.06	55.33	0.085
Error	0.06	1	0.06		

2.2.2. Identification of Significant Factors Using Box–Behnken design (BBD)

Box–Behnken design (BBD) was applied to optimize the selected variables (Sucrose, yeast extract and $FeSO_4$) and to find out the effect of their mutual impact on the production of ergot alkaloids. The yield of ergot alkaloids obtained from extracellular extracts of *Penicillium citrinum* ranged from 13.50 mg/mL to 35.60 mg/mL, respectively. The observed values of ergot were compared with predicted values, as presented in Table 6. Maximum ergot alkaloids yield was observed from extracellular extract of run No. 13 (35.60 mg/mL) and it was also compared with the predicted value (35.60 mg/mL). The lowest value of ergot alkaloids yield was found from run No. 6 (13.50 mg/mL) and it was almost similar to the predicted value (14.16 mg/mL) (Table 6). These values were calculated from the polynomial equation as described in Section 5.5.2.

Table 6. Observed and predicted values of yield using Box–Behnken design (BBD).

Runs	Sucrose (g/100 mL)	Yeast Extract (g/100 mL)	$FeSO_4$ (g/100 mL)	Alkaloids Yield (Observed) mg/ml	Alkaloids Yield (Predicted) mg/ml
1.	41	5	0.06	22.50	21.75
2.	41	39	0.06	16.00	17.42
3.	41	22	0.01	24.55	24.32
4.	41	22	0.11	27.79	27.79
5.	5	5	0.06	18.90	17.94
6.	5	39	0.06	13.50 *	14.16 *
7.	5	22	0.01	20.40	20.06
8.	5	22	0.11	25.20	25.30
9.	23	5	0.01	16.87	16.65
10.	23	39	0.01	17.90	17.56
11.	23	5	0.11	25.20	25.53
12.	23	39	0.11	17.60	16.75
13.	23	22	0.06	35.60	35.60

Where * $p < 0.05$.

ANOVA for BBD Model

The analysis of variance of the three factors (Sucrose, yeast extract and $FeSO_4$) indicated that the ergot alkaloids activity can be well described by the polynomial model with a high coefficient of determination (R^2 = 0.95). The statistical model presented in Table 7 shows that each (sucrose, yeast extract and $FeSO_4$) had a significant impact on the production of ergot alkaloids. It was investigated that yeast extract remarkably influenced the production of ergot alkaloids in extracellular extracts of *Penicillium citrinum* with a value of 290.84 mg/mL. Among the combined interaction effect of these three significant variables (sucrose-yeast extract, sucrose-$FeSO_4$ and yeast extract-$FeSO_4$), the combination of yeast extract-$FeSO_4$ interaction was found to be more significant in extracellular extract (23.52 mg/mL). Lowest yield of ergot alkaloids was obtained in the combination of "Sucrose-$FeSO_4$". Through this statistical model the insignificant interaction coefficients were eliminated, and the final polynomial equation was expressed as follows:

$$Y= -21.244 - 19.656\, x^1 + 3.148\, x^2 + 55.350\, x^3 - 1.791\, x1^2 - 0.02\, x2^2 + 55.350\, x3^2 - 0.019\, x^1x^2 - 0.100\, x^1x^3 + 0.243\, x^2\, x^3 \quad (1)$$

where 'Y' was the predicted response, 'X^1, X^2, X^3' were the values of sucrose, yeast extract and $FeSO_4$, respectively. Through this model it can be assumed that the model accurately represents the data in experimental region. This was confirmed by the residual analysis of the data which is not presented here. The main results of this study are presented in Figures 5–7, which shows the expected ergot alkaloids production and correlation between variables in three dimensional plots.

Table 7. Analysis of variance using BBD.

Variable	Sum of Square	Degree of Freedom	Means Square	F-Value	*p*-Value	*t*-Value
Intercept	175.21	1	175.21	167.58	0.001	−12.55
Sucrose	121.31	1	121.31	11.33	0.002	10.07
Sucrose^2	116.22	1	116.22	119.58	0.002	−10.24
yeast Extract	294.94	1	294.94	259.87	0.000	16.12
yeast Extract^2	308.41	1	308.41	285.67	0.000	−16.63
$FeSO_4$	75.62	1	75.62	76.79	0.004	8.17
$FeSO_4\,^2$	49.91	1	49.91	39.89	0.006	−6.91
Sucrose, yeast Extract	3.16	1	3.16	2.01	0.251	−1.42
Sucrose, $FeSO_4$	0.03	1	0.03	0.04	0.862	−0.18
yeast Extract, $FeSO_4$	24.61	1	24.61	21.01	0.019	−4.58

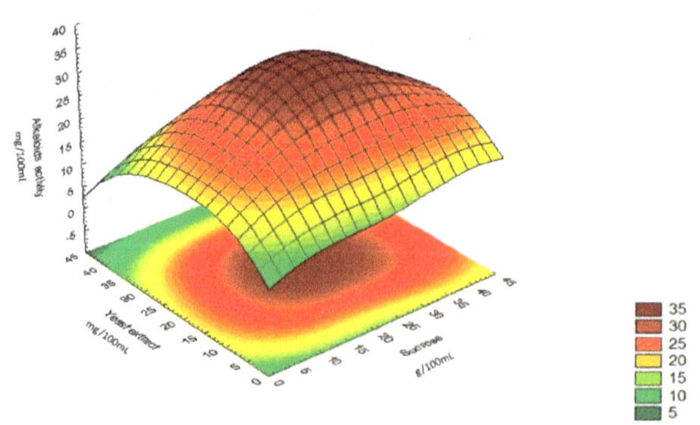

Figure 5. Surface graph showing interactive effect of sucrose and yeast extract on ergot alkaloid production by *Penicillium citrinum*.

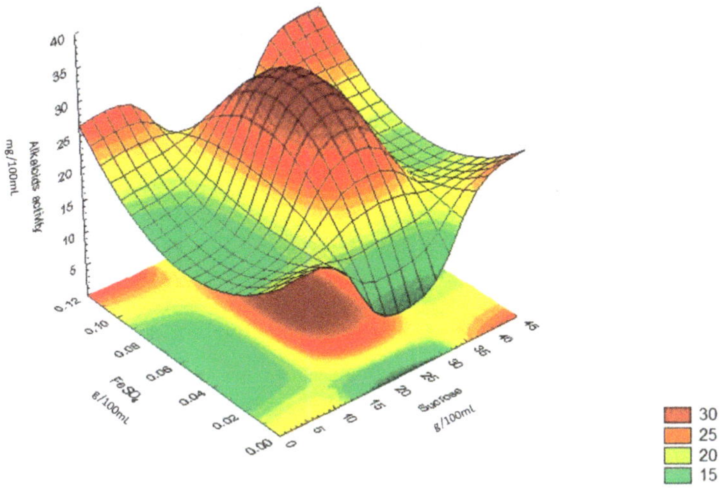

Figure 6. Surface graph showing interactive effect of sucrose and $FeSO_4$ on ergot alkaloid production by *Penicillium citrinum*.

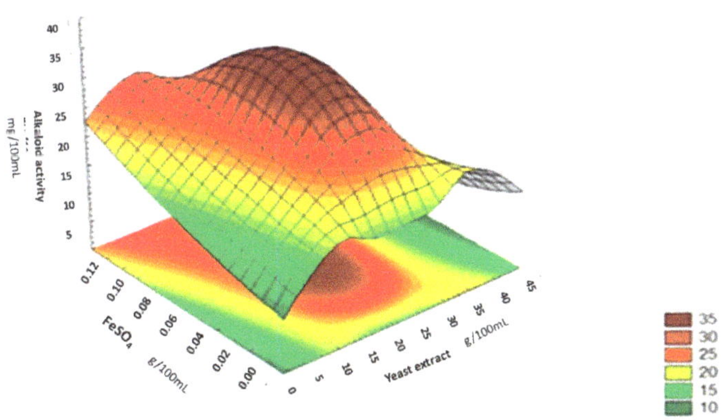

Figure 7. Surface graph showing interactive effect of yeast extract and $FeSO_4$ on ergot alkaloid production by *Penicillium citrinum*.

Figure 5.shows significant additive effect of sucrose and yeast extract on the yield of ergot alkaloids. This two factor impact means that both were dependent on each other for ergot alkaloids production. The shape of the curve and the dark red color in Figure 6 also indicates that maximum alkaloids yield was obtained when 41 g/100 mL of sucrose and 39 g/100 mL of yeast extract was added in the experimental run.

Figure 6 illustrates that increasing the sucrose value at the moderate levels of $FeSO_4$ in the fermentation medium led to maximum ergot alkaloids production. The optimum value deduced from

Figure 6 is in accordance with the mathematically calculated optimum points. The dome shape of the curve shows the significant mutual interaction of sucrose and $FeSO_4$.

In Figure 7, the additive effect of yeast extract and $FeSO_4$ was found and it was observed that maximum ergot alkaloids activity was measured when fermentation medium was supplemented with 0.11 g/100 mL of $FeSO_4$ and 39 g/100 mL of yeast extract. The results obtained as well as predicted by Box–Behnken design showed that a combination of 41 g/100 mL of sucrose, 39 g/100 mL of yeast extract and 0.11 g/100 mL of FeSO4 would favor maximum ergot alkaloids production (290.84 mg/mL).

Regression Analysis for Ergot Alkaloids Production and Comparison between the Observed and Predicted Response

The regression analysis was performed to predict the future response Y (ergot alkaloids yield) corresponding to the experimental data values. The regression analysis explained the difference between the observed and predicted values of the yield (Y). This was calculated by using the following equation:

$$e = Y - Y' \tag{2}$$

where 'e' represents residues, 'Y' is the observed response or yield and 'Y′' is the predicted response or yield of ergot alkaloids. The Figure 8 clearly indicates a non-random pattern among the predicted and observed values for the production of ergot alkaloids by *Penicillium citrinum* (PCUV-4)and a high degree of similarity was observed between the predicted (35 mg/mL)and observed response (35 mg/mL)of ergot alkaloids. It was also found that this model was a better fit model to describe the effect of optimized factors (sucrose, yeast extract and $FeSO_4$) as significant independent variables on the yield of ergot alkaloids.

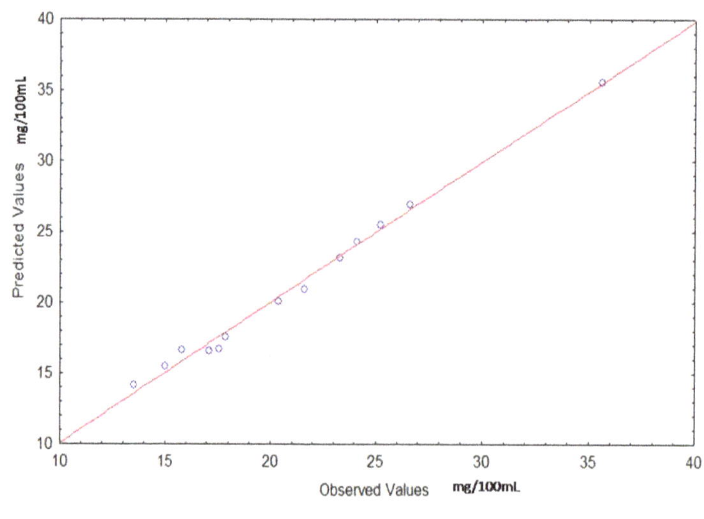

Figure 8. Predicted values of ergot alkaloids production by *Penicillium citrinum* using BBD.

3. Discussion

Fungal organisms are considered as the depository of important secondary metabolites and many them perform significant biologic functions. These fungal species are cosmopolitan in nature and have a potential to fulfill the demand of drugs in the pharmaceutical industry [16]. These amino-alkaloids can also be obtained from higher and lower plants in lysergic acid alkaloids and gliotoxins forms [17].

In this investigation, the wild strain of *Penicillium citrinum* was subjected to physical and chemical mutagens to enhance its capability for ergot alkaloids production. The wild strain of *Penicillium citrinum* was exposed under UV irradiation and it was observed that death of the fungal species is accredited with the harmful effects of UV irradiations. Onyegeme-Okerentet al. [18] described that exposure to UV irradiation can be used for genetic variations in many species of microorganisms. This classical technique was first time used in the 1950s to synthesize maximum amount of penicillin from *Penicillium chrysogenum* Q-176. They also mutated two strains of *Penicillium chrysogenum* (UVP1 and UVP2) after UV treatment for 20 and 25 min. Veerapagu et al. [19] and Moussa [20] also concluded from their studies that with the increase in the amount of irradiations, an increase in the yield of products can be observed, but sometime the production of the fungal product can be decreased due to negative impacts of the irradiations on DNA of the organism. Ethyl methane-sulfonate (EMS) is another powerful carcinogenic mutagen that can cause significant genetic variations in microorganisms. In the present study, *Penicillium citrinum* was exposed in EMS for different time intervals. The lethality of fungal organism was found to be increased with the increase in exposure time and survival percentage of the colonies were found to be decreased (Table 2). El-Bondkly and Abeer [21] also treated *Penicillium roquefortii* with different doses of concentrated EMS and found similar results.

Furthermore, in the present study, the maximum yield of ergot alkaloids was achieved from PCUV-4 and PCEMS-3 mutant strains and their yield was also compared with the yield of wild strains of *Penicillium citrinum* (Table 3). Hamad et al. [22] also found UV irradiation to be more effective than EMS due to strong mutagenic impact of UV light on the DNA of the strain. The study also reported UV mutagenic treatments to be more efficient than the chemical treatments to get strong mutations in the DNA structure of the organism. Nadeem [23] also improved the strain by employing UV, NTG and MMS for the synthesis of alkaline proteases and found the similar results. He concluded that all physical and chemical mutagens can alter the DNA structure for enviable outcomes.

In recent studies, statistical models were reported as effective tools for the optimization of fermentation parameters in biotechnology. There are several studies for the optimization of culture conditions using these sophisticated statistical procedures [24,25]. Therefore, during the present investigation, response surface methodology (RSM) was applied as an empirical statistical technique which can be used for regression analysis of data, received from multi-factorial experimental designs. This technique is very popular for screening and optimization of various parameters of fermentation experiment [3,26]. In this investigation, Plackett–Burman design (PBD) and Box–Behnken design (BBD) were used for optimizing and screening of ingredients used to get the maximum amount of ergot alkaloids (Tables 4 and 5). In the first step of screening the fermentation parameters, sucrose, yeast extract and $FeSO_4$ were found as the most considerable ingredients as analyzed by PBD model. Wu et al. [27] screened some important ingredients of fermentation experiment for synthesizing fumigaclavine C and helvolic acid from *Aspergillus fumigatus* CY018 strain, using the same model of PBD. They also concluded that pH, amount of phosphate and the size of inoculums were considerable ingredients in the production of fumigaclavine C and helvolic acid. Rubina et al. [28] screened various fermentation medium formulations for the synthesis of a thermostable lipase using a recombinant *Escherichia coli* strain BL21. They used a similar Plackett–Burman design and found that glucose, NaCl, temperature and incubation time were the most significant variables influencing lipase production. The R^2 value (0.979) of their study proved that PBD is highly significant model to screen the initial parameters to get the maximum lipase yield. Substantial increases in the yield of many products ranging from five to six fold using similar RSM techniques were reported by several researchers [29,30].

The next step of the present study was to apply Box–Behnken design (BBD) to optimize the yield of the product (ergot alkaloids) during fermentation studies. In this the first step was to identify the independent variables that affect the ergot alkaloids production and the next step was to study their mutual impacts on dependent response. The experiments of BBD were carried out to obtain a quadratic model consisting of 13 runs which are mentioned in Table 6 with their experimental results. Regression analysis and the analysis of variance was also performed for the three significant variables (sucrose, yeast extract and $FeSO_4$). Through BBD, non-significant factors were eliminated, and the reduced model was expressed in a polynomial equation. Through the results of BBD of the present study, it can be assumed that the model accurately represents the data in the experimental region. Therefore, a great enhancement in the yield of ergot alkaloids was achieved using the fewer runs using less expensive statistical method. Lee et al. [31] applied the similar BBD in their experiments and reported an enzyme activity titer of 520 U/L for *B. thermoleovorans* using olive oil in the fermentation medium, which was higher than the activities attained by other thermophilic bacilli.

Statistical designs are significant tools which can also be used to find out the interactive influences of fermentation factors on the process performance. Therefore, during present study, the combined interactions of sucrose, yeast extract and $FeSO_4$ were also analyzed using BBD. It was observed that the combination of yeast extract and $FeSO_4$ is more significant for obtaining the maximum yield of ergot alkaloids from extracts of *Penicillium citrinum* (Table 6). Venil and Lakshmanaperumalsamy [14] applied the similar Box–Behnken design (BBD) for analyzing the interaction effect of temperature, $(NH_4)_2PO_4$ and trace salts on the prodigiosin yield. They also generated a second–order polynomial equation to identify the relationship between the prodigiosin yield and three selected factors. Their fermentation medium was optimized containing 6 g/L of $(NH_4)_2PO_4$ and trace salts (0.6 g/L) and incubation temperature was 30 °C for the enhanced production of the product. The study reported that observed responses of prodigiosin were almost equal to the predicted responses of BBD model as mentioned in the present investigation. In the end of their research, they concluded that the high correlation between the observed and predicted values of their product indicates the validity of BBD model.

Results obtained in this study are comparable to Wang et al. [32] who applied the same BBD model to find out the impact of glucose, peptones and KH_2PO_4 on the cell biomass formation and reported it as an efficient statistical tool. They also obtained a quadratic model and appraised the quadratic results and central points to estimate the pure process variability with chitosanase activity as response. The three components significantly affected the chitosanase activity optimized through the similar BBD model of RSM. Krishnaa et al. [33] optimized the role of pH on the biomass yield of *Borasus flabellifer* in 15 runs of fermentation experiment and concluded that BBD was the best fit model to analyze the maximum runs of experiments in a single batch. The results of the BBD model of this study are similar with Amara [34] who optimized the fermentation conditions for the synthesis of polyhydroxy butyrate by *Bacillus* species and obtained the highest yield of protease. Yasin et al. [35] applied the similar technique of RSM to enhance the flame retardant properties and low mechanical loss to fabric. They upgraded the system parameters of the finishing treatment given to the fabric using Box–Behnken statistical design. They also observed the impacts of fire resistance and other mechanical properties on fabric quality by analyzing the similar regression equation. The R^2 estimation of the responses were above 92% which demonstrated the significance level of relationship between the predicted and experimental values of the experiment. Cheng et al. [36] optimized the nutritional medium composition required for the chitosanase production by *Streptomyces albus* YT2 and identified three significant factors (glucose, peptones and $MgSO_4$) influencing the chitosanase activity by applying similar BBD statistical design. The study reported that chitosanase yield increased from 11.56 U/mL to 39.87 U/mL, which is a 3-fold increase in the yield after using the same BBD model.

Therefore, the present investigation revealed that the combination PBD and BBD for screening and optimization purpose were proved as significant and considerable designs for ergot alkaloids synthesis.

4. Conclusions

Ergot alkaloids are produced by many organisms and the quantity of ergot alkaloids produced by fungi is relatively less, but it can be improved by applying various techniques. Therefore, some physical and chemical mutagens can enhance the yield of the product by improving DNA of the species. In the present study, UV irradiations were proved to be useful mutagenic agent and PCUV-4 mutant was found to be the significant candidate for the production of ergot alkaloids. Ergot alkaloids were in use since many decades due to their pharmaceutical and therapeutic properties. The present investigation also concluded that response surface methodology (RSM) tools such as Plackett–Burman design (PBD) and Box–Behnken design (BBD) are more reliable statistical models to increase the yield of ergot alkaloids in a single step. It was also concluded that these fermentation studies on ergot alkaloids can contribute as an alternative and cost effective methods for the biosynthesis of the important drugs on commercial scales.

5. Materials and Methods

5.1. *Microorganism and Its Maintenance*

Penicillium citrinum was collected from the Department of Botany, GC University, Lahore Pakistan and was grown on malt extract agar (MEA) medium slants. The slants were prepared by dissolving 2 g malt extract and 2 g agar in 100 mL of distilled water in a 250 mL Erlenmeyer flask. The medium was sterilized in autoclaved at 121 °C under 15 lb/inch^2 for 15 min. The mature culture of *Penicillium citrinum* was streaked aseptically to the slants containing 5 mL of MEA medium in a test tube. These inoculated slants were kept in incubator for 5 days at 25 °C so that they may be fully grown. The slants of *Penicillium citrinum* were subcultured after every two weeks and fully grown slants were stored at 4 °C for further analytical studies.

5.2. *Strain Improvement*

5.2.1. Impact of Physical and Chemical Mutagens on *Penicillium citrinum*

The wild strains of *Penicillium citrinum* were mutated using UV irradiation and ethyl methane sulfonate (EMS) for enhancing the yield of ergot alkaloids.

Impact of UV Irradiations

In this step, spore suspension or stock solution of *Penicillium citrinum* was prepared in 50 mL of distilled water. The spore suspension was diluted using serial dilution method and 1 mL of the spore suspension of the last dilution was poured to sterilized Petri plates. The plates containing spore suspensions were placed under 245 nm UV lamp (purifier horizontal clean bench, Thomas Scientific, Zhejiang, China) to induce mutation for 15, 30, 45, 60, 75, 90, 105, 120, 135 and 150 min, respectively. The UV treated (spore suspension containing) Petri plates were placed overnight in dark. After 24 h, malt extract agar (MEA) medium was transferred to UV exposed Petri plates and placed in incubator at 25 °C for 10 days. After 10 days, the percentage of living colonies after mutation was compared with wild (control) colonies of strain. This was done following [37].

Impact of Ethyl Methane Sulfonate (EMS)

The wild strain of *Penicillium citrinum* was also exposed to Ethyl Methane Sulfonate (EMS) for inducing mutation in it. The spore suspension mixture of *Penicillium citrinum* was prepared in 50 mL distilled water and diluted by serial dilution method. After a series of dilutions, from the last dilution, 2.5 mL of spore suspension was transferred to a test tube. Then 1 mL of EMS (0.3 mL of EMS was dissolved in 0.7 mL of double distilled water to make 1 mL solution of EMS) was poured in the test tube containing the spore suspension. The mixture was shaken properly and 1 mL of it was transferred to the sterilized Petri plates. These Petri plates were exposed in EMS solution for 10, 15, 20, 25 and 30 min, respectively. After EMS exposure, malt extract agar (MEA) medium was poured in Petri plates and plates were incubated at 25 °C for 10 days. After 10 days of incubation, the survival percentage of wild and mutant colonies was noted. This method was also done after [37].

5.2.2. Calculation of Survival Percentage of Colonies

Survival percentage of colonies was calculated using the following formula:

$$\text{Survival rate } (\%) = \frac{\text{No. of Mutated Colonies}}{\text{No. of Wild Colonies}} \times 100 \quad (3)$$

5.3. Maintenance of Mutant Strain

The mutant strains of *Penicillium citrinum* were grown on MEA medium slants at 25 °C for 10 days. The fully grown mature slants were kept in refrigerator at 4 °C for further analyses.

5.3.1. Selection of Best UV Mutant for the Production of Ergot Alkaloids

The living colonies of *Penicillium citrinum* after 135 and 150 min of UV exposure were grown on MEA medium slants and named as PCUV-1, PCUV-2, PCUV-3, PCUV-4 and PCUV-5, respectively. These mutants were provided with self-modified, optimized fermentation medium to observe the best producer of ergot alkaloids. The composition of fermentation medium is presented in Table 8.

Table 8. Fermentation medium for the production of ergot alkaloids by wild and mutant strains.

Ingredients.	g/100 mL
NH_4Cl	0.2
Succinic Acid	0.5
Sucrose	5
KH_2PO_4	0.5
Asparagine	0.5
Tryptophan	0.5
yeast Extract	0.5
$MgSO_4$. $7H_2O$	0.03
$FeSO_4$	0.01
$ZnSO_4$	0.002
Fermentation Conditions	
Incubation Time (Days)	21
Inoculum Size (ml)	5
pH	5
Incubation Temperature (°C)	25

5.3.2. Selection of Best EMS Mutant for the Production of Ergot Alkaloids

The fungal colonies grown after 25 min of EMS exposure were streaked on MEA slants and named as PCEMS-1, PCEMS-2 and PCEMS-3, respectively. The self-modified optimized fermentation medium (Table 8) was used to produce the ergot alkaloids by EMS mutants. The yield by mutant strains was also compared with the yield of wild strain.

5.3.3. Response Surface Methodology for Ergot Alkaloids Synthesis

In response surface methodology, statistical models such as Plackett–Burman and Box–Behnken designs were used for the statistical optimization of fermentation ingredients/variables for enhancing the yield of ergot alkaloids [14,38].

5.4. Preparation of Inoculum

Spore suspension of PCUV-4 (best mutant among all UV and EMS treated strains) was prepared by scratching the surface of its colonies and mixing the fungal spores in flask containing 50 mL of distilled water. The number of spores was adjusted at 10^{6-7} spores/mL using hemocytometer.

5.5. Response Surface Methodology

5.5.1. Plackett–Burman design (PBD) for Screening of Fermentation Factors

The Plackett–Burman design (PBD) was used for the screening of fermentation factors (ingredients) which were used to get the highest yield of ergot alkaloids. In this step, "x" variables were screened by formulating "x + 1" fermentation factor [11]. The fermentation factors used in this step were yeast Extract, Succinic acid, Sucrose, Tryptophan, Asparagine, $MgSO_4.7H_2O$, $FeSO_4$, $ZnSO_4$, KH_2PO_4 and pH. First, a numerical factor was given to all the fermentation factors and second, all of these factors were investigated at two different levels such as −1 (low level) and +1 (high level) for the production of ergot alkaloids. The experimental range, levels and design are presented in Tables 9 and 10.

Table 9. Plackett–Burman design (PBD) range and level for screening of factors.

Range and Level		Fermentation Factor
−1	+1	
5	35	Sucrose, X1
5	30	yeast Extract, X2
0.1	1	Succinic acid, X3
0.1	1	Asparagine, X4
0.1	1	Tryptophan, X5
0.1	1	KH_2PO_4, X6
0.25	0.625	$MgSO_4$, X7
0.01	0.1	$FeSO_4$, X8
0.02	0.2	$ZnSO_4$, X9
3	5	pH, X 10

X1, X2 … … X10 are fermentation factors.

Table 10. PBD experimental design for screening of factors.

Runs	Variables (x)									
	X1	X2	X3	X4	X5	X6	X7	X8	X9	X10
	Sucrose	Yeast Extract	Succinic Acid	Asparagine	Tryptophan	$MgSO_4$	KH_2PO_4	$ZnSO_4$	$FeSO_4$	pH
1.	35	5	0.1	0.1	1	0.625	0.1	0.2	0.1	5
2.	35	30	0.1	0.1	0.1	0.625	1	0.02	0.1	5
3.	5	5	0.1	0.1	0.1	0.25	0.1	0.02	0.01	3
4.	5	30	0.1	1	1	0.25	0.1	0.02	0.1	5
5.	35	5	0.1	1	0.1	0.25	1	0.2	0.01	3
6.	35	5	1	1	1	0.25	1	0.02	0.1	3
7.	5	5	1	0.1	1	0.625	1	0.02	0.01	3
8.	5	30	0.1	1	1	0.625	1	0.2	0.01	3
9.	35	30	1	0.1	1	0.25	0.1	0.2	0.01	5
10.	5	30	1	0.1	0.1	0.25	1	0.2	0.1	3
11.	5	5	1	1	0.1	0.625	0.1	0.2	0.1	5
12.	35	30	1	1	0.1	0.625	0.1	0.02	0.01	5

X1, X2 … … X10 are fermentation factors/variables.

The twelve (12) experimental runs were designed as mentioned in Table 10 after Plackett and Burman (1946) [11]. For all of the experiments, three replicates were developed and the average yield (response) of these replicates was measured. The impact of fermentation factors on the yield of ergot alkaloids was measured by formulating the following equation:

$$Y = \beta_o + \sum \beta_{Xi} \quad (4)$$

where, 'Y' is the yield or response, 'β_o' is the intercept, 'β_{Xi}' is the linear coefficient (fermentation factor). The 0.1-N HCl and ammonia solution were used to maintain the pH for all runs designed in PBD. All of these flasks were autoclaved at 121 °C, 15 lb/inch^2 for 15 min and after inoculation flasks were incubated at 25 °C for 21 days.

5.5.2. Identification of Significant Factors Using Box–Behnken Design (BBD)

After screening using PBD model, BBD was employed to find out the most crucial fermentation factors for the production of alkaloids. Here, further optimization of significant fermentation factors and impact of their mutual interaction on the production of ergot alkaloids was observed. This step was done after [12]. In this step, 13 runs were designed, and the significant fermentation factors were studied at three different levels such as low (−1), medium (0) and high (+1). The BBD range and levels are presented in Table 11. All the experiments were done in triplicate and the average of ergot alkaloids yield was considered as the response (Y). The following polynomial equation was formulated to identify the Y as response:

$$Y = \beta_o + \sum \beta_{ii} + X \sum \beta_{ii}\, x\, i^2 + \sum \beta_{ij}\, X_i X_j \quad (5)$$

Table 11. BBD experimental levels.

Level and Range			Fermentation Factor
−1	0	+1	
5	23	41	Sucrose, X1
5	22	39	yeast Extract, X2
0.01	0.06	0.11	$FeSO_4$, X3

In this 'Y' is the predicted response, 'XiXj' is the input variables which influence 'Y'; 'β_o' intercept coefficient, 'β_{ii}' linear coefficient; 'β_{ii} x i^2' quadratic coefficient and 'β_{ij}' interaction of these variables. 'X1, X2 and X3' are fermentation factors. The experimental design of BBD is presented in Table 12. The pH of all runs was maintained at 5.0 using 0.1-N HCl and ammonia solution. The media was autoclaved at 121 °C for 15 min under 15 lb/inch^2 and after inoculation, all flasks were incubated at 25 °C for 21 days.

Table 12. BBD experimental design for optimization of fermentation factors.

Runs	Variables		
	X1	X2	X3
	Sucrose	Yeast Extract	$FeSO_4$
1	41	5	0.06
2	41	39	0.06
3	41	22	0.01
4	41	22	0.11
5	5	5	0.06

Table 12. *Cont.*

Runs	Variables		
	X1	X2	X3
	Sucrose	Yeast Extract	$FeSO_4$
6	5	39	0.06
7	5	22	0.01
8	5	22	0.11
9	23	5	0.01
10	23	39	0.01
11	23	5	0.11
12	23	39	0.11
13	23	22	0.06

5.5.3. Statistical Analyses of RSM

The STATISTICA version 7 (Stat-Ease, Inc., Minneapolis, MN, USA) software was used for statistical analysis and also for developing 3D figures of response surface methodology.

5.6. *Ergot Alkaloids Determination*

After 21 days of incubation the fermented broth was processed and intracellular mass (fungal mycelium) was separated from extracellular liquid (supernatant). The mycelial mass and fermented broth was stored at 4 °C for further analysis.

5.6.1. Ergot Alkaloids in Fermented Broth Extract

The separated fermented broth was centrifuged at 10,000 rpm/min for 10 min at 4 °C and the extract (supernatant) was collected in a separate glass bottle. The extract was purified using rotary evaporator [39]. After purification, 1 mL of purified extract was taken in a test tube and 2 mL of Van Urk reagent was added in it. The reaction mixture was incubated at 37 °C for 30 min in water bath. The optical density (OD) was measured using Spectrophotometer (Hitachi U2900/U2910 double beam) at 590 nm.

5.6.2. Ergot Alkaloids in Extract of Mycelia

The mycelia were placed in oven for drying at 40 °C for 24 h after measuring their initial weights. After 24 h, dried mycelia were soaked in chloroform for 3 h. All the soaked mycelia were crushed and grinded by sonication process at 200 rpm/min for 15 min using Ultrasonic Generator. The paste was homogenized in a homogenizer for 15 minutes. The homogenized pastes were then centrifuged at 10,000 rpm at 4 °C for 10 min. The extract of mycelia was purified using rotary evaporator. The purified extract was assayed using Van Urk reagent and measured at 590 nm in Spectrophotometer (Hitachi U2900/U2910 double beam).

Author Contributions: M.G.S. is the main author of the research study. This is a part of my PhD research thesis. Therefore, the main conceptualization, methodology, research experimentations, interpretations and original draft preparations were done by me. M.N. is the supervisor of my thesis and the major contribution is done by him in designing and determination of the protocols. A.G. did the formal analysis and contributed in writing (review and editing), M.S. done validation and helped in writing (review and editing), H.u.R. helped in the visualization, computing resources and editing of drfat specially the statistical analysis. G.Z.G.; M.U.H.; L.S.; and R.A. contributed in writing (review and editing). R.N. helped in the formal analysis of PBD and BBD. All authors have read and agreed to the published version of the manuscript.

Funding: No outside funding was provided for this research.

Acknowledgments: The authors acknowledge the Pakistan Council of Scientific and Industrial Research, Lahore, Pakistan and the Department of Botany, GC University, Lahore for providing the technical facilities for the smooth conduction of all the experiments of the study.

Conflicts of Interest: There were no conflicts of interest during the research studies.

References

1. Pichersky, E.; Gang, D.R. Genetics and biochemistry of secondary metabolites in plants: An evolutionary perspective. *Trends Plant Sci.* **2000**, *5*, 439–445. [CrossRef]
2. Shahid, M.G.; Nadeem, M.; Baig, S.; Cheema, T.A.; Atta, S.; Ghafoor, G. Screening and optimization of some inorganic salts for the production of ergot alkaloids from *Penicillium* species using surface culture fermentation process. *Pak. J. Pharma. Sci.* **2016**, *29*, 407–414.
3. Shahid, M.G.; Baig, S.; Nadeem, M.; Cheema, T.A.; Nelofar, R.; Saleem, M. Biosynthesis of ergot alkaloids from *Penicillium commune* using response surface methodology (RSM). *Pak. J. Bot.* **2017**, *49*, 1569–1578.
4. Zafar, A.M.; Waseemuddin, S.A.; Azhar, I.; Sualeh, M.; Baig, M.T.; Zoha, S.M.S. Bioactive alkaloids produced by fungi: Updates on alkaloids from the species of the genera *Boletus*, *Fusarium* and *Psilocybe* (Review). *Pak. J. Pharma. Sci.* **2010**, *23*, 349–357.
5. Gulliamon, J.M.; Sabate, J.; Barrio, E.; Cano, J.; Querol, A. Rapid identification of nine Yeast species based on RFLP analysis of ribosomal internal transcribed spacer (ITS) regions. *Arch. Microbiol.* **1998**, *169*, 387–392.
6. Tiwari, K.L.; Jadhav, S.K.; Fatima, A. Culture conditions for the production of thermostable amylase by *Penicillium rugulosum*. *Glob. J. Biotechnol. Biochem.* **2007**, *2*, 21–24.
7. Kozlovsky, A.G.; Zhelifonova, V.P.; Antipova, T.V. Biologically active metabolites of *Penicillium* fungi. *Org. Biomol. Chem.* **2013**, *1*, 11–21.
8. Nina, G.; Neubauer, L.; Tudzynski, P.; Shu-Ming, L. Biosynthesis pathways of ergot alkaloids. *Toxins* **2014**, *6*, 3281–3295.
9. Khurana, S.; Kapoor, M.; Gupta, S.; Kuhad, R.C. Statistical optimization of alkaline xylanase production from *Streptomyces violaceoruber* under submerged fermentation using response surface methodology. *Ind. J. Microbiol.* **2007**, *47*, 144–152. [CrossRef]
10. Trejo, H.M.R.; Lonsane, B.K.; Raimbault, M.; Roussost, S. Spectra of ergot alkaloids produced by *Claviceps purpurea* 1029c in solid-state fermentation system: Influence of the composition of liquid medium used for impregnating sugar-cane pith bagasse. *Proc. Biochem.* **1993**, *28*, 23–27. [CrossRef]
11. Plackett, R.L.; Burman, J.P. The design of optimum multifactorial experiments. *Biometrika* **1946**, *33*, 305–325. [CrossRef]
12. Box, G.E.P.; Behnken, D.W. Some new three level designs for the study of quantitative variables. *Technometrics* **1960**, *2*, 455–475. [CrossRef]
13. Naveena, B.J.; Altaf, M.D.; Bhadriah, K. Selection of medium components by Plackett–Burman design for production of L (+) lactic acid by *Lactobacillus amylophilus* GV6 in SSF using wheatbran. *Bioresour. Technol.* **2005**, *96*, 485–490. [CrossRef]
14. Venil, C.K.; Lakshmanaperumalsamy, P. Application of statistical design to the optimization of culture medium for prodigiosin production by *Serratia marcescens* SWML08. *Malays. J. Microbiol.* **2009**, *5*, 55–61.
15. Shahid, M.G.; Baig, S.; Saleem, M.; Arif, R.; Ghafoor, G.; Liaqat, A. Qualitative and quantitative analysis of ergot alkaloids produced by *Aspergillus niger* through surface culture fermentation process. *Pak. J. Bot.* **2018**, *50*, 2423–2428.
16. Devi, N.N.; Prabakaran, J.J. Bioactive metabolites from an endophytic fungus *Penicillium* sp. isolated from *Centella asiatica*. *Curr. Res. Environ. Appl. Mycol.* **2014**, *4*, 34–43. [CrossRef]
17. Roberts, M.R.; Wink, M. *Alkaloids: Biochemistry, Ecology and Medical Applications*; Plenum: New York, NY, USA, 1998; pp. 135–146.
18. Onyegeme-Okerenta, B.M.; Okochi, V.I.; Chinedu, S.N. Penicillin production by *Penicillium chrysogenum* PCL 501: Effect of UV induced mutation. *Int. J. Microbiol.* **2013**, *12*, 1–10.
19. Veerapagu, M.; Jeya, K.R.; Ponmurugan, K. *Mutational Effect of Penicillium chrysogenum on Antibiotic Production*; Advanced Biotech: Totowa, NJ, USA, 2008; pp. 16–19.
20. Moussa, L.A.A. Effect of some factor including irradiation on the ergot alkaloids production by members of *Penicillium*. *J. Biol. Res.* **2003**, *3*, 65–81.

21. El-Bondkly, A.M.; Abeer, A.K. UV- and EMS- induced mutations affecting synthesis of alkaloids and lipase in *Penicillium roquefortii*. *Arab. J. Biotechnol.* **2007**, *10*, 241–248.
22. Hamad, A.; Haq, I.; Qadeer, M.A.; Javed, I. Screening of *Bacillus licheniformis* mutants for improved production of alpha-amylase. *Pak. J. Bot.* **2001**, *33*, 517–525.
23. Nadeem, M. Biotechnological Production of Alkaline Protease for Industrial Use. Ph.D. Thesis, Department of Zoology, University of the Punjab, New Campus, Lahore, Pakistan, 2009.
24. Rao, J.M.; Kim, C.; Rhee, S. Statistical optimization of medium for the production of recombinant hirudin from *Sacchromyces cerevisiae* using response surface methodology. *Proc. Biochem.* **2000**, *35*, 639–647. [CrossRef]
25. Venil, C.K.; Lakshmanaperumalsamy, P. Applications of response surface methodology in medium optimization for protease production by the new strain of *Serratia marcescens* SWML08. *Pol. J. Microbiol.* **2009**, *58*, 117–124.
26. Mao, X.B.; Eksriwong, T.; Chauvatcharin, S.; Zhong, J.J. Optimization of carbon source and carbon/nitrogen ratio for cordycepin production by submerged cultivation of medicinal mushroom *Cordyceps militaris*. *Proc. Biochem.* **2005**, *40*, 1667–1672. [CrossRef]
27. Wu, Q.; Yong-Chun, S.; Xu, H.; Guo, Y.; Li, J.; Ren-Xiang, T. Medium optimization for enhanced co-production of two bioactive metabolites in the same fermentation by a statistical approach. *J. Asian Nat. Prod. Res.* **2011**, *13*, 1110–1121.
28. Rubina, N.; Ramnan, R.N.; Rahman, R.N.Z.R.; Basri, M.; Ariff, A.B. Sequential optimization of production of a thermostable and organic solvent tolerant lipase by recombinant *Escherichia coli*. *Ann. Microbiol.* **2011**, *61*, 535–544.
29. Liu, Y.; Wang, F.; Tan, T. Cyclic resolution of racemic ibuprofen via coupled efficient lipase and acid base catalysis. *Chirality* **2009**, *21*, 349–353. [CrossRef]
30. Pan, H.; Xie, Z.; Bao, W.; Zhang, J. Optimization of culture conditions to enhance cis-epoxysuccinate hydrolase production in *Escherichia coli* by response surface methodology. *Biochem. Eng. J.* **2008**, *42*, 133–138. [CrossRef]
31. Lee, D.W.; Koh, Y.S.; Kim, B.C.; Choi, H.J.; Kim, D.S.; Suhartono, M.T.; Pyun, Y.R. Isolation and characterization of thermophilic lipase from *Bcillus thermoleovorans* ID-1. *FEMS Microbiol. Lett.* **1999**, *179*, 393–400. [CrossRef]
32. Wang, X.L.; Liu, G.Q. Preliminary select and optimization of submerged fermentation media of *Ganoderma sinense*. *Food Sci. Technol.* **2009**, *34*, 14–16.
33. Krishnaa, D.; Krishnaa, K.S.; Padma, S.R. Response surface modeling and optimization of chromium (Vi) removal from aqueous solution using *Borasus flabellifer* coir powder. *Int. J. Appl. Sci. Eng.* **2013**, *11*, 213–226.
34. Amara, A.A.F. Optimizing PHB and protease production by Box-Behnken design. *J. IILUM Eng.* **2013**, *14*, 15–28. [CrossRef]
35. Yasin, S.; Curti, M.; Behary, N.; Perwuelz, A.; Giraud, S.; Rovero, G.; Guan, J.; Chen, G. Process optimization of eco-friendly flame retardant finish for cotton fabric: A response surface methodology approach. *Surf. Rev. Lett.* **2017**, *24*, 1750114. [CrossRef]
36. Cheng, S.W.; Wang, Y.F.; Hong, B. Statistical optimization of mediumcompositions for chitosanaseproduction by a newly isolated *Streptomyces albus*. *Braz. J. Chem. Eng.* **2012**, *29*, 691–698. [CrossRef]
37. Sreedevi, K.; Venkateswara, R.J.; Lakshmi, N.; Fareedullah, M. Strain improvement of *Aspergillus terreus* for the enhanced production of lovastatin, a HMG-COA reductase inhibitor. *J. Microbiol. Biotechnol. Res.* **2011**, *1*, 96–100.
38. Ryan, K.L.; Christopher, T.M.; Panaccione, D.G. Partial reconstruction of the ergot alkaloid pathway by heterologous gene expression in *Aspergillus nidulans*. *Toxins* **2013**, *5*, 445–455. [CrossRef]
39. Naude, T.W.; Botha, C.J.; Vorster, J.H.; Roux, C.; Van der linde, E.J.; Van der walt, S.L.; Rottinghaus, G.E.; Van Jaarsveld, L.; Lawrence, A.N. *Claviceps cyperi*, a new cause of severe ergotism in dairy cattle consuming maize silage and teff hay contaminated with ergotised *Cyperus esculentus* (nut sedge) on the Highveld of South Africa. *Onderstepoort J. Vet. Res.* **2005**, *72*, 23–37. [CrossRef]

toxins

MDPI

Communication

Using On-Farm Monitoring of Ergovaline and Tall Fescue Composition for Horse Pasture Management

Krista La Moen Lea * and S. Ray Smith

Department of Plant and Soil Sciences, University of Kentucky, N-222C Ag. Science Center North, Lexington, KY 40546, USA; Raysmith1@uky.edu
* Correspondence: Krista.Lea1@uky.edu

Abstract: Central Kentucky horse pastures contain significant populations of tall fescue (*Schedonorus arundinacea* (Schreb.) Dumort) infected with an endophyte (*Epichloë coenophialum* (Morgan-Jones and Gams) Bacon and Schardl) known to produce several ergot alkaloids, with ergovaline in the highest concentration. While most classes of horses are not adversely affected by average levels of ergovaline in pastures, late term pregnant mares have a low tolerance to ergovaline and the related ergot alkaloids. Endophyte-infected tall fescue has been known to cause prolonged gestation, thickened placenta, dystocia, agalactia, and foal and mare mortality. The University of Kentucky Horse Pasture Evaluation Program utilizes ergovaline and endophyte testing, as well as pasture species composition, to calculate ergovaline in the total diet in broodmare pastures. This data is used to develop detailed management recommendations for individual pastures. Application of these recommendations has led to reduced tall fescue toxicity symptoms on these farms, as well as improved pasture management and improved forage quality and quantity.

Keywords: ergovaline; horse pastures; pregnant mares; species composition

Key Contribution: Evaluate tall fescue toxicity potential for pregnant mares by considering ergovaline concentration and pasture species composition.

Citation: Lea, K.L.M.; Smith, S.R. Using On-Farm Monitoring of Ergovaline and Tall Fescue Composition for Horse Pasture Management. *Toxins* **2021**, *13*, 683. https://doi.org/10.3390/toxins13100683

Received: 15 June 2021
Accepted: 23 September 2021
Published: 25 September 2021

1. Introduction

Tall fescue, *Lolium arundinaceum* (Schreb.) Darbysh; *Schedonorus phoenix* (Scop.) Holub, a cool-season, perennial bunch-type grass, dominates over 15 million ha in the southeastern US [1] including approximately 2 million ha in the state of Kentucky. While the grass has good forage quality and palatability, most plants are infected with a common toxic endophyte, *Epichloë coenophialum*. The endophyte and plant form a symbiotic relationship, with the plant provides nutrients and a hospitable environment for the endophyte. The presence of *E. coenophialum* is beneficial to the plant by increasing pest, drought and grazing tolerance [2,3], but is also toxic to grazing livestock. The endemic nature of infected tall fescue in the U.S. traces back to the widespread distribution of the cultivar Kentucky 31+ containing *E. coenophialum* in the 1940s, 1950s, and 1960s [4]. This cultivar continues to be distributed in the U.S. and other countries due to its superior environmental and edaphic adaptation.

A number of ergot alkaloids are present in infected tall fescue, including ergovaline, ergotamine, ergocrystine, and lysergic acid [5]. Ergovaline has been identified as the most prevalent of the ergot alkaloids, representing roughly 80% of total ergot alkaloids produced [6] and is known to be vasoconstrictive, as documented by measurements of the palmar artery and vein [7]. For this reason, ergovaline has historically been the focus of much livestock research and has been correlated with negative physiological reactions in cattle, horses, and small ruminants consuming toxic endophyte-infected tall fescue.

In cattle, reduced average daily gain (ADG) [8], rough hair coat [9], poor breeding efficiency, and elevated body temperature [10] have all been attributed to grazing endophyte-infected tall fescue. As horses are much better at dissipating heat than cattle, most classes of horses are more tolerant to fescue toxicity, though vasoconstriction has been documented in non-pregnant mares [11]. On the other hand, pregnant mares are more sensitive to ergovaline than cattle; mares grazing toxic tall fescue often go several weeks past their due date, resulting in large foals, dystocia, and in some cases foal or mare mortality [12]. These mares can also have thickened or retained placentas [13], further complicating the foaling process. Mares on toxic tall fescue often have limited or no milk production [14,15].

In recent years, novel endophyte tall fescue cultivars have been developed that contain beneficial *Epichloë* endophyte strains. These endophytes provide increased plant persistence and produce non-toxic alkaloids, which confer insect and nematode resistance, but do not cause livestock toxicity. Recent studies have determined that most novel endophyte tall fescue cultivars do not produce significant portions of ergot alkaloids, and mares grazing them are likely to foal normally [16]. However, some cultivars do still produce low levels of ergovaline, below threshold values for cattle, but high enough to potentially cause toxicity in last trimester broodmares [17,18].

The University of Kentucky Horse Pasture Evaluation Program began in 2005 in response to the outbreak of Mare Reproductive Loss Syndrome (MRLS) of 2001/2002. While it was eventually determined that the Eastern Tent Caterpillar was the causal agent for MRLS [19], this event made many horse farm managers acutely aware of the amount of toxic tall fescue in their pastures. The UK Horse Pasture Evaluation Program is a fee-based service offered by University of Kentucky Forage Extension to collect on-farm pasture data and make management recommendations. This program has experienced tremendous growth, in part due to the increasing concerns of tall fescue toxicity on broodmare farms. Here, we will present two case studies where tall fescue toxicity was evaluated on horse farms, and discussion on how ergovaline in total diet is calculated and used to make pasture recommendations.

2. Results

Two case studies will be presented below to showcase how data collected on farms has been used to make management recommendations. Both farms are commercial thoroughbred breeding operations in central Kentucky and have participated in the UK Horse Pasture Evaluation Program for several years. Farm names have been redacted to protect privacy.

2.1. Farm #1

Pastures in Farm #1 were sampled in the late summer of 2020. This farm had significant tall fescue composition, ranging from 23–73% and a farm average of 45%. Endophyte infection was high, averaging 84%, and ergovaline concentration in the fescue ranged from 288 to 977 ppb (parts per billion) across pastures (Table 1). As both ergovaline and tall fescue composition were high, there was minimal ergovaline dilution on this farm. Ergovaline in total diet was over the 200 ppb threshold for late term pregnant mares in 11 of the 15 fields. Although some were just over the threshold, others, such as paddock N, had ergovaline in total diet of 699 ppb, and therefore would be unsafe to hold pregnant mares during their last trimester. In contrast, paddock I showed an ergovaline in total diet of 181 ppb and therefore should be safe for grazing at the time of sampling. Paddock I may show higher ergovaline in the spring, as fescue composition was 32% and endophyte infection was 91%, and may require remediation or additional testing. Total reestablishment was recommended on several fields on this farm.

2.2. Farm #2

Farm #2 had been sampled regularly for a number of years and had followed recommended toxic tall fescue mitigation strategies based on UK Horse Pasture Evaluation Program recommendations. Overall, tall fescue endophyte infection was high at 82% (Table 2), but no fields were of major concern due to two main factors: strong stands of other forages such as Kentucky bluegrass (*Poa pratensis* L.) and orchardgrass (*Dactylis glomerata* L.) which diluted ergovaline consumption, and healthy stands of novel endophyte tall fescue in several pastures. Pasture M1 had been monitored yearly as there was significant tall fescue composition, but ergovaline remained at moderate concentrations and there was a good stand of bluegrass. These combined factors, and the assumption that horses eat randomly in the pasture [20], meant that, in most years, ergovaline in total diet on M1 was low and mares could safely graze. If bluegrass composition declined or ergovaline increased, then this field may become a candidate for total reestablishment. In contrast, Pasture M3 had shown significant toxic tall fescue presence and other weeds issues in past years, therefore a total herbicide kill and reestablishment was recommended. M3 was replanted with a mixture of novel endophyte tall fescue, Kentucky bluegrass, and orchardgrass, and was monitored yearly to watch for toxic fescue encroachment. High endophyte and low ergovaline indicated the presence of a novel endophyte. By following pasture management recommendations, being willing to make changes when needed, and yearly pasture monitoring, this farm has significantly reduced the risk of tall fescue toxicity in broodmares. In the process, they have also reported that their newly established pastures showed improved foal growth rates and the additional forage production allowed hay harvests, thereby reducing purchased feed costs.

Table 1. Data from a representative thoroughbred breeding farm in central Kentucky (Farm #1) showing field acreage, species composition, endophyte infection, ergovaline concentration, and calculated ergovaline in total diet.

Field	TF [1]	BG	OG	WC	WD	NW	BS	Endophyte	Ergovaline	Ergovaline in Total Diet
		---%---							---ppb---	
Pasture 1	23	27	23	0	8	1	19	80	552	178
Pasture 2	33	30	21	0	7	1	7	83	485	189
Pasture 3	46	24	8	0	0	2	15	85	547	321
Pasture 4	73	13	8	0	0	0	6	81	406	314
Pasture 5	57	27	8	0	1	2	4	95	454	283
Pasture 6	38	39	7	0	2	1	13	83	634	287
Pasture 8	43	18	16	0	5	0	15	86	552	306
Paddock E	34	37	20	0	0	0	9	88	542	203
Paddock H	40	9	30	0	8	5	4	94	609	311
Paddock I	32	40	10	0	6	1	10	91	465	181
Paddock J	48	18	20	0	0	1	13	76	288	161
Paddock K	37	17	28	0	0	1	18	95	545	247
Paddock N	59	11	13	0	0	0	14	79	977	699
Paddock T	46	2	11	0	0	2	36	76	415	311
Paddock U	60	2	31	0	0	0	7	67	366	237
Average	45	21	17	0	3	1	13	84	522	282

[1] TF = Tall Fescue, BG = Kentucky Bluegrass, OG = Orchardgrass, WC = White Clover, WD = Weeds, NW = Nimblewill, and BS = Bare Soil and Warm Season Annual Grasses.

Table 2. Data from a thoroughbred breeding farm in central Kentucky (Farm #2) following toxic tall fescue mitigation strategies showing species composition, endophyte infection, ergovaline concentration, and calculated ergovaline in total diet.

Field	TF [1]	BG	OG	WC	WD	NW	BS	Endophyte	Ergovaline	Ergovaline in Total Diet
		---%---							---ppb---	
Pasture M1	22	25	10	1	3	25	4	92	422	162
Pasture M3	44	10	1	1	2	10	17	73	<100 [2]	<100
Field 9	11	19	40	7	2	14	1	90	247	35
Field 10	7	15	20	13	6	29	3	67	289	39
Paddock IP1	15	0	7	3	4	3	58	95	<100	<100
Paddock IP2	67	8	4	0	2	6	12	80	<100	<100
New Paddock	20	7	3	0	2	30	17	90	<100	<100
Field 01	28	11	10	1	12	26	4	75	<100	<100
LC Field 1	13	18	21	23	2	20	1	75	171	31
LC Field 2	4	12	16	6	4	38	17	86	165	18
Farm Average	23	12	13	5	4	20	13	82	259	57

[1] TF = Tall Fescue, BG = Kentucky Bluegrass, OG = Orchardgrass, WC = White Clover, WD = Weeds, NW = Nimblewill, and BS = Bare Soil and Warm Season Annual Grasses. [2] Lower quantification limit is 100 ppb for this method.

3. Discussion

The current literature does not pinpoint an exact threshold of ergovaline in total diet tolerated by pregnant mares, in part due to differences in study designs, feeding methods, pasture sampling, analysis, and reporting. Threshold values set by various extension services range from 0 to 300 ppb. At the University of Kentucky, 200 ppb is the generally accepted threshold value (Table 3), based on work that demonstrated palmar artery constriction around 200 ppb in mares [16]. Pastures testing in the 201–500 ppb range provide a risk when grazed by late term pregnant mares, but some of this risk can be mitigated if steps are taken to reduce ergovaline in total diet, such as mowing to remove seedheads or feeding hay. Pastures with more than 500 ppb in total diet are unlikely to be safe for mares in the last trimester of pregnancy, therefore the recommendation is to remove mares from these pastures.

Table 3. Risk level for late term pregnant mares based on ergovaline concentration in total diet.

Ergovaline in Total Diet	Recommendation for Late-Term Mares
<200 ppb	Low risk—monitor for seasonal fluctuations
201–500 ppb	Risk—take steps to mitigate tall fescue in pasture
>500 ppb	High risk—remove mares from pasture in last 60 days of pregnancy

Following each evaluation, considerable time is spent with each horse farm owner/manager to educate them on the variations in ergovaline concentration in tall fescue. Common sources of variation include season, temperature, precipitation, and pasture management. Ergovaline concentrations are known to vary seasonally, and closely follow the cool season growth curve of the endophyte's host, tall fescue. Figure 1 illustrates seasonal spikes in ergovaline concentration commonly seen in the spring and fall [21].

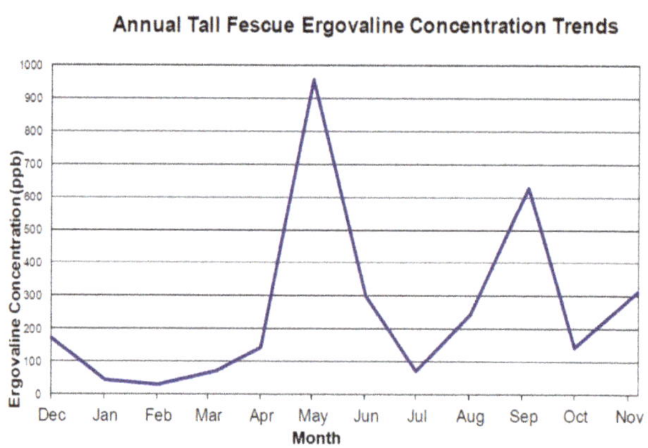

Figure 1. Season variation of ergovaline concentration based on monthly samples during one calendar year from a Kentucky pasture.

Year to year variation is also significant and is not completely understood. Figure 2 provides a bar chart of the range in ergovaline concentration and averages across farms that have participated in this program since 2005. As in Figure 1, a bimodal curve is apparent for average ergovaline concentration, but the range in any given month is also extensive. Annual sampling of pastures suggests that those with higher-than-average ergovaline concentrations one year are likely to be higher than average in subsequent years [22]. Therefore, although ergovaline fluctuates from year to year, the fields with consistently higher ergovaline concentrations should be avoided.

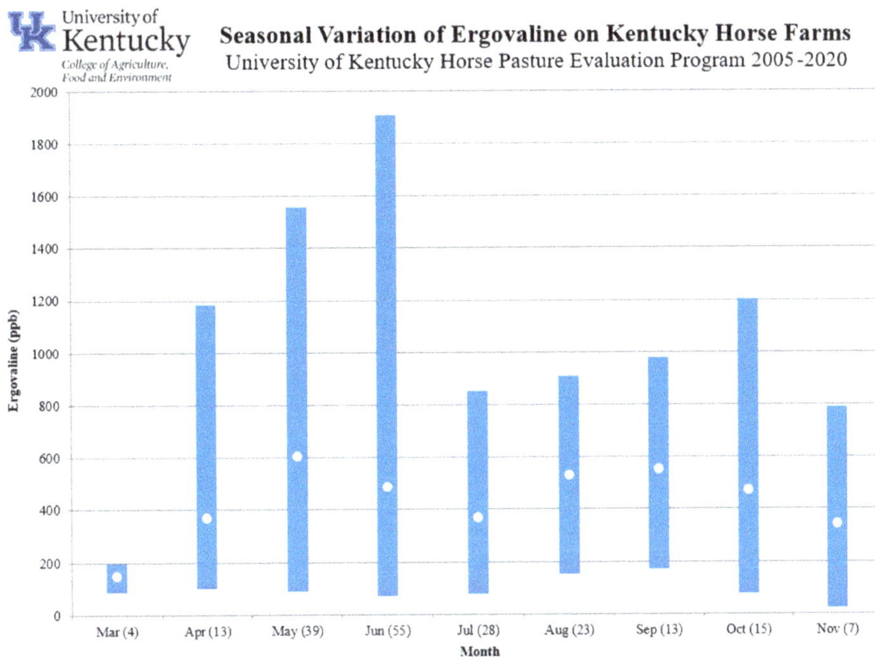

Figure 2. Range (bars) and average (dots) of ergovaline farm averages by month (number of samples) for all program participants since 2005.

Research from Dillard and colleagues [17] found that ergovaline is the most concentrated in the leaf sheath, which is mainly below 10cm. This finding is significant on horse farms that routinely graze or mow closer to the ground than on cattle operations. Horse pastures are typically kept around 15 cm, but small turn-out paddocks are often grazed much closer and frequently have higher ergovaline concentrations compared to larger pastures with taller grazing height.

Endophyte testing was once considered key in understanding and quantifying the toxicity potential in any given pasture. Early studies involving toxic tall fescue simply compared endophyte-infected to endophyte free pastures, such as Putnam [12]. Later studies compared high and low infected pastures to examine toxicity in livestock, such as Aiken, [23]. More advanced analyses, which include ergovaline or total ergot alkaloid quantification, provide a snapshot of toxicity potential in a pasture, but endophyte testing still provides a valid indicator of potential across a season. Endophyte percentage stays relatively unchanged from year to year, and high endophyte percentage suggests that, even if ergovaline or total ergot alkaloids are low at the time of testing, the potential is there for these to dramatically increase and become problematic under different conditions. The University of Kentucky Horse Pasture Evaluation program reports an average of 78% endophyte infection rate across all farms sampled in Kentucky 2005–2020. In addition, when combined with ergovaline analysis at peak times (spring and fall), the presence of a novel endophyte can be confirmed. Novel endophyte tall fescue pastures will have high endophyte percentages, but low ergovaline concentrations.

Average tall fescue presence in sampled horse pastures over the last 15 years is 18%, but the actual composition of tall fescue in individual pastures has ranged from 3 to 73%. Considering this wide range for tall fescue and the fact that ergovaline is not tested on many farms, a more relative risk scale can be used (Table 4). This scale is also useful when ergovaline was tested outside the normal months that late term pregnant mares are on pasture. For example, when tall fescue composition is less than 10% and there are good percentages of other forages, then a pasture would be considered a low risk for tall fescue toxicity. At higher levels of tall fescue, more radical approaches are necessary, unless the fescue is a novel endophyte variety proven by very low ergovaline concentrations.

Table 4. Risk level to late term pregnant mares based on percent tall fescue composition and management recommendations based on the risk level.

Tall Fescue	Risk Level	Management Recommendation
<10%	Very small risk to late term mares.	The only risk would be during severe stress periods (e.g.,—in a hot, dry summer) when the tall fescue may be growing and the KY bluegrass is dormant or when other desirable species are not present.
10–25%	Risk to late term mares is small, but safe pregnancy not guaranteed.	If the last 60–90 days of pregnancy occur in late March/early April or late November/early December, then watch mares carefully. Suppressing tall fescue with herbicide (imazapic) can be considered.
25–50%	Risk to late term mares is significant, especially during stress periods.	Suppressing tall fescue with herbicide (imazapic) can be considered if grazing late term mares in this pasture. Overseed following spraying, but wait for at least two months residue will inhibit the growth of new seedlings.
50–75%	Risk to late term mares is high.	Do not graze pregnant mares during the last 60–90 days of pregnancy. Herbicides used to suppress fescue may result in bare ground and weed growth. Complete reestablishment is recommended if grazing late term mares.
75–100%	Risk to late term mares is very high.	Do not graze pregnant mares during the last 60–90 days of pregnancy.

Final management recommendations are made based on species composition and tall fescue sample results and the other factors that have been discussed. Recommendations may include seeding, spraying, complete renovation, mowing, fertilizing, hay feeding, or removing horses from a pasture. All results and recommendations are discussed with managers to help them find a solution that works for them and their operation.

4. Conclusions

Endophyte and ergovaline testing are key elements of understanding and reducing on-farm risk of tall fescue toxicosis in horses and other livestock. Pasture composition measurements are an additional indicator of toxicosis risk and should always be considered before making management changes. The University of Kentucky Horse Pasture Evaluation Program has found that recommendations based on this data and one-on-one discussions with farm owners/managers has led to improved pasture health, reduced tall fescue toxicity and better forage quality and quantity. Additionally, complete re-establishment has proven successful in mitigating tall fescue toxicity as well as other pasture challenges, such as warm season annual grasses, nimblewill (*Muhlenbergia schreberi* J.F. Gmel.), broadleaf weeds, and bare soil. Since 2005, the UK Horse Pasture Evaluation Program has completed 281 evaluations on over 170 farms in central Kentucky, representing over 67,000 total pasture acres. Recent improvements, including modified sampling methods and digital data collection, have streamlined the evaluation process, providing more consistent information with a rapid turnaround.

5. Materials and Methods

Pastures were sampled between April and November by trained individuals. From 2005 to 2019, visual estimation of a 0.65 × 0.65 m quadrat (made with 3.5 cm diameter PVC pipe), was conducted in 10–20 locations to determine species composition. Pastures less than 2 hectares had 10 sample locations; those 2–4 hectares had 15 sample locations and pastures greater than 4 hectares contained 20 sample locations. Pastures greater than

16 hectares were sub-divided and sampled accordingly. In 2020, pasture sampling was transitioned to the occupancy method as described by Vogel and Masters [24] and Payne [25]. Grids are 0.75 m × 0.75 m and contain 25 smaller squares, 15 cm × 15 cm each, made from metal cattle panels and cut to size with sharp edges ground off and painted for visibility. For each smaller square, the most dominant component was identified. Categories included tall fescue, Kentucky bluegrass, orchardgrass, white clover (*Trifolium repens* L.), broadleaf weeds, nimblewill, and bare soil (which includes warm season annual grasses).

A representative sample of tall fescue plant material was collected from each pasture. The plant material was harvested from 10 to 20 random locations, depending on the size of the pasture, to obtain a minimum total fresh sample weight of at least 300 g. Samples are cut 7–10 cm above the ground and included seedheads, if present. Material was placed on ice in a cooler, and transported to the University of Kentucky Veterinary Diagnostic Laboratory (Lexington, KY, USA. http://vdl.uky.edu/ (accessed on 24 September 2021)) for analysis. The lab flash froze samples in liquid nitrogen and utilized ultra-High Performance Liquid Chromatography with fluorescence detection to quantify ergovaline and its isomer, ergovalinine, as total ergovaline concentration. The full method is described in Lea et al. [26].

Twenty tall fescue tillers were collected throughout each pasture, individual tillers were cut at the soil surface, and placed in a cooler on ice for transport to the University of Kentucky Regulatory Services Lab following the procedure described by Vincelli [27]. Percentage of plants infected with an endophyte was determined using Agrinostics tiller test kit (Watkinsville, GA, USA).

For each pasture sampled, individual grid and total pasture species composition were provided, as well as ergovaline concentration (ppb) and endophyte (%) on a comprehensive datasheet. In addition, "ergovaline in total diet" was calculated using the following formula:

$$\text{Ergovaline in total diet (ppb)} = (\%\text{Tall Fescue}/(\%\text{ Tall Fescue} + \%\text{ Bluegrass} + \%\text{ Orchardgrass} + \%\text{ White Clover})) * \text{ergovaline}.$$

Management recommendations are based on all data collected and may include overseeding, chemical weed control, tall fescue mitigation or complete re-establishment (two rounds of glyphosate is recommended in late summer and early fall, followed by a fall seeding of perennial cool-season grasses). Seeding mixtures typically include Kentucky bluegrass and orchardgrass, and often a novel endophyte tall fescue.

Author Contributions: K.L.M.L. and S.R.S. are intimately involved in all aspects of this program. All authors have read and agreed to the published version of the manuscript.

Funding: This program is partially supported by the University of Kentucky and collects a fee for service from participating farms.

Institutional Review Board Statement: Not applicable.

Informed Consent Statement: Not applicable.

Data Availability Statement: Not applicable.

Conflicts of Interest: The authors declare no conflict of interest.

References

1. Ball, D.M.; Hoveland, C.S.; Lacefield, G.D. Cool season grasses. In *Southern Forages*; Ball, D.M., Hoveland, C.S., Lacefield, G.D., Eds.; International Plant Nutrition Institute: Norcross, GA, USA, 2007; pp. 40–49.
2. Gwinn, K.D.; Gavin, A.M. Relationship between endophyte infestation level of tall fescue seed lots and *Rhizoctonia zeae* seedling disease. *Am. Phytopathol. Soc.* **1992**, *76*, 911–914.
3. Arachevaleta, M.; Bacon, C.W.; Hoveland, C.S.; Radcliffe, D.E. Effect of the tall fescue endophyte on plant response to environmental stress. *Agron. J.* **1989**, *81*, 83–90. [CrossRef]
4. Ball, D.M.; Lacefield, G.D.; Hoveland, C.S. *The Wondergrass: The Story of Tall Fescue in the United States*; Oregon Tall Fescue Commission: Salem, OR, USA, 2019.
5. Klotz, J.L.; Kirch, B.H.; Aiken, G.E.; Bush, L.P.; Strickland, J.R. Effects of selected combination of tall fescue alkaloids on the vasoconstrictive capacity of fescue-naive bovine lateral saphenous veins. *J. Anim. Sci.* **2008**, *86*, 1021–1028. [CrossRef] [PubMed]

6. Belesky, D.P.; Stuedemann, J.A.; Plattner, R.D.; Wilkinson, S.R. Ergopeptine alkaloids in grazed tall fescue. *Agron. J.* **1988**, *80*, 209–212. [CrossRef]
7. Klotz, J.L.; McDowell, K.J. Tall fescue ergot alkaloids are vasoactive in equine vasculature1,2. *J. Anim. Sci.* **2017**, *95*, 5151–5160. [CrossRef] [PubMed]
8. Schuenemann, G.M.; Edwards, J.L.; Hopkins, F.M.; Rohrbach, N.R.; Adair, H.S.; Scenna, F.N.; Waller, J.C.; Oliver, J.W.; Saxton, A.M.; Schrick, F.N. Fertility aspects in yearling beef bulls grazing endophyte-infected tall fescue pastures. *Reprod. Fertil. Dev.* **2005**, *17*, 479–486. [CrossRef] [PubMed]
9. McClanahan, L.K.; Aiken, G.E.; Dougherty, C.T. Influence of rough hair coats and steroid implants on the performance and physiology of steers grazing endophyte-infected tall fescue in the summer. *Prof. Anim. Sci.* **2008**, *24*, 269–276. [CrossRef]
10. Aldrich, C.G.; Paterson, J.A.; Tate, J.L.; Kerley, M.S. The effects of endophyte-infected tall fescue consumption on diet utilization and thermal regulation in cattle. *J. Anim. Sci.* **1993**, *71*, 164–170. [CrossRef] [PubMed]
11. McDowell, K.J.; Moore, E.S.; Parks, A.G.; Bush, L.P.; Horohov, D.W.; Lawrence, L.M. Vasoconstriction in horses caused by endophyte-infected tall fescue seed is detected with Doppler ultrasonography1. *J. Anim. Sci.* **2013**, *91*, 1677–1684. [CrossRef] [PubMed]
12. Putnam, M.R.; Bransby, D.I.; Schumacher, J.; Boosinger, T.R.; Bush, L.; Shelby, R.A.; Vaughan, J.T.; Ball, D.; Brendemuehl, J.P. Effects of the fungal endophyte *Acremonium coenphialum* in fescue on pregnant mares and foal viability. *Am. J. Vet. Res.* **1991**, *52*, 2071–2074. [PubMed]
13. Monroe, J.L.; Cross, D.L.; Hudson, L.W.; Hendricks, D.M.; Kennedy, S.W.; Bridges, W.C. Effect of selenium and endophyte-contaminated fescue on performance and reproduction in mares. *Equine Vet. Sci.* **1988**, *8*, 148–153. [CrossRef]
14. Kosanke, J.L.; Loch, W.E.; Worthy, K.; Ellersieck, M.R. Effect of toxic tall fescue on plasma prolactin and progesterone in pregnant pony mares. In Proceedings of the Tenth Equine Nutrition and Physiology Symposium, Fort Collins, CO, USA, 11–13 June 1987; pp. 663–668.
15. Thompson, F.N.; Caudle, A.B.; Kemppainen, R.J.; Nett, T.M.; Brown, J.; Williams, D.J. Thyroidal and prolactin secretion in agalactic mares. *Theriogenology* **1986**, *25*, 575–580. [CrossRef]
16. McDowell, K.; Taylor, V.; Phillips, T.; Lea, K.; Smith, R.; Aiken, G.; Barrett, M. Pregnant mares grazing a novel endophyte-infected tall fescue foal normally. *J. Equine Vet. Sci.* **2018**, *74*, 56–64. [CrossRef]
17. Dillard, S.L.; Smith, S.R.; Hancock, D.W. Variability of ergovaline and total ergot alkaloid expression among endophytic tall fescue cultivars. *Crop. Sci.* **2019**, *59*, 2866–2875. [CrossRef]
18. Greene, E.; Smith, S.R.; Cotton, K.L.; Davis, D. Comparison of ergovaline concentrations in BarOptima Plus E34 tall fescue and control varieties. In Proceedings of the American Forage and Grassland Council Annual Conference, Covington, KY, USA, 6–8 January 2013.
19. Webb, B.A.; Barney, W.E.; Dahlman, D.L.; DeBorde, S.N.; Weer, C.; Williams, N.M.; Donahue, J.M.; McDowell, K.J. Eastern tent caterpillars (*Malacosoma americanum*) cause mare reproductive loss syndrome. *J. Insect Physiol.* **2004**, *50*, 185–193. [CrossRef]
20. Morrison, J.I.; Smith, S.R.; Aiken, G.E.; Lawrence, L.M. Composition of Horse Diets on Cool—Season Grass Pastures using Microhistological Analysis. *Forage Grazinglands* **2009**, *7*, 1–9. [CrossRef]
21. Allman, R. *Annual Tall Fescue Ergovaline Concentration Trends*; University of Kentucky Forage Extension Program: Lexington, KY, USA, 2007; Data unpublished.
22. Smith, S.R.; Lea, K. Monitoring of horse pasture composition and tall fescue toxicity in Kentucky. In Proceedings of the 73rd Southern Pasture and Forage Crop Improvement Conference, Roanoke, VA, USA, 21–23 May 2019.
23. Aiken, G.; Bransby, D.; McCall, C. Growth of yearling horses compared to steers on high-and low-endophyte infected tall fescue. *J. Equine Vet. Sci.* **1993**, *13*, 26–28. [CrossRef]
24. Vogel, K.P.; Masters, R.A. Frequency grid: A simple tool for measuring grassland establishment. *J. Range Manag.* **2001**, *54*, 653. [CrossRef]
25. Payne, K.M.; Smith, S.R.; Goff, B.M. Enhanced efficiency nitrogen formulations on stockpiled tall fescue production. *Agron. J.* **2020**, *113*, 1596–1606. [CrossRef]
26. Lea, K.; Smith, L.; Gaskill, C.; Coleman, R.; Smith, S.R. Ergovaline stability in tall fescue based on sample handling and storage methods. *Front. Chem.* **2014**, *2*, 76. [CrossRef]
27. Vincelli, R.; Smith, S.R.; Tillery, T. Sampling for the Tall Fescue Endophyte in Pasture or Hay Stands. University of Kentucky Extension Publication PPA-30. 2007. Available online: http://www2.ca.uky.edu/agcomm/pubs/PPA/PPA30/PPA30.pdf (accessed on 24 September 2021).

toxins

MDPI

Article

Ergot and Ergot Alkaloids in Cereal Grains Intended for Animal Feeding Collected in Slovenia: Occurrence, Pattern and Correlations

Janja Babič [1,*], Gabrijela Tavčar-Kalcher [1], Franci Aco Celar [2], Katarina Kos [2], Matjaž Červek [3] and Breda Jakovac-Strajn [1]

1 Institute of Food Safety, Feed and Environment, Veterinary Faculty, University of Ljubljana, Gerbičeva 60, 1000 Ljubljana, Slovenia; gabrijela.tavcar-kalcher@vf.uni-lj.si (G.T.-K.); Breda.JakovacStrajn@vf.uni-lj.si (B.J.-S.)

2 Department of Agronomy, Biotechnical Faculty, University of Ljubljana, Jamnikarjeva 101, 1000 Ljubljana, Slovenia; Franc.Celar@bf.uni-lj.si (F.A.C.); katarina.kos@bf.uni-lj.si (K.K.)

3 Emona RCP, Kavčičeva ulica 72, 1000 Ljubljana, Slovenia; matjaz.cervek@e-rcp.si

* Correspondence: Janja.Babic@vf.uni-lj.si; Tel.: +386-(0)14779233

Received: 25 September 2020; Accepted: 20 November 2020; Published: 21 November 2020

Abstract: This four-year study reports the occurrence of ergot alkaloids (EAs) in cereals intended for animal feeding collected in Slovenia. A total of 517 samples of cereals were analysed using liquid chromatography-tandem mass spectrometry for the presence of EAs. The sample set included wheat, rye, triticale, oat, spelt and barley. The study revealed that 17% of the analysed cereal samples were contaminated with at least one ergot alkaloid. EAs have two epimeric forms: -ine and -inine. The incidence rates of the -ine and -inine forms in the analysed samples were 16% and 15%, respectively. The highest contamination rates were observed in rye (54%), oat (50%) and spelt (30%), where the highest mean concentrations of total EAs were also determined (502 µg/kg, 594 µg/kg and 715 µg/kg, respectively). However, the highest concentrations of total EAs were found in wheat and rye (4217 µg/kg and 4114 µg/kg, respectively). The predominant EAs were ergometrine, ergosine and ergocristinine. The occurrence of six or more ergot alkaloids was observed in 49% of the positive samples. A weak correlation ($p = 0.284$) in the positive samples was found between the mass of sclerotia and the total concentrations of EAs using the Spearman correlation coefficient.

Keywords: ergot alkaloid occurrence; sclerotia; cereals; LC-MS/MS; correlation; survey

Key Contribution: First data on EA occurrence in feed in Slovenia, emphasizing the importance of combining chemical and microscopic methods. Statistically evaluated data presenting several correlations between ergot and ergot alkaloids and individual ergot alkaloids.

1. Introduction

Ergot alkaloids (EAs) are secondary metabolites produced by fungi of the genus *Claviceps*. In Europe, *Claviceps purpurea* is the most widespread *Claviceps* species. The term "*purpurea*" comes from its ability to replace kernels in cereals with hard purple-coloured ergot bodies (sclerotia) containing a variety of alkaloids. *Claviceps purpurea* is known to cause more than 400 plant species, including grasses and economically important cereal grains, such as rye, wheat, triticale, barley, millet and oats, to be infected with the disease known as ergot [1–3]. Sclerotia of *C. purpurea* vary in size (2–20 mm) and are up to ten-fold larger than a normal grain. They have a hard, protective, and black to dark purple cortex outside and a white to grey inside medulla. Sclerotia contain various classes of EAs. More than 50 EAs have been identified [1,4], the most prominent of which are ergometrine (Em), ergotamine (Et), ergosine (Es), ergocristine (Ecr), ergocryptine (Ekr) and ergocornine (Eco), as well as their epimeric forms (-inines), which are biologically less active, but can be converted to the -ine form under various conditions [1,5]. EAs are characterized by the presence of a tetracyclic ergoline ring system and can be classified into four major groups based on the substitution at C^8 [4]: clavine alkaloids and 6,7-secoergolenes, simple lysergic acid derivatives (Em; Figure 1A), ergopeptine alkaloids (Es, Eco, Ekr, Ecr; Figure 1B), and ergopeptam alkaloids (Et, Figure 1B). The chemical structures of the epimeric forms -inine are the same as of the main EAs, but with different configuration at C^8, which is the centre of symmetry. The main EAs are left-hand rotation (R)-isomers, but their epimeric forms -inine are right-hand (S)-isomers [4]. The amount and toxin pattern varies between fungal strains, depending on the host plant and geographical region [1,3]. As stated in [6], the degree of variability in the EA pattern in relation to the fungal species and geographical distribution as well as the host plant is not known at present. More data is needed to identify all factors responsible for the variability in the EA pattern in individual plant species [6]. Currently, the regulations are based on the quantities of ergot sclerotia. In Directive 2002/32/EC on undesirable substances in animal feed [7] and its amendment [8], the maximum content of rye ergot (*Claviceps purpurea*) in feed containing unground cereals has been set at 1000 mg/kg.

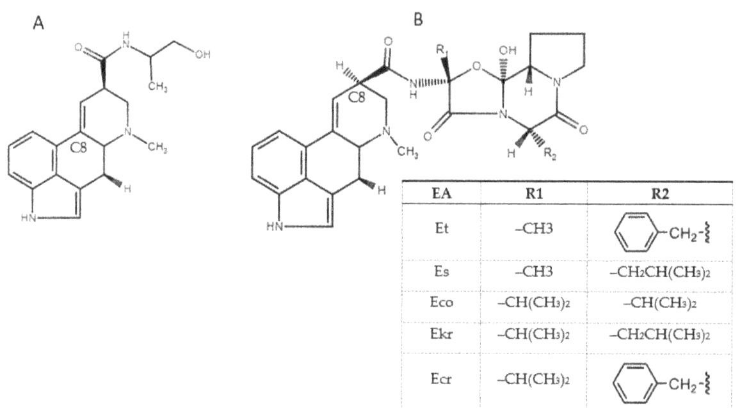

Figure 1. Chemical structures of EAs. (**A**) Ergometrine—Em and (**B**) ergotamine—Et, ergosine—Es, ergocornine—Eco, ergokriptine—Ekr and ergocrystine—Ecr.

However, the physical determination of the contamination rate of cereals by rye ergot is often inaccurate, as the size and mass of the sclerotia may vary considerably, and it is impossible in processed feed and food. Chemical analysis has been suggested for the possible control of potentially contaminated feed and food [6], as various chromatographic methods are available for detecting EAs in feed and food [9,10]. EU Member States were recommended for monitoring data on the presence of EAs in cereals and cereal products and to report their findings to the EFSA [6]. Further, simultaneous determination of the sclerotia content in the samples was recommended to improve the knowledge of the relationship between the content of sclerotia and the level of individual EAs [6].

There are several reports on the presence of EAs in feed from different countries. Ruhland and Tischler [11] reported the occurrence of EAs in 86–100% of feed samples from Germany. The median concentrations in different feeds were 25–96 μg/kg and the maximum concentrations were 149–4883 μg/kg. In a Dutch survey of 184 samples of cereals and compound feed [12], 43% of the cereal samples and 83% of the compound feeds contained EAs with an average concentration of 89 μg/kg and a maximum concentration of 1231 μg/kg. The major detected EAs were Es, Et, Ecr and Ekr. Malysheva et al. [13] presented the occurrence of EAs over a three-year period in a total of 1065 cereal samples originating from 13 European countries. EAs were present in 52% of rye, 27% of wheat and 44% of triticale samples at total EA levels ranging from ≤ 1 µg/kg (≤ LOQ) to 12,340 µg/kg. The most frequently occurring EAs were Es, Ekr and Ecr. An occurrence study in different cereals of the 2012–2014 harvesting season in France was reported by Orlando et al. [14]. The rye and triticale were the most prone to infection. EAs were present in 78% of rye and 54% of triticale samples. In 74% of the samples the most frequent EAs were Et, Es and Ecr. Topi et al. [15] reported EA contamination in Albanian wheat, including 71 samples in the years 2014–2015. In 2014 and 2015, 48.6% and 19.4% of the samples contained EAs, respectively. The EA concentrations were from 10.3 to 975.4 µg/kg. Ecr, Es and Em were the predominant EAs in most samples. The differences in ergot patterns and total EA concentrations are apparently due to the geographical region and environmental conditions during the development of the ergot sclerotia, as mentioned by Krska and Crews in their study [3].

Until now, no data on EA in feed in Slovenia were available. For this reason, a visual and chemical method using LC-MS/MS was introduced to determine ergot and EAs (Em, Et, Es, Ecr, Ekr, Ekr and Eco and their –inine epimers) in cereal grains collected in Slovenia in 2014–2017 and for a correlation study between sclerotia content and EA concentration. The approach of combining visual and chemical methods in the correlation studies will contribute to the knowledge of the relationship between the presence of sclerotia and the content of EAs and the EA pattern in different cereal grains. The four-year study focused on cereals to be used as feed material in Slovenia.

2. Results

2.1. Occurrence of Total Ergot Alkaloids

A total of 517 ground cereal samples were tested in the four-year study of the presence of EAs in various cereals. EAs were determined in 206 wheat, 136 barley, 101 triticale, 35 rye, 23 spelt and 16 oat samples. The samples containing one or more individual EAs at concentrations equal to or above the LOQ (10 µg/kg) were considered positive. The overall occurrence of EAs within each cereal group in the years 2014–2017 is shown in Table 1. Contamination with at least one of the 12 analysed EAs was revealed in 87 of the 517 samples, which amounts to 17% positive rate. The incidence of positive samples was the highest in rye (54%) and oat (50%) samples. In the other cereals, EAs were detected in lower percentages. The total EA concentrations ranged from 14 to 4217 µg/kg. The overall mean concentration level of the positive samples was 448 µg/kg and the median level was 154 µg/kg. The -ine and -inine epimers were present in 95% and 91% of the contaminated samples, respectively. The maximum concentration levels of the –ine and –inine epimers were 2476 µg/kg (wheat, 2014) and 1849 µg/kg (rye, 2017), respectively.

Table 1. Ergot alkaloid occurrence in cereals in the years 2014–2017.

Descriptive Statistics	Wheat	Barley	Triticale	Rye	Spelt	Oat	Total-ines	Total -inines	Total EAs
Number of samples	206	136	101	35	23	16	517	517	517
Number of positive samples	34	6	13	19	7	8	83	79	87
Incidence of positive samples (%)	17	4	13	54	30	50	16	15	17
Min (µg/kg)	14	27	14	25	152	84	10	9.9	14
Max (µg/kg)	4217	1177	2587	4114	2682	2191	2476	1849	4217
Mean (µg/kg)	363	340	417	502	715	594	252	229	448
Median (µg/kg)	102	111	152	154	264	359	84	82	154

The susceptibility of cereals to ergot infection (from most to least) was ranked rye, wheat, triticale, barley, and oats. Rye, as an open pollinator, allows easy access of the fungus into the flowering head, therefore, it is more susceptible to ergot infection than wheat and barley, which are self-pollinators. Oat is rarely affected [9]. However, in our study, a high incidence rate was revealed in oat with a high median concentration, but the maximum concentration of the total EAs was higher in cereals with a higher susceptibility to ergot infection (rye and wheat). The maximum concentration levels of total EAs followed the order wheat (4217 μg/kg, 2014), rye (4114 μg/kg, 2017), spelt (2682 μg/kg, 2014), triticale (2587 μg/kg, 2014), oat (2191 μg/kg, 2016), and barley (1177 μg/kg, 2016).

The distribution according to total EA concentrations in the analysed samples is shown in Figure 2. Within each cereal group, more than 50% of the results were below the LOQ (10 μg/kg). Of the 517 samples analysed, in 430 samples (83%) the concentrations of total EAs were less than the LOQ. Only 17% of the samples contained EAs with concentrations between 14 μg/kg and 4217 μg/kg. The highest percent of samples with an EA concentration below the LOQ was observed in barley (95%), followed by triticale (88%), wheat (84%), spelt (70%) and oat (50%). Of the 35 rye samples analysed, 16 (46%) contained no EA, while the other 19 samples contained very equally distributed EA concentrations between the LOQ and the maximum (4114 μg/kg, Figure 2). The mean concentration of 502 μg/kg was not substantially different from the mean concentration obtained in oat (594 μg/kg) (Table 1). Of the eight contaminated oat samples, five (31%) had an EA concentration between 100 and 500 μg/kg and two had an EA concentration of more than 500 μg/kg, which contributed to a higher mean concentration level in oat samples. Similarly, in all contaminated spelt samples, the concentrations were between 100 μg/kg and 2682 μg/kg, while no samples with concentrations between the LOQ and 100 μg/kg were found. Therefore, the mean concentration was higher in spelt samples than in rye samples (715 μg/kg).

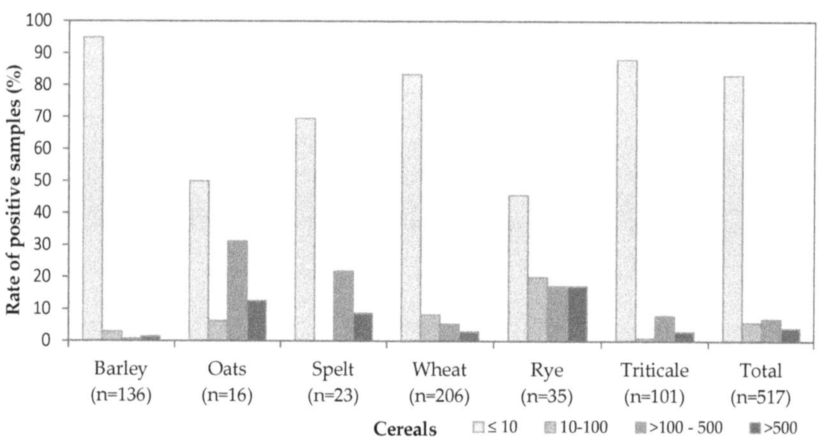

Figure 2. Distribution of the total ergot alkaloid concentrations (μg/kg) in the samples analysed in the years 2014–2017. *n*: number of samples.

The yearly occurrence of total EAs is shown in Figure 3. In the years 2014–2017, EAs were found in 25%, 15%, 10% and 13% of the tested samples, respectively. In 2014, the highest EA incidence was observed in oat, spelt and rye samples, where at least 50% of the samples of each cereal contained one or more EA. The incidences of EAs in spelt and rye samples were similar in 2014 and 2015, while that of oat was lower (33%) in 2015. In 2014, an increased incidence in wheat and triticale was also observed compared to 2015–2017. In 2016 and 2017, the highest incidence of EAs was observed in oat and rye samples, where more than 40% of the samples were contaminated with one or more EA, while contamination of other cereal samples was less than 20%. In 2017, all rye samples contained EAs. The incidence of EAs in spelt, wheat and triticale was the lowest in 2016, with an increased incidence in spelt and triticale in 2017. A review of the observed years indicated that the incidence of EAs in the analysed samples was the highest in 2014, and it decreased in the subsequent years, except for those of rye and oat.

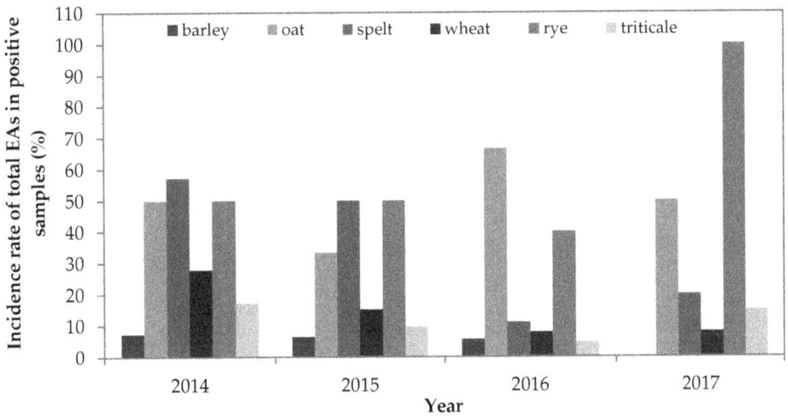

Figure 3. Yearly incidence of total ergot alkaloids (%) in the analysed samples. The samples containing one or more individual EAs at concentrations equal to or above the LOQ (10 μg/kg) were considered positive.

In each cereal group, the differences according to all observed years were also evaluated. The p-values for barley, oat, spelt, rye and triticale were 0.564, 0.880, 0.255, 0.492 and 0.543, respectively. In the mentioned cereal groups, we have not found significant differences in different years; only in the wheat cereal group ($p = 0.005$) was a significant difference observed.

The total EA concentrations in the years 2014–2017 are shown in Figure 4. Since some samples had a very high total EA concentration, the mean concentrations of all cereals are higher than the median concentrations. Due to the higher ratio of results with a concentration of EAs above 500 μg/kg, the median concentration in oat and spelt samples was higher than that of the rye samples. The median concentration, from highest to lowest, was: oat (359 μg/kg), spelt (264 μg/kg), rye (154 μg/kg), triticale (152 μg/kg), barley (111 μg/kg) and wheat (102 μg/kg).

The differences between the mean values of total EA concentrations among the years (2014–2017) were evaluated using the Kruskal–Wallis test. The differences were considered significant if $p < 0.05$. The mean values of total EAs concentrations were significantly different ($p < 0.001$) between the years 2014–2017. Furthermore, the mean EA concentrations in the year 2014 were higher than in 2015, 2016 and 2017 ($p < 0.001$), in the year 2015 they were lower than in 2016 ($p = 0.123$) and 2017 ($p = 0.507$) and in the year 2016 they were lower than 2017 ($p = 0.452$).

Statistical differences with $p < 0.05$ in the mean values of total EA concentrations were found between different cereal groups. The total mean EA concentration in barley was higher than in all other cereal groups (oat, spelt, wheat, rye and triticale) with $p < 0.001$. The total mean EA concentration in spelt was higher than in triticale ($p = 0.047$) and barley ($p < 0.001$), but lower than in the other cereals (p-values between 0.082 and 0.317). The total mean EA concentrations were higher in oat than in wheat, barley and triticale ($p < 0.001$), while it was lower in oat than in spelt ($p = 0.167$) and rye ($p = 0.582$). No significant difference was found between wheat and triticale ($p = 0.478$), but the total mean EA concentration in wheat was higher than in barley, oat and rye ($p < 0.001$). In addition, the total mean EA concentration in rye was higher than in triticale, wheat and barley ($p < 0.001$).

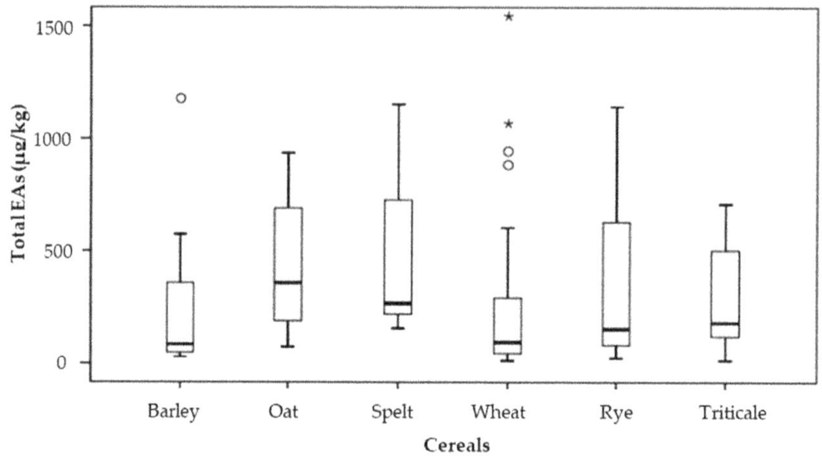

Figure 4. Box-plot graph summarizing the total content of 12 ergot alkaloids (as shown in the legend of Figure 7) in the positive samples (containing at least one EA with a concentration of 10 µg/kg or more) for each cereal species in the years 2014–2017. Median concentrations are indicated by horizontal lines in the boxes encompassing the 25th–75th percentiles. Outliers in the 95th percentiles are indicated as empty dots (○) and stars (*).

2.2. *Co-Occurrence of Ergot Alkaloids*

The co-occurrence of EAs in positive samples of each cereal and in all positive samples together is shown in Figure 5. Of the 87 positive samples, eight (9.2%) and 10 (11.5%) samples contained only one or two EAs, respectively. Three or more and six or more EAs were present in 79.3% and 49.4% of all samples, respectively.

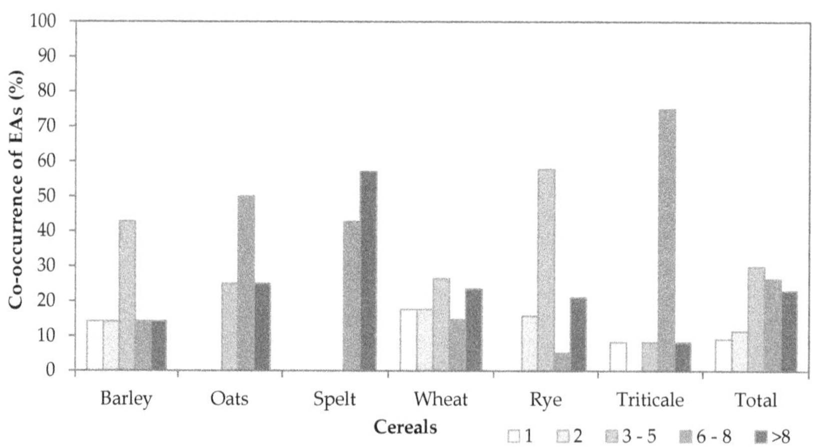

Figure 5. Co-occurrence of EAs (as indicated in the legend of Figure 7). Only positive samples are considered (containing at least one EA with a concentration of 10 µg/kg or more).

A difference in the co-occurrence of EAs was found between oat, spelt and triticale on one side and barley, wheat and rye on the other side. In oat, spelt and triticale, the positive samples were contaminated with at least three EAs, but in 75–100% of the samples six or more EAs were present.

In barley, wheat and rye, more than 50% of the positive samples contained fewer than six EAs. In the spelt samples, all positive samples contained six or more EAs per sample; in oat and triticale samples, the co-occurrence exceeded 75%; while in barley, rye and wheat, it was lower than 40% (Figure 5).

In spelt, 43% of the positive samples contained six to eight co-occurring EAs, while 57% contained more than eight EAs. Six to eight or more EAs were also found in oat (50%) and triticale (75%) samples, whereas the co-occurrence of more than eight EAs was found in 25% or fewer certain cereal samples. The co-occurrence of three to five EAs was the highest in rye (58%) and barley (43%), followed by wheat (27%) and oat (25%), while the lowest co-occurrence was observed in triticale (8%). An equal distribution of co-occurrences over a complete range from one to more than eight EAs was observed in wheat samples. The co-occurrence of three to five EAs and more than eight EAs was 27% and 24%, respectively, while the other co-occurrences were approximately 15%.

A statistically significant difference ($p < 0.05$) between the mean value of the number of co-occurring EAs in positive samples and the type of cereal ($p = 0.035$) was found. The results are presented in Figure 6. The mean of EA number in spelt samples was greater than in barley ($p = 0.046$), triticale ($p = 0.043$), rye ($p = 0.010$) and wheat samples ($p = 0.019$). All other co-occurrences of EAs in cereals were not significantly different ($p = 0.053$–0.812).

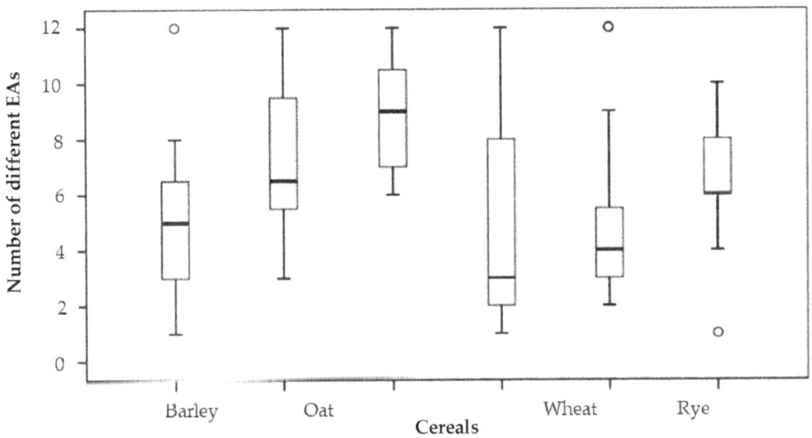

Figure 6. Number of different ergot alkaloids in cereals. Mean values are indicated by horizontal lines in the boxes encompassing the 25th–75th percentiles. Outliers in the 95th percentiles are indicated as empty dots (○).

2.3. Individual Ergot Alkaloids

In 87 positive samples, all EA forms were found, but some EA forms were observed more often than others (Figure 7, dark bars). The occurrence of individual EAs in the positive samples was evaluated as the ratio (%) between the number of an individual EA and the number of the positive samples. The most frequently occurring alkaloids in the positive samples were Em, Es, ergosinine (Esn), Ecr and ergocristinine (Ecrn). Es and Esn were present in 57 (66%) and 58 (67%) of the samples, respectively, while Ecr and Ecrn were found in 57 (66%) and 56 (64%) of the 87 positive samples. Em had a higher rate of occurrence (63%) than the corresponding epimer Emn, which was found in only 29% of the positive samples.

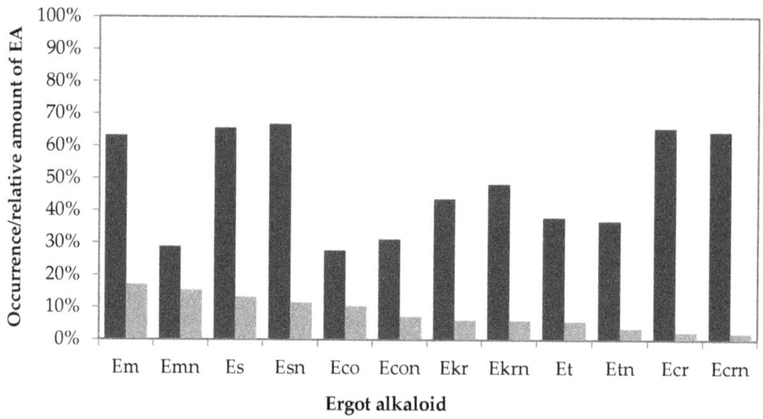

Figure 7. Occurrences (dark bars, % of 87 positive samples) and relative amounts (light bars, % of the sum of individual EA concentration compared to the sum of total EA concentration (39100 µg/kg)) of the individual EAs in the positive cereal samples. Legend: Em—ergometrine, Emn—ergometrinine, Es—ergosine, Esn—ergosinine, Eco—ergocornine, Econ—ergocorninine, Ekr—ergocryptine, Ekrn—ergocryptinine, Et—ergotamine, Etn—ergotaminine, Ecr—ergocristine and Ecrn—ergocristinine.

The relative EA amounts (%) of individual EAs present in the positive samples are shown in Figure 7 with light bars. The relative amounts were calculated as the ratio of the sum of the individual EA concentration to the sum of the total EA concentration in the positive samples. The relative amounts of -ine epimers were higher than those of -inine epimers. The predominant EA was Em (17%) and its epimer Emn (15.3%), followed by Es (13%), Esn (11.5%), Eco (10.4%) and Econ (7.2%). The most abundant EAs were the Em/Emn and Es/Esn epimers.

The occurrence of individual EAs in the positive samples varied depending on the cereal group (Figure 8). All twelve individual EAs were found in all observed cereal groups, but with different EA patterns and different relative EA amounts. Em, Es, Esn and Ecr, Ecrn were determined in 58–100% of all positive samples for a specific cereal. Em was most frequently found in spelt (86%), oat (75%) and triticale (67%), while in barley, wheat, and rye it was slightly lower (58%). Emn was present in 24–38% of the positive samples in all cereals, with the highest occurrence in oat and the lowest in wheat. Es and Esn were present in spelt in all positive samples (100%) and in 92% and 83% of the positive triticale samples, respectively. The epimers Es and Esn were present in 75% of oat and 71% of barley but in less than 70% of wheat and rye. More barley, spelt, wheat and triticale samples were contaminated by Ecr than by its epimer, Ecrn, which was higher in occurrence in oat and rye. Eco, Ekr, Et and their corresponding epimers (Econ, Ekrn and Etn) were infrequently present in the positive samples. Et and Etn were found in more than 50% of the positive oat and spelt samples, Ekr and Ekrn were found in oat, spelt and triticale samples, while more often Eco was found only in spelt. The occurrence of Et ranged from 26% in rye to 62.5% in oat, while the Etn level ranged from 14% in barley to 62.5% in oat. Ekr and Ekrn were present in 86% of the positive samples of spelt, while their lowest level was observed in barley (28.5%).

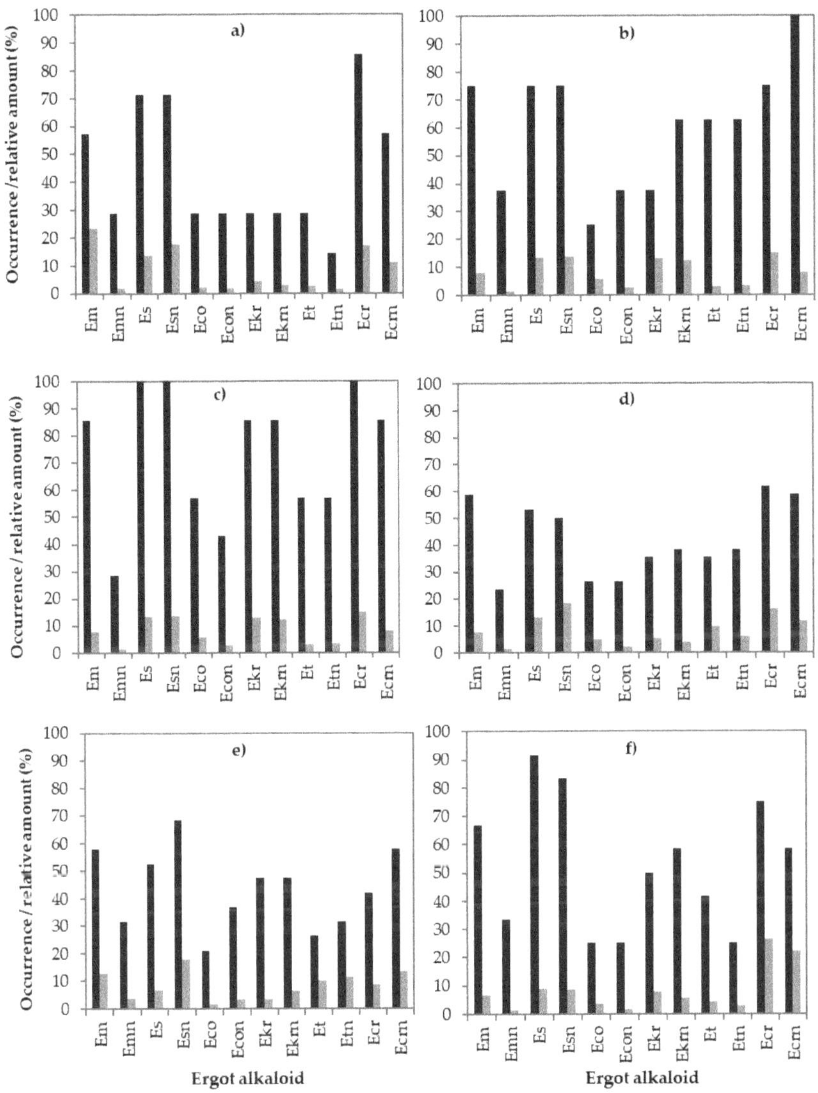

Figure 8. Occurrences (dark bars, % of 87 positive samples) and relative amounts (light bars, % of the sum of individual EA concentration compared to the sum of total EA concentration (39,100 µg/kg)) of the individual EAs in different cereals: (**a**) barley, (**b**) oat, (**c**) spelt, (**d**) wheat, (**e**) rye and (**f**) triticale. Only positive samples are considered (containing at least one EA with a concentration of 10 µg/kg or more).

The relative amounts of individual EAs differed between cereal groups. In all samples, both epimeric forms, –ine and –inine, were found, however, they did not always appear together. In most cases, –ine forms were present in a higher ratio than their epimers (–inine form) with one exception (the epimers Es and Esn). In all cereals, Esn rather than Es was the predominant alkaloid at a ratio of 8.5–25%. If a higher occurrence of an individual EA was found in a positive sample, usually a higher ratio was observed.

2.4. Ergot (Sclerotia) Content and Analytical Correlation Study

Of the 432 samples analysed in the years 2014–2016, the EAs were found in 76 samples. However, with microscopic inspection, sclerotia were found in only 52 positive samples. In the 76 samples, the mean and maximum concentrations of the total EAs were 439 μg/kg and 4217 μg/kg, respectively. The mean and median sclerotia weights in the samples were 134 mg and 36.5 mg, respectively. The average sclerotia content per kg of cereal was 0.14 g/kg. The most abundant individual EAs in the samples were Em, Es, Esn, Ecr and Ecrn (58–67%), followed by Ekr, Ekrn, Et and Etn (40–49%). The contributions of Eco, Econ and Emn were 29% or less.

A Spearman's rank-order correlation was run to determine the relationship between the mass of sclerotia (mg) or sclerotia content (g/kg) and the total EAs concentrations (μg/kg) as well as individual EA concentrations (Table S1). The results were evaluated as statistically significant at the level $p < 0.01$ or 0.05 and the strength of the correlation was determined by comparing calculated correlation coefficient values (r) with Evans guidelines (Table S2) [16]. There was a weak, positive, statistically significant correlation between the mass of sclerotia or sclerotia content and the total EA concentrations ($r = 0.284$, $p = 0.008$; $r = 0.315$, $p = 0.003$). Furthermore, the concentrations of all individual EAs were significantly different with $p < 0.01$ or 0.05 and were weakly or moderately correlated according to the correlation coefficient values shown in Table S2. There was a strong, positive, statistically significant ($r \geq 0.65$, $p < 0.001$) correlation between each EA pair –ine and –inine.

In a further experiment, a visual screening before chemical analysis of cereal samples collected in the year 2017 was performed. A total of 85 cereal samples were analysed. The results for the visual inspection as well as the chemical determination are shown in Table 2. In 65 samples (76.4%, denoted A and D in Table 2), the results of the visual inspection for the presence and absence of sclerotia, matches with the result of the chemical analysis. In 20 of the positive samples (denoted B), sclerotia were found with an absence of EAs. Cases where no sclerotia were found although alkaloids appeared to be present were none. The average total EA concentration in cereals containing sclerotia was 512 μg/kg and the median concentration was 139 μg/kg. In 11 samples, where the masses of sclerotia and the EAs were determined, the most frequently present individual EAs were Es, Ecr and Em. Compared to the present study, much higher average levels were found in cereals in the study by Mulder et al. [12].

Table 2. The number of the positive and negative samples for the visual screening and for the chemical analysis. *n*: number of samples.

Combined Results ($n = 85$)	Sample Group	n	%
Visually positive/chemically positive	A	11	12.9
Visually positive/chemically negative	B	20	23.5
Visually negative/chemically positive	C	0	0.0
Visually negative/chemically negative	D	54	63.5

3. Discussion

3.1. Occurrence of Total Ergot Alkaloids

In this study, 517 samples of cereals were analysed for EA content. There are only a few available surveys on the occurrence of EAs focusing on feed. In Table 3, an overview of the results from different studies is presented. Considering all samples in a specific cereal group, the incidence of positive samples in Slovenian cereals was the highest in rye (54%), followed by oat and spelt (50% and 30%), while wheat and triticale were lower (17% and 13%). Remarkably, the incidence of positive samples in rye was almost identical to the values reported in Europe (52%) and the Netherlands (50%) [12,13] but differed from the results of Ruhland and Tischler [11], who observed EA contamination in all 15 analysed rye samples and also a high incidence rate for other grains from Germany with levels from 86% to 93%. The incidence of EAs in wheat samples in our study was much lower than those in studies from Germany, the Netherlands and the USA [11,12,17], but close to the results from Europe [13]. In the study, more than half of the wheat samples originating from Germany, Poland, Switzerland and the Czech Republic did not contain EAs, which is also similar to our results. Out of 206 Slovenian wheat samples, 83% were not contaminated with EAs. The incidence of EAs in triticale samples was similar to those reported in the studies of Malysheva et al. [13] and Mulder et al. [12], where 44% and 33.3% of 27 and 45 analysed samples, respectively, were contaminated. The limited number of samples in the European study did not allow a robust analysis as conducted by Malysheva et al. [13] and could also be the reason for a slightly higher incidence rate of positive samples in European samples than in Slovenian samples, where results were obtained from 101 analysed samples. The highest incidence rate in triticale samples was observed in 14 German samples (93%) and the lowest was observed in our samples (13%), as shown in Table 3. The literature results for the incidence of EAs in barley are scarce. Tittlemier et al. [18] analysed a total of 25 barley samples, but the incidence rate was not provided. In our study, most of the 136 analysed barley samples were uncontaminated. In 4% of the positive samples, the maximum and median concentrations were 1177 μg/kg and 111 μg/kg, respectively.

Topi et al. [15] reported on the presence of EAs in 71 wheat samples collected during the years 2014 and 2015. The maximum concentration of EAs was 975.4 μg/kg, which was lower than our result (4217 μg/kg). A similar maximum content of total EAs was observed for American wheat samples (4760 μg/kg), while in other countries (Canada, Europe, Germany, the Netherlands) the maximum concentrations in wheat samples were from 529 μg/kg to 1236 μg/kg (Table 3), but median concentrations were from <1 ug/kg (< LOQ) to 195 μg/kg. The highest median concentration level was observed in Albanian wheat (226.7 μg/kg). The median concentration in Slovenian wheat samples was between the values obtained in the samples collected in European countries (< LOQ) and in Albania (226.7 μg/kg). Of the 35 rye samples analysed, the median and maximum concentrations of 154 μg/kg and 4114 μg/kg were determined, respectively. A similar maximum concentration (4850 μg/kg) was reported by Meister and Batt [19]. Malysheva and co-workers [13] reported a much higher maximum concentration (12,340 μg/kg), which was found in a rye sample from Switzerland, while the results for the samples from the other countries mentioned in their work were below 500 μg/kg. Other studies performed by Mulder et al. [12] and Ruhland and Tischler [11], demonstrated lower maximum concentrations. In Dutch and German rye, it was 1231 μg/kg and 1067 μg/kg, respectively. The higher concentration in rye samples was expected because it is particularly susceptible to infection by *C. purpurea*, as rye is a cross-pollinator with opened florets [9].

In the German study [11] and triticale samples from different countries of Europe [13], the maximum concentration was the same (1103 μg/kg), with a median concentration of 25 μg/kg and < LOQ. In the European study, in 73% of all analysed samples, the results were below the LOQ. The statistical approach of Malysheva and co-workers [13] included not only positive samples but also the results of all samples. The median concentration in their study was below the LOQ for wheat and triticale samples and equalled the LOQ (1 μg/kg) for rye samples. The concentration considering only positive samples was 152 μg/kg. This result was higher than those in positive samples from the Netherlands (63.9 μg/kg) and Germany (25 μg/kg), where the number of analysed samples was much lower. If the same approach as that of Malysheva et al. [13] had been used in our study, the median concentration would also have been < LOQ, because in our case the ratio of the results below the LOQ was very high (83%) Additionally, in our study, we also presented the occurrence of total EAs in spelt and oat. There are no data in the literature to compare to these results. The maximum levels of total EAs in oat and spelt samples were similar (2191 μg/kg and 2682 μg/kg, respectively) as in triticale samples (2587 μg/kg). The median concentration was higher in oat samples, as in spelt samples. Since the number of samples was low, the data represent a rough evaluation of the EA occurrence in the two cereals in the years 2014–2017.

Table 3. The occurrence of ergot alkaloids (total) in the cereals. Summary of available data.

Country	Year of Sampling	Feed	Number of Samples	Positive Sample Rate (%)	Median (µg/kg)	Max (µg/kg)	LOD/LOQ (µg/kg)	Method of Analysis	Reference
Slovenia	2014–2017	Wheat	206	17	102 *	4217	3/10	LC-MS/MS	This study
		Barley	136	4	111 *	1177			
		Triticale	101	13	152 *	2587			
		Rye	35	54	154 *	4114			
		Spelt	23	30	264 *	2682			
		Oat	16	50	359 *	2191			
Albania	2014–2015	Wheat	71	33.8	226.7 *	975.4	3/10	LC-MS/MS	[15]
Germany	2013	Rye	60	67	237 *	4850	–	LC-MS/MS	[19]
Canada	2010–2012	Wheat	117	–	195	666	-/2	LC-MS/MS	[18]
		Barley	25	–	47	584			
		Rye	1	–	–	149			
Europe	2009–2012	Rye	157	52	1	12,340	-/1	LC-MS/MS	[13]
		Wheat	137	27	<LOQ	701			
		Triticale	27	44	<LOQ	1103			
USA	2011	Wheat	10	70	85 *	4760	–	LC-MS/MS	[17]
The Netherlands	2007–2010	Rye	69	50.7	121 *	1231	2/10	LC-MS/MS	[12]
		Triticale	45	33.3	63.9 *	297			
		Wheat	18	38.9	56 *	529			
		Other cereals	4	25.0	961 *	961			
Germany	2005–2007	Rye	15	100	96 *	1067	5/10	HPLC-FLD	[11]
		Triticale	14	93	25 *	1103			
		Wheat	21	86	29 *	1236			
		Other grains	14	93	44 *	140			

* Notes: only positive samples considered.

3.2. Co-Occurrence and Individual Ergot Alkaloids

In the contaminated samples of the four-year survey, all EAs were found. In the analysed samples, the number of co-occurring EAs was usually higher than three EAs. The results are shown in Figure 6. Only one EA was present in 9% of all samples, with the highest occurrence in barley, wheat and triticale samples. However, in all oat, spelt and rye samples, more than one EA was found. The co-occurrence of two EAs was revealed in barley, wheat, rye samples, but not in triticale samples. Three to five EAs co-occurred mostly in rye (58%) and barley (43%) samples but were not found in spelt samples, which usually contained six or even more than eight EAs. Similar high co-occurrences of more than six EAs were observed in triticale samples, with a lower co-occurrence of more than eight EAs. A difference in the co-occurrences of EAs was found between oat, spelt and triticale positive samples and between barley, wheat and rye positive samples. In oat, spelt and triticale, the positive samples were contaminated with at least three EAs, but in 75–100% of the samples, six or more EAs were present. Furthermore, no EA can be used to predict the total EA concentrations, because none of the EAs were determined in all contaminated samples in all cereals.

In most analysed samples, the –ine and –inine forms of EAs were found together. The occurrence of a single EA was observed for Emn and Et in barley. The distribution of EAs in positive samples is evident in Figures 7 and 8. The predominant EAs were Es and Ecr with the corresponding epimers and Em. The less frequently occurring EAs were Emn, Eco, and Et with the corresponding epimers, as they were present in 21–42% of the positive samples. However, the least frequently occurring EAs were Eco and the corresponding epimers Econ and Emn, which were found in less than 30% of the positive samples. A scarce amount of data is available on the EA pattern in grains or feed. Topi et al. [15] reported the ergot pattern for wheat samples harvested from 2014 to 2015. In the year 2014, the most frequently occurring EAs were Em and Es, which were present in 76.5% and 70.6% of the positives, respectively. The least frequently occurring EAs were Econ and Ekrn in 5.9% and 11.8% of the positive samples, respectively. The predominant EAs in the Slovenian wheat samples were Ecr and its epimer Ecrn, which were present in 61.8% and 58.8% of the positives, respectively. In our wheat samples, Es and Em were also found, as in Albanian wheat, but less frequently. They were present in 58.8% and 52.9% of the positives, respectively. In the Albanian samples from the year 2015 [15], the predominant EA was Ecr, which was present in 71.4% of the positive samples. Ecr and Ecrn were also the predominant alkaloids in wheat samples in Canada and France, followed by Et [14,18]. In contrast, in wheat samples from Belgium, only Em and Emn were detected, while the study of European samples concluded that Es, Ecr, and Ekr were the predominant EAs in most samples [13]. In our study, in more than 70% of the barley samples, the predominant EAs were Es and Ecr and their epimers, which were observed also in oat, spelt and triticale samples and were slightly less predominant in wheat and rye. Additionally, Em was present in more than 57% of all cereal positive samples, while Emn was the predominant EA in oat, rye, and triticale, but in other cereals its level was below 30%. Ekr and Ekrn were usually found in less than 40% of the positive samples for a specific cereal, with the highest occurrence in spelt samples. The least frequently present EAs in barley positive samples were Eco, Et and Ekr and the corresponding epimers, which were present in less than 30% of the positives. A similar low occurrence of Eco, Et, and Ekr was observed in wheat. In rye and triticale, only the Eco and Et occurrences were less than 40%, but Ekr was higher (47.4–58.3%). Overall, we agree with the conclusion of Malysheva et al. [13] and Orlando et al. [14] that the EA patterns and alkaloid levels found in the samples depend on the cereal type. For the cereals we analysed, such as barley, oat, spelt, and triticale, no literature data is available for a results comparison.

The geographical region and environmental conditions, the weather condition shortly before and during the flowering period of the cereal, and data on cereal cultivars and *Claviceps* species infecting the individual sample should be taken into account [3,20] as reasons for the differences in the EA pattern and EA contamination. These data were not collected in the study and, therefore, no conclusions can be drawn about the correlation of EA pattern and EA contamination on geographical and climatic conditions or *Claviceps* species. However, the available data from the meteorological stations show that the annual average temperatures in all years (2014–2017) were above the long-term average (1981–2010), in 2014 even by more than 1.6 °C. The fluctuations in annual average temperatures depend on both the year and the region. Annual precipitation in the period 2014–2017 fluctuated significantly compared to the long-term average (1981–2010), both by year and by individual regions of Slovenia. In 2014 there was significantly more precipitation (from 5% to more than 40% more) than the long-term average (1981–2010). In 2016 and 2017, some regions also received more precipitation than the long-term average, but the deviation was slightly smaller (0–20%). 2015 was drier than the long-term average in most regions.

3.3. Sclerotia Mass and Analytical Correlation Study

The visual screening method is a great supporting warning tool for ergot contamination, but only in combination with a chemical method for determination of EAs because, in some cases, the ergot bodies are present in samples as dust and with the visual method a false negative result could be produced, which is well apparent from our results in Table 2. With a chemical analysis, we also determined the presence of EAs in the samples where the results for visual screening were negative.

The objective of this study was to determine correlations between the EA concentrations of the individual EA and the total concentrations of EA as well as between the sclerotia contents and the EA concentrations. The results are shown in Table S1. Most EAs correlated with each other, but with no specific predominant individual EA, which could be used as a marker for determining the total EA concentration based on the content of a specific single EA in the sample. The results were in accord with the information available in the literature [12,21]. In contrast, a strong linear relationship between the concentration of ergot alkaloids and the presence of ergot sclerotia was reported by Tittlemier et al. [18] in Canadian wheat and other cereals and by Orlando et al. [14] in French cereals.

The results indicate that a higher sclerotia mass does not necessarily mean a higher total EA concentration in the sample. A fluctuation of the total EA concentrations in the samples with sclerotia must be expected because of the variability of the sclerotia mass and size in the samples. Furthermore, the fluctuation of the total EA concentration also varies between samples with equal sclerotia masses. The concentration of ergot body in cereals is for an animal restricted to 1000 mg/kg in feed materials and compound feed containing unground cereals [7,8], but the sclerotia content in unground cereals could not be used as an indicator of the EA contamination level of cereals intended as animal feed, because of the fluctuation of total EA concentration. Both methods, visual and chemical, should be used to monitor the EA in feed as recommended in [6]. Because of the diversity of EA patterns in cereals, the total EA concentration could be used to set the maximum level of EAs in cereals intended for animal feeding.

4. Conclusions

In this work, the occurrence of EAs in cereals in Slovenia was presented for the first time. The study contributed to the data on the EA content in feed material in Southeastern Europe, which are still scarce. The combination of visual and chemical methods improved the knowledge about the relationship between the content of sclerotia and the level of individual EA. The presented data outlined the importance of establishing maximal permitted levels of EA in feed and food. The maximum permitted level based on the presence of sclerotia [7,8] is not sufficient because a weak correlation between content of ergot sclerotia and the concentrations levels of individual and total EAs was observed. All individual EAs correlated with each other, but a specific EA that could be used as a marker for the determination of EAs in cereals was not found. As several cereal samples are contaminated with EAs, farmers still need to manage ergot to reduce the risk of occurrence of EA by using preventive methods to manage EAs concentrations. It is recommended that the official control of cereals continues with both visual and chemical methods.

5. Materials and Methods

5.1. Sample Collection

The samples (wheat, barley, triticale, rye, oat, and spelt) were collected in all regions of Slovenia in the years 2014–2017. The samples were taken according to Commission Regulation (EC) No. 152/2009 [22] and its amendment [23]. A total of 517 samples (173 samples in 2014, 123 in 2015, 136 in 2016 and 85 in 2017) were collected and tested for the presence of EAs and for sclerotia using visual and mass analyses. For the preparation of the final samples, the collected samples were divided into equal amounts using a divider according to Commission Regulation (EC) No. 152/2009 [22] and its amendment [23].

5.2. Standards and Chemicals

The EA standards Em, Es, Et, Eco, Ekr, Ecr and their -inine epimers, as well as MycoSep 150 Ergot columns were purchased from Romer (Biopure, Tulln, Austria). Stock standard solutions and the mixed working standard solutions were prepared in acetonitrile and stored in amber glass vials at −20 °C. The certified purity of individual standard substances was between 95.1% ± 4.9% and 99.0% ± 1.0%. The concentrations of the –ine and –inine stock standard solutions were 100 and 25 μg/mL, respectively. However, the exact concentrations of individual EA stock standard solutions obtained by reconstituting the content of ampoules in 5 mL of acetonitrile were provided by the producer. Acetonitrile (Honeywell, NC, USA) and ammonium carbonate (Merck, Darmstadt, Germany) were of analytical or LC-MS-grade purity. Deionised water was prepared using a Milli-Q system (Millipore, Bedford, MA, USA). The extraction solution was a mixture of acetonitrile and a solution of ammonium carbonate (0.2 g of ammonium carbonate per 1 L of deionised water) in the ratio of 84:16 (*v*/*v*). For the final sample solution, the two components were mixed in the ratio of 50:50 (*v*/*v*). Mobile phase component A was deionised water containing 0.2 g of ammonium carbonate per litre and component B was acetonitrile containing 0.1% of formic acid per litre.

5.3. Analytical Procedure

For the determination of EAs, a procedure described by Topi et al. [15] was used. It was based on the procedures of Mulder et al. [12]; Diana Di Mavungu et al. [21]; Kokkonen and Jestoi [25]; Crews et al. [26]; and Krska et al. [27]. Sample preparation included the extraction of EAs from ground cereals (20 g) with 100 mL of extraction solution. After extraction, 4 mL of the extract was transferred to a glass tube and purified by passing it through a Mycosept 150 Ergot column (Romer, Biopure, Tulln, Austria). Afterwards, 1 mL of the extract was evaporated to dryness under a vacuum at + 60 °C (Büchi Labortechnik AG, Flawil, Switzerland) and redissolved in 500 μL of a mixture of acetonitrile and ammonium carbonate in the ratio of 50:50. Chemical analysis of the EAs was performed with an LC-MS/MS. Chromatographic separation was performed with an Acquity UPLC (Waters, MS, USA) on a 2.7 μm Ascentis Express Phenyl-hexyl column, 2.1 × 100 mm (Supelco, Bellefonte, PA, USA). Two components of the mobile phase were mixed in a gradient mode. The starting composition of the eluent was 95% A and 5% B. The portion of component B was linearly increased to 25% within 1 min and further increased to 60% within the next 7 min. In the next 0.1 min, the proportion of component B was returned to 5% and then held for 5 min. The mobile phase flow rate was 100 μL/min and the column temperature was 30 °C. MS/MS analysis was performed using an ESI+ and a triple quadrupole mass spectrometer operated in MRM mode. The capillary voltage was 3.5 kV, the desolvation temperature was 500 °C, and the ion source temperature was 150 °C. The retention times of single EAs and monitored transitions are provided in Table 4. Quantification was performed using matrix-matched calibration. However, no internal standard was used.

Table 4. Retention times of ergot alkaloids and monitored transitions.

Ergot Alkaloid	Retention Time (min)	Precursor Ion (m/z)	Product Ion 1 (m/z)	Product Ion 2 (m/z)
Ergometrine	4.38	326.22	223.17	208.07
Ergometrinine	4.74	326.20	180.18	223.16
Ergosine	6.56	548.35	208.06	268.15
Ergosinine	6.75	548.41	223.16	277.19
Ergocornine	7.08	562.35	208.05	268.20
Ergocorninine	7.51	562.35	223.09	277.18
Ergocryptine	7.47	576.35	223.09	208.05
Ergocryptinine	7.99	576.35	223.09	305.17
Ergotamine	6.97	582.35	208.04	223.08
Ergotaminine	7.24	582.35	223.08	297.10
Ergocristine	7.84	610.35	208.05	223.09
Ergocristinine	8.47	610.35	305.17	223.09

5.4. Method Validation

The validation was performed using cereals and compound feed samples spiked with each EA at the concentration levels of 10, 200 and 500 µg/kg. The method validation procedure and validation results were presented in our previous work [15]. The recovery rates ranged from 100% to 124%. The intra-day and inter-day precisions of EAs expressed as the relative standard deviation (RSD_r and RSD_R, respectively) were from 5.8% to 17% and from 7% to 35%, respectively. The results are shown in Table S3 in the supplementary section. Essentially, the limit of detection (LOD) and the limit of quantification (LOQ) of single EAs were estimated as concentrations resulting in a signal-to-noise ratio of 3:1 and 10:1, respectively. However, as the limit of quantification (LOQ) of single EAs, the lowest tested concentration of 10 µg/kg was accepted according to the needs of the clients.

5.5. Ergot (Sclerotia) Content and Analytical Correlation Study

In the years 2014–2016, the collected samples were divided into two reduced samples using a divider. One reduced sample was used for chemical analysis of EAs and the second for sclerotia visual analysis and sclerotia mass determination. The chemical determination of EAs was conducted with an LC-MS/MS as described above. Sclerotia were visually identified in each sample by their purplish/black colour and their cylindrical shape with round ends using the microscopic method recommended by the International Association of Feedingstaff Analysis (IAG) [28]. Sclerotia were removed from a sample and weighed, and the ergot concentration was calculated. The sclerotia mass (mg) was determined only in cereal samples contaminated by EAs ($n = 76$).

In the year 2017, all samples were visually inspected for sclerotia before chemical analysis. Sclerotia were removed from a sample, visually identified as mentioned above and weighed, and the ergot concentration was calculated. Afterwards, sclerotia were returned to the original sample. It was homogenized and ground for the determination of EAs.

5.6. Statistical Evaluation

The statistical analysis was performed using SPSS Statistics 23 (IBM, Armonk, NY, USA) [29]. All calculations were performed using a nonparametric approach because of the small samples. The differences in the occurrence of EAs among the years as well as among the different types of cereals were tested using the Kruskal–Wallis test and Mann–Whitney test using the medium-bound approach for left-censored results. The correlations between the mass or ergot concentration in cereals and occurrences of individual or total EAs in positive samples were calculated using the Spearman's correlation coefficient. For all tests, $p < 0.05$ was considered statistically significant.

Supplementary Materials: The following are available online at http://www.mdpi.com/2072-6651/12/11/730/s1, Table S1. Analytical correlations in positive samples; Table S2. Guidelines for interpreting the strength of the correlation (Evans, 1996); Table S3. Performance characteristics of the analytical procedure for the determination of ergot alkaloids.

Author Contributions: F.A.C., K.K., M.Č. and B.J.-S. contributed to the study conception and design. B.J.-S. was the head of the project V4-1403. J.B. wrote the manuscript. Sample collection was performed by F.A.C., K.K. and M.Č., J.B. and G.T.-K. performed LC-MS/MS analysis, validation of the method and interpretation of results. J.B. prepared data for statistical evaluation. Statistical data analysis was performed by G.T.-K. Supervision was performed by B.J.-S. All authors read and approved the final manuscript.

Funding: This research was funded within the framework of the project V4-1403: Emerging toxic substances in Slovenian feed, by the Ministry of Agriculture, Forestry and Food and the Slovenian Research Agency and with support by the Slovenian Research Program P4-0092 and by the METROFOOD-RI included in ESFRI Roadmap. The authors would like to thank American Journal Experts for proofreading the English language in the manuscript.

Acknowledgments: The authors would like to thank Katarina Pavšič-Vrtač and Irena Indihar for laboratory support.

Conflicts of Interest: The authors declare no conflict of interest.

References

1. European Food Safety Authority (EFSA). Scientific opinion on ergot alkaloids in food and feed. EFSA Panel on Contaminants in the Food Chain (CONTAM). *EFSA J.* **2012**, *10*, 2798–2956.
2. Haarmann, T.; Rolke, Y.; Giesbert, S.; Tudzynski, P. Ergot: From witchcraft to biotechnology. *Mol. Plant Pathol.* **2009**, *10*, 563–577. [CrossRef] [PubMed]
3. Krska, R.; Crews, C. Significance, chemistry and determination of ergot alkaloids: A review. *Food Addit. Contam. Part A* **2008**, *25*, 722–731. [CrossRef] [PubMed]
4. Flieger, M.; Wurst, M.; Shelby, R. Ergot alkaloids—Sources, structures, and analytical methods. *Folia Microbiol.* **1997**, *42*, 3–30. [CrossRef] [PubMed]
5. Buchta, M.; Cvak, L. Ergot alkaloids and other metabolites of the genus Claviceps. In *Ergot: The Genus Claviceps*; Kren, V., Cvak, L., Eds.; Harwood Academic Publishers: Amsterdam, The Netherlands, 1999; pp. 173–200.
6. European Commission. Recommendation No 2012/154/EU of 15 March 2012 on the monitoring of the presence of ergot alkaloids in feed and food. *Off. J. Eur. Union* **2012**, *77*, 20–21.
7. European Commission. Directive 2002/32/EC of the European Parliament and of the Council of 7 May 2002 on undesirable substances in animal feed. *Off. J. Eur. Union* **2002**, *140*, 10–21.
8. European Commission. Regulation (EU) No 574/2011of 16 June 2011 amending Annex I to Directive 2002/32/EC of the European Parliament and of the Council as regards maximum levels for nitrite, melamine, *Ambrosia* spp. and carry-over of certain coccidiostats and histomonostats and consolidating Annexes I and II thereto. *Off. J. Eur. Union* **2011**, *159*, 7–24.
9. Coufal-Majewski, S.; Standford, K.; McAllister, T.; Blakley, B.; McKinnon, J.; Chaves, A.V.; Wang, Y. Impact of cereal ergot in food animal production. *Front. Vet. Sci.* **2016**, *3*, 1–13. [CrossRef]
10. Crews, C. Analysis of ergot alkaloids. *Toxins* **2015**, *7*, 2024–2050. [CrossRef]
11. Ruhland, M.; Tischler, J. Determination of ergot alkaloids in feed by HPLC. *Mycotoxin Res.* **2008**, *24*, 73–79. [CrossRef]
12. Mulder, P.P.J.; Van Raamsdonk, L.W.D.; Van Egmond, H.J.; Van Der Horst, T.; De Jong, J. *Ergot Alkaloids and Sclerotia in Animal Feeds. Dutch Survey 2007–2010*; Technical Report No. 2012.005; RIKILT, Wageningen UR: Wageningen, The Netherlands, 2012; pp. 1–54.
13. Malysheva, S.V.; Larionova, D.A.; Di Mavungu, J.D.D.; De Saeger, S. Pattern and distribution of ergot alkaloids in cereals and cereal products from European countries. *World Mycotoxin J.* **2014**, *7*, 217–230. [CrossRef]
14. Orlando, B.; Maumené, C.; Piraux, F. Ergot and ergot alkaloids in French cereals: Occurrence, pattern and agronomic practices for managing the risk. *World Mycotoxin J.* **2017**, *10*, 327–337. [CrossRef]
15. Topi, D.; Strajn, B.J.; Vrtač, K.P.; Kalcher, G.T. Occurrence of ergot alkaloids in wheat from Albania. *Food Addit. Contam. Part A* **2017**, *34*, 1333–1343. [CrossRef] [PubMed]
16. Evans, J.D. *Straightforward Statistics for the Behavioral Sciences*; Brooks/Cole Publishing: Pacific Grove, CA, USA, 1996.

17. Wegulo, S.N.; Carlson, M.P. Ergot of small grain cereals and grasses and its health effects on humans and livestock. *Ext. Circ.* **2011**, *1880*, 1–7.
18. Tittlemier, S.A.; Drul, D.; Roscoe, M.; McKendry, T. Occurrence or ergot and ergot alkaloids in western Canadian wheat and other cereals. *J. Agric. Food Chem.* **2015**, *63*, 6644–6650. [CrossRef]
19. Meister, U.; Batt, N. Fusarium toxins and ergot alkaloids in rye of the federal state Brandenburg harvested 2013. In Proceedings of the 36th Mycotoxin Workshop, Göttingen, Germany, 16–18 June 2014; p. 131.
20. Miedaner, T.; Geiger, H.H. Biology, genetics, and management of ergot (*Claviceps* spp.) in rye, sorghum, and pearl millet. *Toxins* **2015**, *7*, 659–678. [CrossRef]
21. Grusie, T.; Cowan, V.; Singh, J.; McKinnon, J.; Blakley, B. Correlation and variability between weighing, counting and analytical methods to determine ergot (*Claviceps purpurea*) contamination of grain. *World Mycotoxin J.* **2017**, *10*, 209–218. [CrossRef]
22. European Commission. Regulation No. 152/2009 of 27 January 2009 laying down the methods of sampling and analysis for the official control of feed. *Off. J. Eur. Union* **2009**, *54*, 1–130.
23. European Commission. Regulation No 691/2013 of 19 July 2013 amending Regulation (EC) No. 152/2009 as regards methods of sampling and analysis. *Off. J. Eur. Union* **2013**, *197*, 1–12.
24. Di Mavungu, J.D.; Larionova, D.; Malysheva, S.V.; Van Peteghem, C.; De Saeger, S. Survey on Ergot Alkaloids in Cereals Intended for Human Consumption and Animal Feeding. Scientific Report to EFSA 2011. Available online: https://efsa.onlinelibrary.wiley.com/doi/abs/10.2903/sp.efsa.2011.EN-214 (accessed on 2 April 2019).
25. Kokkonen, M.; Jestoi, M. Determination of ergot alkaloids from grains with UPLC-MS/MS. *J. Sep. Sci.* **2010**, *33*, 2322–2327. [CrossRef]
26. Crews, C.; Anderson, W.A.C.; Rees, G.; Krska, R. Ergot alkaloids in some rye-based UK cereal products. *Food Addit. Contam. Part B* **2009**, *2*, 79–85. [CrossRef] [PubMed]
27. Krska, R.; Stubbings, G.; Macarthur, R.; Crews, C. Simultaneous determination of six major ergot alkaloids and their epimers in cereals and foodstuffs by LC–MS–MS. *Anal. Bioanal. Chem.* **2008**, *391*, 563–576. [CrossRef] [PubMed]
28. Method for the Determination of Ergot (Claviceps purpurea Tul.) in Animal Feedingstuff, IAG-Method A4. 2008. Available online: http://www.iag-micro.org/files/iag-a4_ergot.pdf (accessed on 22 April 2020).
29. International Business Machines Corporation. *IBM SPSS Statistics for Windows*, version 23.0; International Business Machines Corporation: Armonk, NY, USA, 2015; Available online: https://www-01.ibm.com/support/docview.wss?uid=swg21476197 (accessed on 12 March 2019).

Publisher's Note: MDPI stays neutral with regard to jurisdictional claims in published maps and institutional affiliations.

toxins

MDPI

Review

Use of Integrative Interactomics for Improvement of Farm Animal Health and Welfare: An Example with Fescue Toxicosis

Ryan S. Mote and Nikolay M. Filipov *

Interdisciplinary Toxicology Program, Department of Physiology and Pharmacology, University of Georgia, Athens, GA 30602, USA; ryan.mote25@uga.edu

* Correspondence: filipov@uga.edu

Received: 8 July 2020; Accepted: 24 September 2020; Published: 1 October 2020

Abstract: Rapid scientific advances are increasing our understanding of the way complex biological interactions integrate to maintain homeostatic balance and how seemingly small, localized perturbations can lead to systemic effects. The 'omics movement, alongside increased throughput resulting from statistical and computational advances, has transformed our understanding of disease mechanisms and the multi-dimensional interaction between environmental stressors and host physiology through data integration into multi-dimensional analyses, i.e., integrative interactomics. This review focuses on the use of high-throughput technologies in farm animal research, including health- and toxicology-related papers. Although limited, we highlight recent animal agriculture-centered reports from the integrative multi-'omics movement. We provide an example with fescue toxicosis, an economically costly disease affecting grazing livestock, and describe how integrative interactomics can be applied to a disease with a complex pathophysiology in the pursuit of novel treatment and management approaches. We outline how 'omics techniques have been used thus far to understand fescue toxicosis pathophysiology, lay out a framework for the fescue toxicosis integrome, identify some challenges we foresee, and offer possible means for addressing these challenges. Finally, we briefly discuss how the example with fescue toxicosis could be used for other agriculturally important animal health and welfare problems.

Keywords: integrative interactomics; integrome; metabolomics; microbiome; tall fescue; *Epichloë coenophiala*; fescue toxicosis

Key Contribution: This review presents the benefits of integrating multi-'omics data sets (i.e., integrative interactomics) to more comprehensively evaluate farm animal health and adverse toxic insults, while providing the ability to gain molecular insights into systemic responses to numerous internal/external stressors. Based on the fescue toxicosis example herein, we highlight an initial framework for integrative interactomics that allows for in-depth analysis of complex systemic pathophysiology, addition/subtraction of external variables to assess pertinent effects, while steering the field towards developing scalable, translatable, and reproducible integrative interactomics studies.

1. Introduction

As world population continues to climb [1], improving production efficiency without sacrificing rigorous quality or safety standards is crucial for sustaining the agricultural supply chain [2]. Significant geopolitical and extreme weather events pose additional negative pressures, while consistent economic growth drives agricultural demand, leading to new challenges for producers and scientists alike [2]. Other extenuating circumstances, like increased risk of a global pandemic in our highly connected world, i.e., the Coronavirus disease 2019 (COVID-19) pandemic the world is currently faced with [3], also pose a threat to the global food supply. In the face of these challenges, maintenance of a sustainable, efficient, and adaptable agricultural supply chain is essential. Scientific advances have begun to increase the efficiency of outlining pathophysiological responses by generating high-volume, high-quality datasets quickly, providing novel biological targets for historical and future maladies that may have been missed when using more classical, low-throughput techniques [4]. Thus, recent scientific and technological advances, in conjunction with classical techniques, will continue to improve the ability of the agricultural sector to adapt to both predictable and unforeseen future challenges.

High-throughput technologies are changing the way we investigate complex biological systems by generating large data sets coupled with advanced computational and bioinformatics tools. This includes recent efforts towards combining multi-level data matrices (e.g., genomics, proteomics, metabolomics, etc.). The main focus of the current review is on the application of these technologies to agriculturally important, mycotoxin-induced diseases.

Next-generation sequencing (NGS) can generate high-quality sequencing data regardless of sample origin, increasing our understanding of how genotype influences the ability of organisms to adapt to internal and external stressors. For example, NGS can screen plants for genes associated with agronomic benefits [5]. From a farm animal perspective, NGS has outlined the relationship between ruminant microbes and animal productivity [6,7], allowing investigation of complex ecosystems derived from NGS' culture-independency. Improvements in sequencing technologies have also provided new insights into ecological shifts that occur in the host microbiota after different dietary/treatment regiments [8,9]. Culturomics, a new field centered on cultivating ruminant microorganisms, utilizes NGS and is driven by understanding the biochemical potential of individual microbes [10].

High-resolution mass spectrometry propelled the investigation of the proteome and metabolome [11]. Metabolomics, the study of all low molecular weight compounds within a biological matrix [12], can be used to examine biochemical interactions [13] or to further our understanding of food quality, processing and safety [14]. Much like NGS, advances in the metabolomics computational framework have increased the utility and capacity of metabolomics by improving detection and quantification of both prominent and low abundance metabolites without sacrificing quality [15–18]. Notably, metabolomics scalability has many potential uses; one of the most promising is for biomonitoring of environmental exposures, as suggested for deployed troops [19]. Although computational metabolomics has been proposed for evaluation of the human exposome for a detailed review on the exposome; see: [20], these advances are translatable for improving food safety/quality and, as detailed here, agricultural animal health/wellness.

One important multi-'omics development is computational tools that integrate different 'omics data types (e.g., genome, proteome, metabolome, etc.) into single, interpretable outputs, allowing an evaluation of the integrome through integrative interactomics [21]. Traditionally, interactomics is a broad, yet well-established field focused on mRNA-protein [22] and/or protein-protein [23] interactions within a large spectrum of biological systems and disorders [24–26]. To contrast this view of the interactome, herein we have defined the integrome as the integration of multi-'omics data sets through integrative interactomics, allowing for the description of systemic biological interactions and multi-organ-integrative responses to internal or external variables. The proposed generic framework for the integrome is presented in Figure 1, where grazing animals are used as the example. This initial iteration of the integrome has multiple components. First, molecular characteristics of dietary components are influenced by environmental factors, and a composition of dietary, environmental, and toxin-related exposures determine what animals are exposed to on a daily basis (Figure 1). In response to oral exposures, the animal's enteric physiology shifts to adapt to dietary, environmental and toxicological pressures by altering the microbiome, microbial metabolites, and diet-derived metabolites. The enteric immunoprofile, metabolite/microbe interactions with gut associated lymphoid tissues (GALT), and potential effects on gut wall integrity further effect downstream (patho)physiology. Overall, these enteric physiological shifts dictate what metabolites are present within the gastrointestinal (GI) tract and are absorbed into systemic circulation. Further, the enteric immune and enteroendocrine systems signal and influence downstream physiological processes in peripheral tissues (Figure 1). Evaluating metabolites that enter circulation and their fate (e.g., metabolism, receptor interactions, etc.) in the larger biological system is a reflection on how diet and environmental exposures influence downstream host physiology (Figure 1). Finally, microbes and metabolites excreted from the animal, along with grazing and other environmental stresses, may influence, for example in pasture production settings, plant physiology, creating a feedback loop between the plant, dietary/environmental exposures and (patho)physiological changes within the animal (Figure 1). Notably, the excrements (fecal matter, urine) are easily accessible tools and can be used for investigating novel biomarkers of exposure or effect. This framework evaluates systemic, multi-organ responses to dietary and environmental stressors, while considering effects on the entirety of the diet-environment-animal interactions. Overall, technological and computational advances have opened new opportunities to evaluate global, systemic biological interactions. For a detailed review of the computational methods associated with multi-'omics data integration, the mathematical concepts behind these methods, and their application towards toxicological risk assessment, which is outside the scope of the current paper, the readers are directed to the following reviews [27–30].

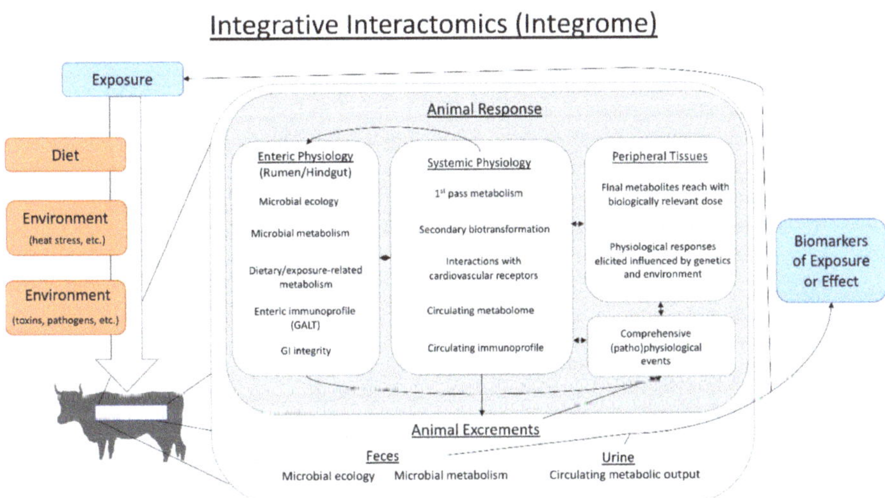

Figure 1. A general, adaptable schematic of the integrome as evaluated through integrative interactomics based on dietary/toxic exposures from an animal agricultural perspective. Animals are exposed to combinations of dietary components, ambient environmental conditions, and external toxicants or pathogens from the environment. Multi-'omics data are collected and integrated within multi-compartment animal responses, which consists of evaluation of enteric (rumen and hindgut herein) and systemic physiological changes, understanding effects on peripheral tissues, and how this results in (patho)physiological effects of interest. Further, animal excrements are an ideal, readily accessible biological matrix that can be utilized to identify biomarker(s) of exposure and/or effect, including those associated with pathophysiological changes. Finally, evaluating how components of the excrements feed back into what animals are exposed to through their diet is also important to the integrome. Overall, this model can help outline global changes that occur in any situation through systematic evaluation of complex biological processes and interactions.

2. Technological Advances Increase Throughput for Toxicology Screening

The field of toxicology stands to benefit from integrative interactomics [30]. For example, the Toxicology in the 21st century (Tox21) joint venture in the United States is focused on increasing throughput for toxicity screenings (https://tox21.gov/). Drawing conclusions about how individual compounds or mixtures interact within a biological system (e.g., cell, organism) is difficult, but using top-down strategies (i.e., using high-volume global analyses to understand molecular signatures, such as 'omics-based analyses) can provide valuable biological insights missed by bottom-up strategies. Moreover, top-down strategies can evaluate adaptive and pathophysiological responses in plants [31], humans [32], and model organisms [33], which is an important step in understanding the toxicity mechanisms and consequences of environmental exposures. Examples in this regard are low-dose chronic exposures to heavy metal and pesticide mixtures [34,35]. Although these approaches have been previously applied in different contexts, they have specific utility for livestock and large animal physiology/toxicology. Recent work highlighted the bovine genome-microbiome relationship [36], bovine microbiota response to dietary additives [37], and the use of host-microbiota relationship to establish more efficient production systems [38].

Mycotoxin exposures are a major livestock concern. Aflatoxins perturb the immune system [39] and rumen motility [40], while *Fusarium* mycotoxins (e.g., T-2 toxin) adversely impact livestock reproduction [41]. Reproductive and growth impairments, discussed in detail later, are also associated with *Epichloë* mycotoxins [42]. Integrative interactomics provides the opportunity to identify novel, mutli-level evidence of direct/indirect mycotoxin effects that, when coupled with traditional methods, can be exploited therapeutically.

3. Maintaining Animal Productivity and Wellness in Adverse Environments, Toxicity Included

Improving and maintaining farm animal productivity and wellness is critical. NGS has provided insights into the importance of host genetics on the bovine microbiota [38,43], which seems to be stable after rumen transplant in dairy cattle [44], and the association between microbial genes in the rumen and feed efficiency [45–47]. Further, how the microbiota is influenced by diets/dietary additives has been evaluated [48,49]. As shown recently, the ruminant microbiota is influenced mostly by diet, only partially by the host, and a core microbiota exists across geographies [50,51]. Moreover, rumen microbial communities are influenced by age and environment [52], indicating the developing microbiota could be important for diseases that are either heritable and/or have pre- or postnatal contributing factors. Overall, evaluating the enteric microbiota after exposure to dietary additives or environmental contaminants, as well as after disease onset, is important, since understanding the adaptive microbiota response to these adverse events can help provide a complete picture of the microbiota's role in health and disease.

Metabolomics can also provide important insights into livestock productivity. For example, metabolomics-based methods were used to associate rumen fluid metabolites with feed efficiency [53], assess biochemical variability in bovine milk [54], and identify differences in the bovine milk metabolome during early and late gestation [55]. Metabolomics is used widely to search for biomarker signatures of feed efficiency and disease states in cattle. Livestock metabolomics is not major focus of this review; for an in-depth review, refer to Goldansaz, et al. [56].

How fungal mycotoxins/metabolites influence rumen microbial ecology in animals consuming contaminated feedstuffs is of major interest [57]. While ruminants are considered less susceptible to mycotoxins than monogastrics, ruminant mycotoxicoses do occur and are problematic for animal health and welfare, for a detailed review on ruminant mycotoxicosis, see [58]. Aflatoxin B1 (AB1) is a well-studied mycotoxin; its effects and/or potential biomarkers for exposure in dairy cattle plasma [59,60], as well as rumen fluid and milk [61], have been investigated with metabolomics-based methods. The non-ruminant (porcine) colonic microbiota, namely its functional biodiversity, is impacted by *Fusarium* mycotoxins [62]; however, despite this, and the utility of NGS and metabolomics in evaluating animal productivity, integrating these separate data sets towards understanding systemic biological responses and the livestock integrome is yet-to-be widely done.

As a whole, animal agriculture would benefit greatly from the development, characterization, and utilization of integrative interactomics-based approaches, and this has been proposed for human health as well [63]. Although not in livestock, using an integrative metagenomics and metabolomics approach, Lu, et al. [64] reported that arsenic exposure in drinking water significantly perturbs the host (mouse) microbiota, metabolome, and multiple microbiota-related metabolites. A similar approach was also able to identify a novel relationship between the microbiota, certain metabolites and colorectal cancer, indicating a potential utility of microbial metabolites as diagnostic or therapeutic targets [65]. Finally, Tang, et al. [66] were able to identify microbiota and metabolites that were significantly influenced by diet, indicating that dietary constituents not only influence the gut microbiota, but have systemic effects as well. Similar investigations integrating microbiota and metabolomics data from multiple biological compartments in livestock could provide significant insights into the mechanisms by which dietary, environmental, and mycotoxin exposures influence animal performance.

New data from our lab [67–69], among other recent reports [38,70,71], indicate that developing integrative interactomics-based approaches to complement other, more traditional methods for evaluating the livestock integrome could increase our understanding about the development and adaptive versus pathophysiological responses to one of the costliest diseases to the United States beef industry, fescue toxicosis. The rest of this review will focus on introducing what we currently know about fescue toxicosis development and pathophysiology and how newly generated 'omics data could be integrated to help fill crucial knowledge gaps related to this disease.

4. Fescue Toxicosis

4.1. Tall Fescue and Epichloë Coenophiala

Tall fescue, *Lolium arundinaceum*, is a cool season perennial grass that is well adapted to the Southeastern United States, covering approximately 14 million hectares of land. Tall fescue has many compelling agronomic attributes, such as persistence under drought and grazing stresses, resistance to herbivores, nematodes, and insects, and greater potential for nutrient uptake [72]. Ultimately, these attributes give it a competitive advantage over other cultivars and stem from an infection of the plant with the endophytic fungus *Epichloë coenophiala* [72].

E. coenophiala grows intercellularly and systemically above ground [73] and is vertically transmitted through tall fescue seed heads [74]. Interestingly, the agronomic benefits of *E. coenophiala* infection appear to be heritable, with both the plant and endophyte genotype influencing offspring persistence [75]. Endophyte derived bioactive secondary metabolites are considered responsible for the increased persistence of *E. coenophiala*-infected tall fescue. While some metabolites are beneficial, one group, the ergot alkaloids, have been shown to be detrimental to grazing livestock, inducing the development of fescue toxicosis [76]. Ergot alkaloids are the most abundant class of bioactive secondary metabolites produced by *E. coenophiala* and are the ones of greatest economic concern in the US [42,77] and worldwide [78,79].

4.2. Ergot Alkaloids: Presence, Biosynthesis, and Monoaminergic Activities

The ergot alkaloid profile of tall fescue is variable and consists of lysergic acid derivatives and ergopeptine alkaloids. Both share the same pharmacophore (D-lysergic acid), but classification is derived from functional groups bonded to the primary 8th carbon [80]. Lysergic acid derivatives have a carboxamide functional group, with varying chemical entities bound here used for classification. The ergopeptines have a tripeptide moiety bound to carbon 8 and this is used for ergopeptine nomenclature. For example, the major ergopeptine alkaloid derived from *E. coenophiala*, ergovaline, has L-alanine, L-proline, and L-valine as constituents, whereas ergotamine, a *Claviceps purpurea* ergopeptine, has L-alanine, L-proline, and L-phenylalanine as constituents [77].

Studying the activities of ergot alkaloids in biological systems is difficult because the pharmacophore (i.e., the D-lysergic acid backbone) bears structural similarities with dopamine (DA), norepinephrine (NE), and serotonin [5HT; 80]. This allows ergot alkaloids to have systemic interactions, eliciting serotonergic, dopaminergic, and adrenergic activities that can influence numerous physiological functions [80].

4.3. Ergot Alkaloid Metabolism in Ruminant Animals

The pharmacokinetics of bromocryptine, a synthetic ergot alkaloid, have been extensively studied, but the toxicokinetics of the natural alkaloids of endophyte-infected tall fescue are not well characterized. That said, ergot alkaloids and their metabolites (i.e., lysergic acid) have been detected in the serum after direct injection [81], urine [82,83], bile [83], fluids of the rumen and abomasum [84], subcutaneous fat [85], and liver and kidney tissues [86]; outside the urine, in many instances, the levels are minute. The amount of ergot alkaloids detected decrease through sequential sampling along the GI tract, with one study finding 50–60% recovered from abomasal contents and 5% in the feces [87]. Further, as much as 93–96% of ergot alkaloids are excreted in the urine, although the main metabolite detected in the urine is lysergic acid, suggesting extensive metabolism [83]; lysergic acid is also the main metabolite that crosses gastric tissues [84]. Parent ergopeptine alkaloids are not transported across gastric tissues in vivo and are likely metabolized in the rumen, indicating that downstream processes may contribute to fescue toxicosis development [88]. Overall, these data indicate that ergot alkaloids undergo complex pre-systemic metabolism in the rumen. That said, the effects of ergot alkaloids and their metabolites on microbial populations and microbial metabolism, the potential downstream perturbations caused by this (e.g., microbial-mediated global metabolic effects, immune system effects, etc.), and when the animals transition from an adaptive response to a pathophysiological state need to be investigated; top-down, integrative interactomics-based approaches can help begin this journey.

4.4. Biomarkers of Ergot Alkaloid Exposure and Effect

Biomarkers are reproducible, quantifiable biological constructs that result from biological activities and can be used to identify different medical or disease states [89]. Biomarkers need to have high reproducibility across studies or trials while being able to encapsulate the entirety of what it is they are being used to identify. Generally speaking, there are two types of common biomarkers: (1) biomarkers of exposure and (2) biomarkers of effect [90]. Biomarkers of exposure are biological indicators that exposure to a particular substance or compound of interest has occurred, whereas biomarkers of effect indicate that, generally, a pathological effect or disease state is present within a given system. While many biomarkers for fescue toxicosis have been proposed, evidence suggests most of the proposed biomarkers are biomarkers of exposure. For example, it has previously been shown that if ergot alkaloids cross gastric tissues into systemic circulation, they can interact with dopamine receptor subtype 2 (DRD2) receptors on anterior pituitary lactotrophs to decrease serum prolactin [91]. Decreases in serum prolactin have been proposed as a biomarker for ergot alkaloid exposure, but prolactin levels are known to be influenced by photoperiod [92], acute ambient temperature changes [93], and different forms of stress [94,95], which could all lead to reproducibility problems. Detection of total ergot alkaloids in the urine is an accurate, less variable method to assess ergot alkaloid exposure when compared to previous methods, such as decreased serum prolactin [82,83]. This is because ninety-four percent of ergot alkaloid excretion occurs in the urine, appearing as early as 12 h post-exposure, and concentrations are exposure level- and duration-dependent [82]. While urinary alkaloids are of great utility as a sensitive and reproducible biomarker of ergot alkaloid exposure and correlate with reduction in average daily weight gains [ADG; 82], this approach does not speciate individual ergot alkaloids or their metabolites. Speciating ergot alkaloids and metabolites, in plasma, urine, or rumen fluid, might help identify how ruminants metabolize ergot alkaloids and other metabolic processes that may be affected [82,84], which is where integrating multi-'omics data sets will be beneficial.

4.5. Adverse Effects of Toxic Tall Fescue Grazing and Ergot Alkaloids on Livestock

Despite notable stand persistence, which sparked initial interest in tall fescue as a pasture cultivar, reports regarding the detrimental impact of tall fescue grazing in livestock began as early as the 1940's [96–98]. While cattle grazing tall fescue exhibit numerous signs, such as fescue foot, thermoregulatory impairment, and decreased feed intake, the most economically costly are lowered weight gains and reproductive insufficiencies [99].

Decreased muscle accretion and weight gains are common findings in steers grazing toxic tall fescue. This is a major concern in the beef industry, as these effects can go unnoticed despite their financial impacts [99]. Weight gains are decreased, in part, by lower feed intake and shortened grazing periods; these signs are exacerbated under hot and humid environmental conditions [i.e., heat stress; 99]. The molecular mechanism(s) that lead to decreased food intake are unknown, but current investigations are attempting to elucidate these mechanisms. Considering the dopaminergic and serotonergic activities of ergot alkaloids, it is plausible that one mechanism could be through alterations of gut motility. Strickland, et al. [100] were some of the first to suggest that interaction between ergopeptine alkaloids and enteric receptors [101] could potentially influence gut motility and feed intake. Though the authors noted the interactions between ergot alkaloids and enteric DRD2 receptors, we now know enteric serotonergic activities of ergot alkaloids are another plausible source for altered gut motility induced by toxic tall fescue exposure [101,102]. In support, increases in the baseline tonus of the rumen/reticulum and increased amplitude of reticular contractions follow intravenous infusion of ergovaline, the major ergopeptine in tall fescue [103]. Of note, increased rumen fill that could not be explained by increased dry matter intake, indicate ruminal passage rates may be decreased [104–107] in the absence of changes in energy balance (e.g., energy intake, O_2 consumption, CO_2 production, heat production, etc.) or feed digestibility [105]. Moreover, voluntary dry matter intake can be inhibited by increases in rumen fill and restricted flow of digesta to the lower gastrointestinal tract [108]. This indicates that, although digestibility and basal metabolic rates may not be affected, changes in gastrointestinal physiology post ergot alkaloid exposure, among other specific metabolic changes (i.e., certain metabolic pathways), could contribute to decreased muscle accretion and weight gains associated with fescue toxicosis through molecular changes. Although these data refer to direct actions of ergot alkaloids, most ergot alkaloids are degraded in the rumen [88]. Therefore, studies investigating the mechanisms behind the decreased weight gains and altered feeding behaviors need to consider: (1) ergot alkaloid metabolites that could directly influence host physiology and (2) indirect effects of ergot alkaloids and their metabolites.

5. The Case for Integrative Interactomics in Fescue Toxicosis Studies

Much is known about specific effects of toxic tall fescue grazing and ergot alkaloid exposure in the bovine, but there are still gaps of knowledge about fescue toxicosis pathophysiology. While there is a need for studying specific effects and/or responses, there is also an urgent need to examine the entire system as one integrated set of responses that contribute to decreased animal performance. First, from a plant perspective, evaluating global effects of *E. coenophiala* infection on important plant physiological characteristics (e.g., rhizosphere, phyllosphere, and endosphere microbiota, the plant metabolome, root exudates, etc.) will provide deeper insights into endophyte-derived molecular changes in the plant that animals are exposed to. Further, top-down strategic approaches may help identify global plant responses, outside of fungal ergot alkaloids, that could contribute to fescue toxicosis pathophysiology. If we assume ergot alkaloids, or the ergot alkaloid pharmacophore, enter the bloodstream and reach target receptors, their receptor promiscuity would result in complex physiological changes. The complexity of studying fescue toxicosis is derived from having to understand the plant-endophyte relationship growing in different geographical regions alongside direct and indirect effects of the entire tall fescue plant, endophyte, and endophyte-derived compounds on the grazing animal. However, an integrative interactomics-based approach would provide multi-level global analysis of different compartments between the plant and animal, providing a breadth of data that begins to address the true complexity of the disease by providing unique integrated information about multi-levels effects of *E. coenophiala*. When these are coupled with specific analyses of production parameters (i.e., weight gain) and/or physiological measurements (e.g., respiration rates), integrative interactomics will help with the identification and prioritization of therapeutic and management strategies of the disease. The complex relationship between the soil, plant, endophyte, and animal is the ideal case for studying the integrome through such systems biology-based approaches. The rest of this review will focus on how systems biology-based approaches have been used previously and demonstrate why integrative interactomics should be used to evaluate the fescue toxicosis integrome and use it to improve disease management alongside current methodologies.

5.1. *Effects of Epichloë Coenophiala Infection on the Plant*

Felitti, et al. [109] performed transcriptome analyses on *Neotyphodium* (*Epichloë*) *coenophiala*, *Neotyphodium lolii*, and *Epichloë festucae*, and were able to compare functional changes in expressed sequence tag (EST), a subsequence of cDNA, resulting from growing these fungi on different media, indicating endophyte selection may result in functional in planta changes. Another study found that the endophytes that infect tall fescue plants play a significant role in modulating the rhizosphere and root exudate and how those secretions are modified by soil microbes, indicating a potential utility for endophyte selection to manipulate soil qualities [110]. In support, Rojas, et al. [111] noted that *E. coenophiala* infection significantly impacts the soil and rhizosphere fungal community population, with similar effects being seen for all endophytes tested. From another perspective, Rasmussen, et al. [112] have published a review summarizing their work, where they've assessed the effects of *Neotyphodium lolii* infection on the *Lolium perenne* plant metabolome. One difficulty of this approach is assessing what changes in the plant metabolome may be associated with biotransformation of endophytic metabolites, but Rasmussen and colleagues outlined a potential network of metabolites that are highly integrated between the endophyte and the plant, indicating complex plant-endophyte cross talk [112]. In support, a meta-analysis indicated that *Epichloë* endophytes promote resistance to insects through certain alkaloids and jasmonic acid pathway metabolites Bastias, et al. [113]. Altogether, these data indicate that endophyte infection significantly alters the rhizosphere and plant microbiota and metabolome, via an incredibly complex plant-endophyte relationship. However, no study has combined plant data in the same study that assesses animal responses to toxic tall fescue grazing.

5.2. *Effects of E. coenophiala-Infected Tall Fescue on the Animal*

5.2.1. Toxic Tall Fescue Effects along the Bovine Alimentary Tract

The fate of ergovaline along the enteric tract was previously investigated [88]; it was undetectable in rumen fluid or urine in both in vitro and in vivo experiments. Only lysergic acid was able to cross rumen and omasal tissues, indicating that parent ergopeptine alkaloids likely do not make it into circulation at significant quantities through enteral absorption without undergoing metabolism/biotransformation. That said, one possible route that has yet to be thoroughly explored is possible ergot alkaloid absorption through the lymphatic system. The initial parts of the lymphatic system rely on peristalsis as one mechanism to fill and empty [114]. Considering ergot alkaloids can influence peristalsis through interactions with receptors in the enteric smooth muscle [100–103], ergot alkaloids could be absorbed through the lymphatic networks of the intestinal smooth muscle [115] while also influencing the ability of the initial lymphatic system to fill and empty. However, this possibility needs to be explored further. Overall, these data, combined with Section 4.3., point towards indirect mechanisms through which ergot alkaloids induce fescue toxicosis. Thus, understanding direct, indirect, and systemic effects of toxic tall fescue, ergot alkaloids and their metabolites (e.g., lysergic acid) is crucial to deciphering fescue toxicosis pathogenesis and progression.

To date, limited studies exist examining the effects of *E. coenophiala* on the bovine microbiota. The fungus *Aspergillus terreus* was consistently found along the enteric tract in steers that exhibited signs of fescue toxicosis (e.g., fescue foot), indicating enteric fungi may play a role in fescue toxicosis etiology or be useful as a biomarker [116]. Clavine alkaloids, which are ergot alkaloid precursors, appear to have antibiotic-like properties [117,118]. Previous work has also shown that ewes inoculated with an antibiotic resistant *Escherichia coli O157:H7* and fed a high endophyte-infected, versus low endophyte-infected seed shed a greater number of antibiotic resistance genes [119]. Recently, Harlow, et al. [120] investigated rumen bacteria that may degrade ergopeptine alkaloids by evaluating hyper-ammonia producing bacteria (HAB) and tryptophan-utilizing bacteria in vitro; they found that all tested HAB (e.g., *Clostridium* spp.) degraded ergovaline, with varying degrees of effect. Another recent study [71] assessed the effects of tall fescue seed and red clover isoflavones on global rumen microbial populations in vitro and found limited effects of tall fescue seed on volatile fatty acids. Interestingly, significant effects on bacteria aligned within the *Ruminococcaceae*, *Coriobacteriaceae*, and *Erysipelotrichaceae* families, which is similar to results from our lab where we saw E+ exposure altered the abundances of operational taxonomic units (OTUs) aligned to these families in vivo [67]. Most importantly, we found that toxic tall fescue grazing steers had a unique fecal microbiota structure that could provide insights into easily accessible biomarkers of exposure or effect. In another study, we found similar effects on the fecal microbiota ecology, but also that the fecal microbiota in Angus steers is more easily influenced by heat stress in toxic tall fescue grazing steers versus those on non-toxic tall fescue pastures [69]. This could have multiple origins, one being impaired gut wall integrity. Of note, Alrashedi [121] found increased ruminal *Firmicutes* and decreased fecal *Bacteroidetes* in ewes grazing high and medium endophyte-infected tall fescue pastures, with ruminal *Prevotella* and fecal *Coriobacteriaceae* operational taxonomic units (OTUs) associating with body weight changes. Finally, Koester, et al. [122] reported significantly decreased bacterial and fungal fecal diversity and richness in Angus cattle that had low tolerance to toxic tall fescue exposure. In this study, the *Neocallimastigaceae* family was increased in high tolerant steers, whereas the genus *Thelebolus* was increased in lower tolerance steers. Interestingly, this study also reported farm-to-farm variability in the fecal microbiota of Angus cattle, indicating that putative fecal biomarkers should be systematically evaluated to ensure cross-study validation. Overall, these data affirm that ergot alkaloids/toxic tall fescue significantly alter gut physiology and microbiota along the enteric tract. However, more in vivo analyses need to be performed to understand the importance of the microbiota in fescue toxicosis adaptation/development and whether any of the potentially adverse effects could be targeted therapeutically at gut level.

5.2.2. Metabolic Effects of Toxic Tall Fescue Grazing

For a comprehensive review on the overarching physiological changes associated with fescue toxicosis, the authors would direct readers to Strickland, et al. [123]. Briefly, most cellular components of the blood are not affected by toxic tall fescue grazing, but decreased erythrocyte size and hemoglobin content have been reported [124]. One consistent finding in studies analyzing the effects of toxic tall fescue on blood parameters is decreased circulating cholesterol and triglyceride levels [86,124]. Interestingly, bolus injection with ergotamine decreases plasma insulin, while increasing plasma glucagon concentrations within 1 h post-injection [125]. The impact of ergot alkaloids on plasma cortisol, another metabolism regulating hormone, levels are variable [86,126]. Cattle consuming toxic tall fescue have large variations in circulating norepinephrine and epinephrine levels, which may be associated with altered behavior [127,128]. Tall fescue affects circulating metabolites and hormones and both intra- and extra-hepatic enzyme activities [86], indicating global metabolic effects. Jackson, et al. [129] analyzed specific blood analytes in steers on either high or low endophyte-infected tall fescue pastures, but we first utilized untargeted high-resolution metabolomics to assess global biochemical changes that result from toxic tall fescue grazing [68]. Both plasma and urine metabolomes, namely amino acid and glycerophospholipid metabolism [68], were among the major effects. Interestingly, some metabolic effects are modulated by hot and humid environmental conditions [69], highlighting the complexity of fescue toxicosis pathophysiology. While this work is a starting point, to expand upon these initial results, substantial additional efforts that assess global metabolic changes in response to *E. coenophiala* infection or exposure, in the fescue plant and grazing animal, are needed.

5.2.3. Understanding the Fescue Toxicosis Integrome

Integrating microbiota and metabolomics data will help outline the relationship between the microbiota, global metabolism, and other parameters such as animal performance. For example, it is well known that shifts in the microbiota influence the presence of microbial metabolites that, in turn, modulate other downstream processes. Toxic tall fescue grazing alters metabolism; however, the microbiome, metabolome and other important physiological parameters have generally been investigated in isolation and not in an integrative manner despite their interconnectedness. The fescue toxicosis integrome is complex; although the plant-endophyte-animal relationship has not been evaluated, we have begun to move in this direction. We successfully integrated the plasma and urine metabolomes with fecal microbiota data to interrogate changes in the microbiota-metabolome relationship that occur after toxic tall fescue exposure [69]. Within this study, the bidirectional relationship between the microbiota and metabolome [130,131], and how this shifts after placement on toxic versus non-toxic tall fescue pastures, was investigated. To do this, we used sparse partial least squares regression (sPLS) to integrate the datasets followed by a differential network analysis to reveal a highly correlated network of fecal OTUs and plasma/urine metabolites. We then identified the OTUs that were correlated with parameters of interest (e.g., average daily weight gains) and found a subnetwork that provides potentially valuable therapeutic targets. As highlighted in Figure 2, understanding the plant-endophyte relationship and how changes in the plant influence bovine physiology through integrating plant and animal data provides a framework for understanding how additional stressors or geographical/cultivar change the disease.

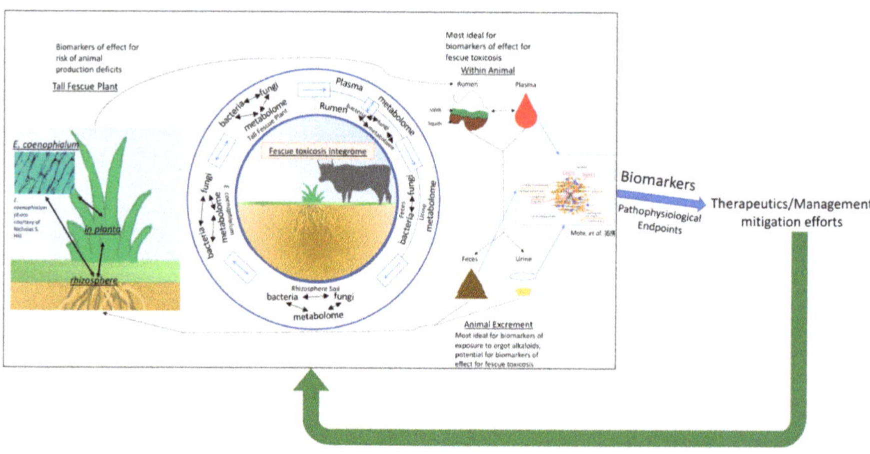

Figure 2. A schematic of the fescue toxicosis integrome. The integrome begins by understanding the complex relationship between the rhizosphere microbiota and metabolome, tall fescue plant microbiota and metabolome, and how these are influenced by the endophyte and grazing stresses. Next, endophyte-related changes are integrated into the animal physiology through microbiota or metabolic changes in the rumen that are reflected by metabolic changes in the plasma prior to microbiota and metabolome effects at the excretion level (feces and urine). The integrome will also allow for understanding the potential for feedback regulation of plant physiology and plant-endophyte relationship by components of the animal excrements, alongside grazing stresses. Of note, changes on the plant side have the potential to be used as biomarkers that have the utility of predicting the risk of animal production deficits of fescue toxicosis. Further, while blood tissues and rumen samples are most suitable for finding biomarkers of effects, animal excrements are ideal for identifying biomarkers of exposure (possibly biomarkers of effects as well) because of their ease of access. One example from our

previous paper [69] shows how top-down strategies, while centering of global effects, can also identify biological features (i.e., OTUs and metabolic features) that associate with pathophysiological effects (e.g., animal weight gains), highlighting their utility as biomarkers. Biomarkers and pathophysiological endpoints can be used, in conjunction with other methods, to identify potential new and/or improved fescue toxicosis mitigation approaches. They will be then evaluated for their efficacy at multiple levels and the outcomes of the evaluation will inform stakeholders on best ways to improve disease management. Similar approach can be adapted to other conditions, including other economically important toxicoses.

6. Where Is This Approach for Fescue Toxicosis Research Headed?

Currently, it is increasingly apparent that indirect, monoamine receptor-independent mechanisms play an important role in fescue toxicosis pathogenesis. To evaluate these, we propose a framework that can be used for interrogating the fescue toxicosis integrome. Understanding the soil-plant-endophyte relationship and how altering this relationship could then influence the toxic fescue grazing beef is a necessary first step (Figure 2). Next, integrating these data with the animal response data (e.g., the rumen/fecal microbiota and metabolomes) will pinpoint some important features that could be targeted therapeutically at the plant level, or as other (i.e., non-ergot alkaloid) potential biomarkers that predict animal production deficit risk (Figure 2). This approach can also point to where pathophysiological changes begin, what other processes may be directly/indirectly influenced within the animal and explain systemic effects of fescue toxicosis (Figure 2). Finally, what the animal is exposed to, as a result of toxic tall fescue grazing will influence host physiology, and ultimately, what that animal excretes. Consequently, grazing animal excrements and grazing stresses will affect the tall fescue plant/endophyte physiology. The integrome considers how these changes, in combination with other environmental factors, influence what the animals are exposed to (Figure 2). Animal biological matrices, such as tissues or blood, have a higher barrier to accessibility, but still provide utility for identifying biomarkers of effects resultant from sensitivity to physiological (whether GI or systemic) changes (Figure 2). The urine or fecal matter provide unique opportunity to identify biomarkers (metabolites or microbes [OTUs]) of exposure to ergot alkaloids and toxic tall fescue, as they are easily accessed (Figure 2). These two readily accessible biological matrices would be ideal for biomarkers of effect as well, if a urine/fecal change is reflective of systematic effects. We used a similar approach to begin investigating the integrome, where we integrated plasma/urine high-resolution metabolomics and fecal bacterial microbiota data to investigate the microbiota-metabolome interactions [69]. Ultimately, we identified one small cluster of metabolites anchored by bacterial OTUs of interest. One *Ruminococcaceae*, one *Clostridium*, and one *Peptococcaceae* candidate genus *rc4-4* were highlighted within this network, as they were highly correlated with physiological endpoints of interest (i.e., animal weight gains, urinary ergot alkaloids), plasma (e.g., 4,8-dihydroxyquinoline) and urine (e.g., 18-oxocortisol) metabolites, and were present in the networks under both thermoneutral and hot and humid environmental conditions. Therefore, these metabolites could be useful as biomarkers of a hindgut microbiota associated with signs of fescue toxicosis pathophysiology (i.e., reduced animal productivity) under most ambient environmental conditions [69]. Expanding upon this by integrating the foregut and plant information will be an important step forward towards deciphering the fescue toxicosis integrome.

One of the other most important issues moving forward will be ensuring analytical and computational pipelines are comparable. For example, not only should researchers include sequencing depth and coverage for NGS-based genomics; it was recently shown that differing preprocessing (i.e., Mothur vs QIIME) methods of NGS data can influence the abundance profiles of less common microbes of the rumen [132]. Further, recent comparisons between pipelines leading to amplicon sequence variants (ASVs) and OTUs seem to indicate some differences in specificity between the two, with different OTU-based pipelines producing some non-overlapping spurious OTUs [133]. For metabolomics, differences in analytical platforms, computational methods, and database matching algorithms for metabolomics could also result in metabolic feature variability [134–137]. Finally, while still a less visible, yet budding field, multi-'omics integration methodologies, such as xMWAS [21], can vary, and their continual development and improvement will increase consistency. Overall, while this integrative approach will be a great complement to classical techniques and provide additional context for ongoing fescue toxicosis research, much progress still needs to be made to ensure its proper implementation; the current framework could be helpful as an initial guideline.

7. Applicability of Integrative Interactomics across Agriculture

While the latter portion of this review focused on developing a framework for using integrative interactomics to help interrogate fescue toxicosis' pathophysiology from a multi-level perspective, this concept is widely applicable across the agricultural sector. The influence of the microbiota on animal production and in numerous disease states, as well as use of metabolomics to identify potential biomarkers of interest, is increasing. For example, Kumar, et al. [138] proposed similar systems-biology based techniques as a potential route for smarter agricultural innovation based on the idea that these techniques help link genotype and phenotype. Shameer, et al. [139] took it a step further by emphasizing using these techniques to develop sustainable agricultural systems, an important and timely endeavor. From an animal perspective, as stated before, these systems biology techniques have been used to assess animal productivity, more so in dairy cattle [140,141]. Although no study has comprehensively evaluated what we have termed the integrome, the idea of integrating 'omics data sets has been previously proposed as a potential tool from a meat science perspective [142]. To provide an analogous example of how this could be useful as a complementary approach to classical methods for fescue toxicosis, Johne's Disease, a bovine disease that culminates with weight loss, is used. Recent investigation found that Johne's Disease is associated with profound fecal microbiota dysbiosis [143]. Using this framework could help investigators to detect downstream processes that are affected by gastrointestinal dysbiosis induced by Johne's Disease, select ones that track with wasting and muscle loss, and identify possible new therapeutic target(s). Moreover, it is well known that many mycotoxins are detrimental to livestock productivity, whether through reproductive inefficiencies or by reducing weight gains [144]; this framework is readily applicable towards any mycotoxin, highlighting the translatability of this approach across farm animal exposures and diseases.

8. Conclusions

In this review, the case for integrative interactomics as the next step in toxicology and animal agriculture is presented. It should be clear that studying the integrome will be a widely adaptable field moving forward as we continue to improve methodologies. For fescue toxicosis, the proposed framework of the fescue toxicosis integrome is an important step forward. As high-throughput techniques to evaluate different aspects of how toxic tall fescue grazing influences livestock physiology become a mainstay, providing a framework can help align the field towards a cohesive, integrative approach to this complex problem. Taking such an approach will allow identification true fescue toxicosis pathophysiological outcomes, which can help address some study-to-study variations. Using the outcomes from integrative interactomics analyses in conjunction with key production and/or pathophysiology indices will lead to therapeutically valuable target(s) identification and, ultimately, better disease management. Overall, adopting an integrome-based framework, like the one proposed here, has the potential to benefit animal agriculture moving forward.

Funding: The research used as the foundation for this review was funded in part by National Institute of Food and Agriculture (NIFA) Agriculture and Food Research Initiative (AFRI) Grants #67030-25004 and 67015-31301 to NMF. We would like to thank the Interdisciplinary Toxicology Program, the Department of Physiology and Pharmacology, and the Graduate School of the University of Georgia for partial support to RSM. The support from the Lalita and Raghubir Sharma Distinguished Professorship in Toxicology to NMF is acknowledged.

Conflicts of Interest: The authors declare no conflict of interest.

References

1. United Nations, DESA. *World Population Prospects 2019*; United Nations, Department of Economic and Social Affairs: New York, NY, USA, 2019.
2. USDA OCE. *USDA Agricultural Projections to 2028*; USDA Office of the Chief Economist Outlook Board: Washington, DC, USA, 2019; p. 108.
3. Li, J.Y.; You, Z.; Wang, Q.; Zhou, Z.J.; Qiu, Y.; Luo, R.; Ge, X.Y. The epidemic of 2019-novel-coronavirus (2019-nCoV) pneumonia and insights for emerging infectious diseases in the future. *Microbes Infect.* **2020**, *22*, 80–85. [CrossRef] [PubMed]
4. Capalbo, S.M.; Antle, J.M.; Seavert, C. Next generation data systems and knowledge products to support agricultural producers and science-based policy decision making. *Agric. Syst.* **2017**, *155*, 191–199. [CrossRef] [PubMed]
5. Unamba, C.I.; Nag, A.; Sharma, R.K. Next Generation Sequencing Technologies: The Doorway to the Unexplored Genomics of Non-Model Plants. *Front. Plant Sci.* **2015**, *6*, 1074. [CrossRef] [PubMed]
6. Matthews, C.; Crispie, F.; Lewis, E.; Reid, M.; O'Toole, P.W.; Cotter, P.D. The rumen microbiome: A crucial consideration when optimising milk and meat production and nitrogen utilisation efficiency. *Gut Microbes* **2019**, *10*, 115–132. [CrossRef]
7. Henderson, G.; Cox, F.; Ganesh, S.; Jonker, A.; Young, W.; Global Rumen Census, C.; Janssen, P.H. Rumen microbial community composition varies with diet and host, but a core microbiome is found across a wide geographical range. *Sci. Rep.* **2015**, *5*, 14567. [CrossRef]
8. Loor, J.J.; Elolimy, A.A.; McCann, J.C. Dietary impacts on rumen microbiota in beef and dairy production. *Anim. Front.* **2016**, *6*, 22–29. [CrossRef]
9. Wolff, S.M.; Ellison, M.J.; Hao, Y.; Cockrum, R.R.; Austin, K.J.; Baraboo, M.; Burch, K.; Lee, H.J.; Maurer, T.; Patil, R.; et al. Diet shifts provoke complex and variable changes in the metabolic networks of the ruminal microbiome. *Microbiome* **2017**, *5*, 60. [CrossRef]
10. Zehavi, T.; Probst, M.; Mizrahi, I. Insights Into Culturomics of the Rumen Microbiome. *Front. Microbiol.* **2018**, *9*, 1999. [CrossRef]
11. Blum, B.C.; Mousavi, F.; Emili, A. Single-platform 'multi-omic' profiling: Unified mass spectrometry and computational workflows for integrative proteomics-metabolomics analysis. *Mol. Omics* **2018**, *14*, 307–319. [CrossRef]
12. Griffiths, W.J. *Metabolomics, Metabonomics, and Metabolic Profiling*; Royal Society of Chemistry: London, UK, 2008.
13. Bundy, J.G.; Davey, M.P.; Viant, M.R. Environmental metabolomics: A critical review and future perspectives. *Metabolomics* **2008**, *5*, 3–21. [CrossRef]
14. Cevallos-Cevallos, J.M.; Reyes-De-Corcuera, J.I.; Etxeberria, E.; Danyluk, M.D.; Rodrick, G.E. Metabolomic analysis in food science: A review. *Trends Food Sci. Technol.* **2009**, *20*, 557–566. [CrossRef]
15. Yu, T.; Jones, D.P. Improving peak detection in high-resolution LC/MS metabolomics data using preexisting knowledge and machine learning approach. *Bioinformatics* **2014**, *30*, 2941–2948. [CrossRef] [PubMed]
16. Go, Y.M.; Walker, D.I.; Liang, Y.; Uppal, K.; Soltow, Q.A.; Tran, V.; Strobel, F.; Quyyumi, A.A.; Ziegler, T.R.; Pennell, K.D.; et al. Reference Standardization for Mass Spectrometry and High-resolution Metabolomics Applications to Exposome Research. *Toxicol. Sci.* **2015**, *148*, 531–543. [CrossRef] [PubMed]
17. Yu, T.; Park, Y.; Johnson, J.M.; Jones, D.P. apLCMS–adaptive processing of high-resolution LC/MS data. *Bioinformatics* **2009**, *25*, 1930–1936. [CrossRef] [PubMed]
18. Uppal, K.; Soltow, Q.A.; Strobel, F.H.; Pittard, W.S.; Gernert, K.M.; Yu, T.; Jones, D.P. xMSanalyzer: Automated pipeline for improved feature detection and downstream analysis of large-scale, non-targeted metabolomics data. *BMC Bioinform.* **2013**, *14*, 15. [CrossRef]
19. Mallon, T.M.; Krahl, P.K.; Haines, K.M., Jr.; Walker, D.I.; Thatcher, T.; Woeller, C.F.; Thakar, J.; Hopke, P.K.; Gaydos, J.C.; Smith, M.R.; et al. Use of Biomarkers to Assess Environmental Exposures and Health Outcomes in Deployed Troops. *J. Occup. Environ. Med.* **2019**, *61* (Suppl. 12), S1–S4. [CrossRef]
20. Niedzwiecki, M.M.; Walker, D.I.; Vermeulen, R.; Chadeau-Hyam, M.; Jones, D.P.; Miller, G.W. The Exposome: Molecules to Populations. *Annu. Rev. Pharmacol. Toxicol.* **2019**, *59*, 107–127. [CrossRef]
21. Uppal, K.; Ma, C.; Go, Y.M.; Jones, D.P.; Wren, J. xMWAS: A data-driven integration and differential network analysis tool. *Bioinformatics* **2018**, *34*, 701–702. [CrossRef]

22. Koster, T.; Marondedze, C.; Meyer, K.; Staiger, D. RNA-Binding Proteins Revisited—The Emerging Arabidopsis mRNA Interactome. *Trends Plant Sci.* **2017**, *22*, 512–526. [CrossRef]
23. Maccarrone, G.; Bonfiglio, J.J.; Silberstein, S.; Turck, C.W.; Martins-de-Souza, D. Characterization of a Protein Interactome by Co-Immunoprecipitation and Shotgun Mass Spectrometry. In *Multiplex Biomarker Techniques: Methods and Applications*; Guest, P.C., Ed.; Springer: New York, NY, USA, 2017; pp. 223–234. [CrossRef]
24. Ghadie, M.A.; Coulombe-Huntington, J.; Xia, Y. Interactome evolution: Insights from genome-wide analyses of protein-protein interactions. *Curr. Opin. Struct. Biol.* **2018**, *50*, 42–48. [CrossRef]
25. Shukla, E.; Chauhan, R. Host-HIV-1 Interactome: A Quest for Novel Therapeutic Intervention. *Cells* **2019**, *8*, 1155. [CrossRef] [PubMed]
26. Park, D.I.; Turck, C.W. Interactome Studies of Psychiatric Disorders. In *Reviews on Biomarker Studies in Psychiatric and Neurodegenerative Disorders*; Guest, P.C., Ed.; Springer International Publishing: Cham, Switzerland, 2019; pp. 163–173. [CrossRef]
27. Subramanian, I.; Verma, S.; Kumar, S.; Jere, A.; Anamika, K. Multi-omics Data Integration, Interpretation, and Its Application. *Bioinform. Biol. Insights* **2020**, *14*. [CrossRef] [PubMed]
28. Ulfenborg, B. Vertical and horizontal integration of multi-omics data with miodin. *BMC Bioinform.* **2019**, *20*, 649. [CrossRef] [PubMed]
29. Bersanelli, M.; Mosca, E.; Remondini, D.; Giampieri, E.; Sala, C.; Castellani, G.; Milanesi, L. Methods for the integration of multi-omics data: Mathematical aspects. *BMC Bioinform.* **2016**, *17*, S15. [CrossRef]
30. Canzler, S.; Schor, J.; Busch, W.; Schubert, K.; Rolle-Kampczyk, U.E.; Seitz, H.; Kamp, H.; von Bergen, M.; Buesen, R.; Hackermüller, J. Prospects and challenges of multi-omics data integration in toxicology. *Arch. Toxicol.* **2020**, *94*, 371–388. [CrossRef]
31. Zhao, L.; Zhang, H.; White, J.C.; Chen, X.; Li, H.; Qu, X.; Ji, R. Metabolomics reveals that engineered nanomaterial exposure in soil alters both soil rhizosphere metabolite profiles and maize metabolic pathways. *Environ. Sci. Nano* **2019**, *6*, 1716–1727. [CrossRef]
32. Walker, D.I.; Lane, K.J.; Liu, K.; Uppal, K.; Patton, A.P.; Durant, J.L.; Jones, D.P.; Brugge, D.; Pennell, K.D. Metabolomic assessment of exposure to near-highway ultrafine particles. *J. Expo. Sci. Environ. Epidemiol.* **2019**, *29*, 469–483. [CrossRef]
33. Houten, S.M.; Chen, J.; Belpoggi, F.; Manservisi, F.; Sanchez-Guijo, A.; Wudy, S.A.; Teitelbaum, S.L. Changes in the Metabolome in Response to Low-Dose Exposure to Environmental Chemicals Used in Personal Care Products during Different Windows of Susceptibility. *PLoS ONE* **2016**, *11*, e0159919. [CrossRef]
34. Van Meter, R.J.; Glinski, D.A.; Purucker, S.T.; Henderson, W.M. Influence of exposure to pesticide mixtures on the metabolomic profile in post-metamorphic green frogs (Lithobates clamitans). *Sci. Total Environ.* **2018**, *624*, 1348–1359. [CrossRef]
35. Grison, S.; Fave, G.; Maillot, M.; Manens, L.; Delissen, O.; Blanchardon, E.; Dublineau, I.; Aigueperse, J.; Bohand, S.; Martin, J.C.; et al. Metabolomics reveals dose effects of low-dose chronic exposure to uranium in rats: Identification of candidate biomarkers in urine samples. *Metabolomics* **2016**, *12*, 154. [CrossRef]
36. Golder, H.M.; Thomson, J.M.; Denman, S.E.; McSweeney, C.S.; Lean, I.J. Genetic Markers Are Associated with the Ruminal Microbiome and Metabolome in Grain and Sugar Challenged Dairy Heifers. *Front. Genet.* **2018**, *9*, 62. [CrossRef] [PubMed]
37. Wang, B.; Ma, M.P.; Diao, Q.Y.; Tu, Y. Saponin-Induced Shifts in the Rumen Microbiome and Metabolome of Young Cattle. *Front. Microbiol.* **2019**, *10*, 356. [CrossRef] [PubMed]
38. Myer, P.R. Bovine Genome-Microbiome Interactions: Metagenomic Frontier for the Selection of Efficient Productivity in Cattle Systems. *Msystems* **2019**, *4*, e00103–e00119. [CrossRef]
39. Patterson, D.S.P.; Shreeve, B.J.; Roberts, B.A.; Berrett, S.; Brush, P.J.; Glancy, E.M.; Krogh, P. Effect on calves of barley naturally contaminated with ochratoxin A and groundnut meal contaminated with low concentrations of aflatoxin R1. *Res. Vet. Sci.* **1981**, *31*, 213–218. [CrossRef]
40. Cook, W.O.; Richard, J.L.; Osweiler, G.D.; Trampel, D.W. Clinical and pathologic changes in acute bovine aflatoxicosis: Rumen motility and tissue and fluid concentrations of aflatoxins B1 and M1. *Am. J. Vet. Res.* **1986**, *47*, 1817–1825.
41. Placinta, C.M.; D'Mello, J.P.F.; Macdonald, A.M.C. A review of worldwide contamination of cereal grains and animal feed with Fusarium mycotoxins. *Anim. Feed Sci. Technol.* **1999**, *78*, 21–37. [CrossRef]
42. Klotz, J.L. Activities and Effects of Ergot Alkaloids on Livestock Physiology and Production. *Toxins* **2015**, *7*, 2801–2821. [CrossRef]

43. Li, F.; Li, C.; Chen, Y.; Liu, J.; Zhang, C.; Irving, B.; Fitzsimmons, C.; Plastow, G.; Guan, L.L. Host genetics influence the rumen microbiota and heritable rumen microbial features associate with feed efficiency in cattle. *Microbiome* **2019**, *7*, 92. [CrossRef]
44. Weimer, P.J.; Stevenson, D.M.; Mantovani, H.C.; Man, S.L. Host specificity of the ruminal bacterial community in the dairy cow following near-total exchange of ruminal contents. *J. Dairy Sci.* **2010**, *93*, 5902–5912. [CrossRef]
45. Delgado, B.; Bach, A.; Guasch, I.; Gonzalez, C.; Elcoso, G.; Pryce, J.E.; Gonzalez-Recio, O. Whole rumen metagenome sequencing allows classifying and predicting feed efficiency and intake levels in cattle. *Sci. Rep.* **2019**, *9*, 11. [CrossRef]
46. Lima, J.; Auffret, M.D.; Stewart, R.D.; Dewhurst, R.J.; Duthie, C.A.; Snelling, T.J.; Walker, A.W.; Freeman, T.C.; Watson, M.; Roehe, R. Identification of Rumen Microbial Genes Involved in Pathways Linked to Appetite, Growth, and Feed Conversion Efficiency in Cattle. *Front. Genet.* **2019**, *10*, 701. [CrossRef]
47. Jewell, K.A.; McCormick, C.A.; Odt, C.L.; Weimer, P.J.; Suen, G. Ruminal Bacterial Community Composition in Dairy Cows Is Dynamic over the Course of Two Lactations and Correlates with Feed Efficiency. *Appl. Environ. Microbiol.* **2015**, *81*, 4697–4710. [CrossRef] [PubMed]
48. Belanche, A.; Kingston-Smith, A.H.; Griffith, G.W.; Newbold, C.J. A Multi-Kingdom Study Reveals the Plasticity of the Rumen Microbiota in Response to a Shift From Non-grazing to Grazing Diets in Sheep. *Front. Microbiol.* **2019**, *10*, 122. [CrossRef] [PubMed]
49. Khafipour, E.; Li, S.; Tun, H.M.; Derakhshani, H.; Moossavi, S.; Plaizier, J.C. Effects of grain feeding on microbiota in the digestive tract of cattle. *Anim. Front.* **2016**, *6*, 13–19. [CrossRef]
50. Mao, S.; Zhang, M.; Liu, J.; Zhu, W. Characterising the bacterial microbiota across the gastrointestinal tracts of dairy cattle: Membership and potential function. *Sci. Rep.* **2015**, *5*, 16116. [CrossRef]
51. de Oliveira, M.N.; Jewell, K.A.; Freitas, F.S.; Benjamin, L.A.; Totola, M.R.; Borges, A.C.; Moraes, C.A.; Suen, G. Characterizing the microbiota across the gastrointestinal tract of a Brazilian Nelore steer. *Vet. Microbiol.* **2013**, *164*, 307–314. [CrossRef]
52. O'Hara, E.; Kenny, D.A.; McGovern, E.; Byrne, C.J.; McCabe, M.S.; Guan, L.L.; Waters, S.M. Investigating temporal microbial dynamics in the rumen of beef calves raised on two farms during early life. *FEMS Microbiol. Ecol.* **2020**, *96*. [CrossRef]
53. Artegoitia, V.M.; Foote, A.P.; Lewis, R.M.; Freetly, H.C. Rumen Fluid Metabolomics Analysis Associated with Feed Efficiency on Crossbred Steers. *Sci. Rep.* **2017**, *7*, 2864. [CrossRef]
54. Boudonck, K.J.; Mitchell, M.W.; Wulff, J.; Ryals, J.A. Characterization of the biochemical variability of bovine milk using metabolomics. *Metabolomics* **2009**, *5*, 375–386. [CrossRef]
55. Klein, M.S.; Almstetter, M.F.; Schlamberger, G.; Nurnberger, N.; Dettmer, K.; Oefner, P.J.; Meyer, H.H.; Wiedemann, S.; Gronwald, W. Nuclear magnetic resonance and mass spectrometry-based milk metabolomics in dairy cows during early and late lactation. *J. Dairy Sci.* **2010**, *93*, 1539–1550. [CrossRef]
56. Goldansaz, S.A.; Guo, A.C.; Sajed, T.; Steele, M.A.; Plastow, G.S.; Wishart, D.S. Livestock metabolomics and the livestock metabolome: A systematic review. *PLoS ONE* **2017**, *12*, e0177675. [CrossRef] [PubMed]
57. Leng, R.A. Biofilm compartmentalisation of the rumen microbiome: Modification of fermentation and degradation of dietary toxins. *Anim. Prod. Sci.* **2017**, *57*, 2188–2203. [CrossRef]
58. Gallo, A.; Giuberti, G.; Frisvad, J.C.; Bertuzzi, T.; Nielsen, K.F. Review on Mycotoxin Issues in Ruminants: Occurrence in Forages, Effects of Mycotoxin Ingestion on Health Status and Animal Performance and Practical Strategies to Counteract Their Negative Effects. *Toxins* **2015**, *7*, 3057–3111. [CrossRef] [PubMed]
59. Ogunade, I.; Jiang, Y.; Pech Cervantes, A. DI/LC-MS/MS-Based Metabolome Analysis of Plasma Reveals the Effects of Sequestering Agents on the Metabolic Status of Dairy Cows Challenged with Aflatoxin B1. *Toxins* **2019**, *11*, 693. [CrossRef]
60. Ogunade, I.; Jiang, Y.; Adeyemi, J.; Oliveira, A.; Vyas, D.; Adesogan, A. Biomarker of Aflatoxin Ingestion: (1)H NMR-Based Plasma Metabolomics of Dairy Cows Fed Aflatoxin B(1) with or without Sequestering Agents. *Toxins* **2018**, *10*, 545. [CrossRef]
61. Wang, Q.; Zhang, Y.; Zheng, N.; Guo, L.; Song, X.; Zhao, S.; Wang, J. Biological System Responses of Dairy Cows to Aflatoxin B1 Exposure Revealed with Metabolomic Changes in Multiple Biofluids. *Toxins* **2019**, *11*, 77. [CrossRef]

62. Piotrowska, M.; Slizewska, K.; Nowak, A.; Zielonka, L.; Zakowska, Z.; Gajecka, M.; Gajecki, M. The effect of experimental fusarium mycotoxicosis on microbiota diversity in porcine ascending colon contents. *Toxins* **2014**, *6*, 2064–2081. [CrossRef]
63. Shaffer, M.; Armstrong, A.J.S.; Phelan, V.V.; Reisdorph, N.; Lozupone, C.A. Microbiome and metabolome data integration provides insight into health and disease. *Transl. Res.* **2017**, *189*, 51–64. [CrossRef]
64. Lu, K.; Abo, R.P.; Schlieper, K.A.; Graffam, M.E.; Levine, S.; Wishnok, J.S.; Swenberg, J.A.; Tannenbaum, S.R.; Fox, J.G. Arsenic exposure perturbs the gut microbiome and its metabolic profile in mice: An integrated metagenomics and metabolomics analysis. *Environ. Health Perspect.* **2014**, *122*, 284–291. [CrossRef]
65. Yang, Y.; Misra, B.B.; Liang, L.; Bi, D.; Weng, W.; Wu, W.; Cai, S.; Qin, H.; Goel, A.; Li, X.; et al. Integrated microbiome and metabolome analysis reveals a novel interplay between commensal bacteria and metabolites in colorectal cancer. *Theranostics* **2019**, *9*, 4101–4114. [CrossRef]
66. Tang, Z.-Z.; Chen, G.; Hong, Q.; Huang, S.; Smith, H.M.; Shah, R.D.; Scholz, M.; Ferguson, J.F. Multi-Omic Analysis of the Microbiome and Metabolome in Healthy Subjects Reveals Microbiome-Dependent Relationships Between Diet and Metabolites. *Front. Genet.* **2019**, *10*, 454. [CrossRef] [PubMed]
67. Mote, R.S.; Hill, N.S.; Skarlupka, J.H.; Turner, Z.B.; Sanders, Z.P.; Jones, D.P.; Suen, G.; Filipov, N.M. Response of Beef Cattle Fecal Microbiota to Grazing on Toxic Tall Fescue. *Appl. Environ. Microbiol.* **2019**, *85*, e00032-19. [CrossRef] [PubMed]
68. Mote, R.S.; Hill, N.S.; Uppal, K.; Tran, V.T.; Jones, D.P.; Filipov, N.M. Metabolomics of fescue toxicosis in grazing beef steers. *Food Chem. Toxicol.* **2017**, *105*, 285–299. [CrossRef] [PubMed]
69. Mote, R.S.; Hill, N.S.; Skarlupka, J.H.; Tran, V.T.; Walker, D.I.; Turner, Z.B.; Sanders, Z.P.; Jones, D.P.; Suen, G.; Filipov, N.M. Toxic tall fescue grazing increases susceptibility of the Angus steer fecal microbiota and plasma/urine metabolome to environmental effects. *Sci. Rep.* **2020**, *10*, 2497. [CrossRef]
70. Melchior, E.A.; Myer, P.R. Fescue toxicosis and its influence on the rumen microbiome: Mitigation of production losses through clover isoflavones. *J. Appl. Anim. Res.* **2018**, *46*, 1280–1288. [CrossRef]
71. Melchior, E.A.; Smith, J.K.; Schneider, L.G.; Mulliniks, J.T.; Bates, G.E.; McFarlane, Z.D.; Flythe, M.D.; Klotz, J.L.; Goodman, J.P.; Ji, H.; et al. Effects of red clover isoflavones on tall fescue seed fermentation and microbial populations in vitro. *PLoS ONE* **2018**, *13*, e0201866. [CrossRef]
72. Clay, K.; Schardl, C. Evolutionary origins and ecological consequences of endophyte symbiosis with grasses. *Am. Nat.* **2002**, *160* (Suppl. 4), S99–S127. [CrossRef]
73. Christensen, M.J.; Bennett, R.J.; Ansari, H.A.; Koga, H.; Johnson, R.D.; Bryan, G.T.; Simpson, W.R.; Koolaard, J.P.; Nickless, E.M.; Voisey, C.R. Epichloe endophytes grow by intercalary hyphal extension in elongating grass leaves. *Fungal Genet. Biol.* **2008**, *45*, 84–93. [CrossRef]
74. Schardl, C.L.; Young, C.A.; Pan, J.; Florea, S.; Takach, J.E.; Panaccione, D.G.; Farman, M.L.; Webb, J.S.; Jaromczyk, J.; Charlton, N.D.; et al. Currencies of mutualisms: Sources of alkaloid genes in vertically transmitted epichloae. *Toxins* **2013**, *5*, 1064–1088. [CrossRef]
75. Missaoui, A.; Hill, N. Use of accelerated aging as a surrogate phenotyping approach to improve endophyte survival during storage of tall fescue seed. *Field Crop. Res.* **2015**, *183*, 43–49. [CrossRef]
76. Guerre, P. Ergot alkaloids produced by endophytic fungi of the genus Epichloe. *Toxins* **2015**, *7*, 773–790. [CrossRef] [PubMed]
77. Young, C.A.; Charlton, N.D.; Takach, J.E.; Swoboda, G.A.; Trammell, M.A.; Huhman, D.V.; Hopkins, A.A. Characterization of Epichloe coenophiala within the US: Are all tall fescue endophytes created equal? *Front. Chem.* **2014**, *2*, 95. [CrossRef] [PubMed]
78. Maruo, V.M.; Bracarense, A.P.; Metayer, J.P.; Vilarino, M.; Oswald, I.P.; Pinton, P. Ergot Alkaloids at Doses Close to EU Regulatory Limits Induce Alterations of the Liver and Intestine. *Toxins* **2018**, *10*, 183. [CrossRef]
79. Philippe, G. Lolitrem B and Indole Diterpene Alkaloids Produced by Endophytic Fungi of the Genus Epichloe and Their Toxic Effects in Livestock. *Toxins* **2016**, *8*, 47. [CrossRef] [PubMed]
80. Berde, B.; Stürmer, E. *Introduction to the Pharmacology of Ergot Alkaloids and Related Compounds as a Basis of Their Therapeutic Application*; Springer: Berlin/Heidelberg, Germany, 1978; pp. 1–28. [CrossRef]
81. Moubarak, A.S.; Piper, E.L.; Johnson, Z.B.; Flieger, M. HPLC Method for Detection of Ergotamine, Ergosine, and Ergine after Intravenous Injection of a Single Dose. *J. Agric. Food Chem.* **1996**, *44*, 146–148. [CrossRef]
82. Hill, N.S.; Thompson, F.N.; Stuedemann, J.A.; Dawe, D.L.; Hiatt, E.E., 3rd. Urinary alkaloid excretion as a diagnostic tool for fescue toxicosis in cattle. *J. Vet. Diagn. Investig.* **2000**, *12*, 210–217. [CrossRef]

83. Stuedemann, J.A.; Hill, N.S.; Thompson, F.N.; Fayrer-Hosken, R.A.; Hay, W.P.; Dawe, D.L.; Seman, D.H.; Martin, S.A. Urinary and biliary excretion of ergot alkaloids from steers that grazed endophyte-infected tall fescue. *J. Anim. Sci.* **1998**, *76*, 2146–2154. [CrossRef]
84. Hill, N.S.; Thompson, F.N.; Stuedemann, J.A.; Rottinghaus, G.W.; Ju, H.J.; Dawe, D.L.; Hiatt, E.E., 3rd. Ergot alkaloid transport across ruminant gastric tissues. *J. Anim. Sci.* **2001**, *79*, 542–549. [CrossRef]
85. Realini, C.E.; Duckett, S.K.; Hill, N.S.; Hoveland, C.S.; Lyon, B.G.; Sackmann, J.R.; Gillis, M.H. Effect of endophyte type on carcass traits, meat quality, and fatty acid composition of beef cattle grazing tall fescue. *J. Anim. Sci.* **2005**, *83*, 430–439. [CrossRef]
86. Zbib, N.; Repussard, C.; Tardieu, D.; Priymenko, N.; Domange, C.; Guerre, P. Toxicity of endophyte-infected ryegrass hay containing high ergovaline level in lactating ewes. *J. Anim. Sci.* **2015**, *93*, 4098–4109. [CrossRef]
87. Westendorf, M.L.; Mitchell, G.E., Jr.; Tucker, R.E.; Bush, L.P.; Petroski, R.J.; Powell, R.G. In vitro and in vivo ruminal and physiological responses to endophyte-infected tall fescue. *J Dairy Sci.* **1993**, *76*, 555–563. [CrossRef]
88. Ayers, A.W.; Hill, N.S.; Rottinghaus, G.E.; Stuedemann, J.A.; Thompson, F.N.; Purinton, P.T.; Seman, D.H.; Dawe, D.L.; Parks, A.H.; Ensley, D. Ruminal Metabolism and Transport of Tall Fescue Ergot Alkaloids. *Crop Sci.* **2009**, *49*, 2309–2316. [CrossRef]
89. Strimbu, K.; Tavel, J.A. What are biomarkers? *Curr. Opin. HIV AIDS* **2010**, *5*, 463–466. [CrossRef] [PubMed]
90. Mayeux, R. Biomarkers: Potential uses and limitations. *NeuroRx* **2004**, *1*, 182–188. [CrossRef] [PubMed]
91. Hurley, W.L.; Convey, E.M.; Leung, K.; Edgerton, L.A.; Hemken, R.W. Bovine prolactin, TSH, T and T concentrations as affected by tall fescue summer toxicosis and temperature. *J. Anim. Sci.* **1980**, *51*, 374–379. [CrossRef]
92. Karg, H.; Schams, D. Prolactin release in cattle. *J. Reprod. Fertil.* **1974**, *39*, 463–472. [CrossRef]
93. Smith, V.G.; Hacker, R.R.; Brown, R.G. Effect of alterations in ambient temperature on serum prolactin concentration in steers. *J. Anim. Sci.* **1977**, *44*, 645–649. [CrossRef]
94. Bryant, G.D.; Linzell, J.L.; Greenwood, F.C. Plasma prolactin in goats measured by radioimmunoassay: The effects of teat stimulation, mating behavior, stress, fasting and of oxytocin, insulin and glucose injections. *Hormones* **1970**, *1*, 26–35. [CrossRef]
95. Yayou, K.; Ito, S.; Yamamoto, N.; Kitagawa, S.; Okamura, H. Relationships of stress responses with plasma oxytocin and prolactin in heifer calves. *Physiol. Behav.* **2010**, *99*, 362–369. [CrossRef]
96. Cunningham, I. Tall fescue grass is poison for cattle. *N. Z. J. Agric.* **1948**, *77*, 519.
97. Cunningham, I.J. A Note on the Cause of Tall Fescue Lameness in Cattle. *Aust. Vet. J.* **1949**, *25*, 27–28. [CrossRef]
98. Jacobson, D.R.; Miller, W.M.; Seath, D.M.; Yates, S.G.; Tookey, H.L.; Wolff, I.A. Nature of Fescue Toxicity and Progress toward Identification of the Toxic Entity. *J. Dairy Sci.* **1963**, *46*, 416–422. [CrossRef]
99. Paterson, J.; Forcherio, C.; Larson, B.; Samford, M.; Kerley, M. The effects of fescue toxicosis on beef cattle productivity. *J. Anim. Sci.* **1995**, *73*, 889–898. [CrossRef] [PubMed]
100. Strickland, J.R.; Oliver, J.W.; Cross, D.L. Fescue toxicosis and its impact on animal agriculture. *Vet. Hum. Toxicol.* **1993**, *35*, 454–464. [PubMed]
101. Snider, M.A.; Harmon, D.L.; Klotz, J.L. Pharmacologic assessment of bovine ruminal and mesenteric vascular serotonin receptor populations. *J. Anim. Sci.* **2018**, *96*, 1570–1578. [CrossRef] [PubMed]
102. Trotta, R.J.; Harmon, D.L.; Klotz, J.L. Interaction of ergovaline with serotonin receptor 5-HT2A in bovine ruminal and mesenteric vasculature. *J. Anim. Sci.* **2018**, *96*, 4912–4922. [CrossRef]
103. McLeay, L.M.; Smith, B.L. Effects of ergotamine and ergovaline on the electromyographic activity of smooth muscle of the reticulum and rumen of sheep. *Am. J. Vet. Res.* **2006**, *67*, 707–714. [CrossRef]
104. Foote, A.P.; Kristensen, N.B.; Klotz, J.L.; Kim, D.H.; Koontz, A.F.; McLeod, K.R.; Bush, L.P.; Schrick, F.N.; Harmon, D.L. Ergot alkaloids from endophyte-infected tall fescue decrease reticuloruminal epithelial blood flow and volatile fatty acid absorption from the washed reticulorumen. *J. Anim. Sci.* **2013**, *91*, 5366–5378. [CrossRef]
105. Koontz, A.F.; Kim, D.H.; McLeod, K.R.; Klotz, J.L.; Harmon, D.L. Effect of fescue toxicosis on whole body energy and nitrogen balance, in situ degradation and ruminal passage rates in Holstein steers. *Anim. Prod. Sci.* **2015**, *55*, 988–998. [CrossRef]

106. Koontz, A.F.; Bush, L.P.; Klotz, J.L.; McLeod, K.R.; Schrick, F.N.; Harmon, D.L. Evaluation of a ruminally dosed tall fescue seed extract as a model for fescue toxicosis in steers. *J. Anim. Sci.* **2012**, *90*, 914–921. [CrossRef]
107. Koontz, A.F.; Kim, D.H.; Foote, A.P.; Bush, L.P.; Klotz, J.L.; McLeod, K.R.; Harmon, D.L. Alteration of fasting heat production during fescue toxicosis in Holstein steers. *J. Anim. Sci.* **2013**, *91*, 3881–3888. [CrossRef] [PubMed]
108. Allen, M.S. Physical constraints on voluntary intake of forages by ruminants. *J. Anim. Sci.* **1996**, *74*, 3063–3075. [CrossRef] [PubMed]
109. Felitti, S.; Shields, K.; Ramsperger, M.; Tian, P.; Sawbridge, T.; Webster, T.; Logan, E.; Erwin, T.; Forster, J.; Edwards, D.; et al. Transcriptome analysis of Neotyphodium and Epichloe grass endophytes. *Fungal Genet. Biol.* **2006**, *43*, 465–475. [CrossRef] [PubMed]
110. Guo, J.; McCulley, R.L.; McNear, D.H., Jr. Tall fescue cultivar and fungal endophyte combinations influence plant growth and root exudate composition. *Front. Plant. Sci.* **2015**, *6*, 183. [CrossRef]
111. Rojas, X.; Guo, J.; Leff, J.W.; McNear, D.H., Jr.; Fierer, N.; McCulley, R.L. Infection with a Shoot-Specific Fungal Endophyte (Epichloe) Alters Tall Fescue Soil Microbial Communities. *Microb. Ecol.* **2016**, *72*, 197–206. [CrossRef]
112. Rasmussen, S.; Parsons, A.J.; Newman, J.A. Metabolomics analysis of the Lolium perenne–Neotyphodium lolii symbiosis: More than just alkaloids? *Phytochem. Rev.* **2009**, *8*, 535–550. [CrossRef]
113. Bastias, D.A.; Martinez-Ghersa, M.A.; Ballare, C.L.; Gundel, P.E. Epichloe Fungal Endophytes and Plant Defenses: Not Just Alkaloids. *Trends Plant Sci.* **2017**, *22*, 939–948. [CrossRef]
114. Miller, M.J.; McDole, J.R.; Newberry, R.D. Microanatomy of the intestinal lymphatic system. *Ann. N. Y. Acad. Sci.* **2010**, *1207* (Suppl. 1), E21–E28. [CrossRef]
115. Unthank, J.L.; Bohlen, H.G. Lymphatic pathways and role of valves in lymph propulsion from small intestine. *Am. J. Physiol. Gastrointest. Liver Physiol.* **1988**, *254*, G389–G398. [CrossRef]
116. Futrell, M.C.; Farnell, D.R.; Poe, W.E.; Watson, V.H.; Coats, R.E. Fungal Populations in the Rumen Associated with Fescue Toxicosis. *J. Environ. Qual.* **1974**, *3*, 140–143. [CrossRef]
117. Eich, E.; Eichberg, D.; Muller, W.E. Clavines. New antibiotics with cytostatic activity. *Biochem. Pharmacol.* **1984**, *33*, 523–526. [CrossRef]
118. Eich, E.; Eichberg, D.; Schwarz, G.; Clas, F.; Loos, M. Antimicrobial activitiy of clavines. *Arzeimittelforschung* **1985**, *35*, 1760–1762.
119. Looper, M.L.; Edrington, T.S.; Flores, R.; Burke, J.M.; Callaway, T.R.; Aiken, G.E.; Schrick, F.N.; Rosenkrans, C.F., Jr. Influence of dietary endophyte (Neotyphodium coenophialum)-infected tall fescue (Festuca arundinacea) seed on fecal shedding of antibiotic resistance-selected Escherichia coli O157:H7 in ewes. *J. Anim. Sci.* **2007**, *85*, 1102–1108. [CrossRef] [PubMed]
120. Harlow, B.E.; Goodman, J.P.; Lynn, B.C.; Flythe, M.D.; Ji, H.; Aiken, G.E. Ruminal tryptophan-utilizing bacteria degrade ergovaline from tall fescue seed extract. *J. Anim. Sci.* **2017**, *95*, 980–988. [CrossRef] [PubMed]
121. Alrashedi, S. Effect of Endophyte-Infected Tall Fescue Toxins on Growth Performance and the Microbial Community in the Rumen and Feces in Pregnant Ewes. University of Arkansas. 2017. Available online: http://scholarworks.uark.edu/etd/2532 (accessed on 2 April 2020).
122. Koester, L.; Poole, D.; Serão, N.; Schmitz-Esser, S. Beef cattle that respond differently to fescue toxicosis have distinct gastrointestinal tract microbiota. *PLoS ONE* **2020**, *15*, e0229192. [CrossRef]
123. Strickland, J.; Aiken, G.E.; Spiers, D.; Fletcher, L.R.; Oliver, J. Physiological Basis of Fescue Toxicosis. In *Tall Fescue for the Twenty-First Century*; American Society of Agronomy: Madison, WI, USA, 2009; Volume 53, pp. 203–227. [CrossRef]
124. Oliver, J.W.; Schultze, A.; Rohrbach, B.W.; Fribourg, H.; Ingle, T.; Waller, J. Alterations in hemograms and serum biochemical analytes of steers after prolonged consumption of endophyte-infected tall fescue. *J. Anim. Sci.* **2000**, *78*, 1029–1035. [CrossRef]
125. Browning, R., Jr.; Gissendanner, S.J.; Wakefield, T., Jr. Ergotamine alters plasma concentrations of glucagon, insulin, cortisol, and triiodothyronine in cows. *J. Anim. Sci.* **2000**, *78*, 690–698. [CrossRef]
126. Coufal-Majewski, S.; Stanford, K.; McAllister, T.; Blakley, B.; McKinnon, J.; Chaves, A.V.; Wang, Y. Impacts of Cereal Ergot in Food Animal Production. *Front. Vet. Sci.* **2016**, *3*, 15. [CrossRef]

127. Schmidt, S.P.; Hoveland, C.S.; Clark, E.M.; Davis, N.D.; Smith, L.A.; Grimes, H.W.; Holliman, J.L. Association of an endophytic fungus with fescue toxicity in steers fed Kentucky 31 tall fescue seed or hay. *J. Anim. Sci.* **1982**, *55*, 1259–1263. [CrossRef]
128. Patterson, J.; Looper, M.C.; Williamson, B.; Rosenkrans, C. Effects of fescue cultivar and heat shock protein haplotype on growth and fertility of crossbred beef heifers. *Anim. Sci. Ark. Anim. Sci.* **2011**, *597*, 57–59.
129. Jackson, J.J.; Lindemann, M.D.; Boling, J.A.; Matthews, J.C. Summer-Long Grazing of High vs. Low Endophyte (Neotyphodium coenophialum)-Infected Tall Fescue by Growing Beef Steers Results in Distinct Temporal Blood Analyte Response Patterns, with Poor Correlation to Serum Prolactin Levels. *Front. Vet. Sci.* **2015**, *2*, 77. [CrossRef] [PubMed]
130. Amar, J. Microbiota-Host Crosstalk: A Bridge Between Cardiovascular Risk Factors, Diet, and Cardiovascular Disease. *Am. J. Hypertens.* **2018**, *31*, 941–944. [CrossRef] [PubMed]
131. Goncalves, S.M.; Lagrou, K.; Duarte-Oliveira, C.; Maertens, J.A.; Cunha, C.; Carvalho, A. The microbiome-metabolome crosstalk in the pathogenesis of respiratory fungal diseases. *Virulence* **2017**, *8*, 673–684. [CrossRef] [PubMed]
132. Lopez-Garcia, A.; Pineda-Quiroga, C.; Atxaerandio, R.; Perez, A.; Hernandez, I.; Garcia-Rodriguez, A.; Gonzalez-Recio, O. Comparison of Mothur and QIIME for the Analysis of Rumen Microbiota Composition Based on 16S rRNA Amplicon Sequences. *Front. Microbiol.* **2018**, *9*, 3010. [CrossRef] [PubMed]
133. Prodan, A.; Tremaroli, V.; Brolin, H.; Zwinderman, A.H.; Nieuwdorp, M.; Levin, E. Comparing bioinformatic pipelines for microbial 16S rRNA amplicon sequencing. *PLoS ONE* **2020**, *15*, e0227434. [CrossRef]
134. Lee, M.Y.; Hu, T. Computational Methods for the Discovery of Metabolic Markers of Complex Traits. *Metabolites* **2019**, *9*, 66. [CrossRef]
135. Schrimpe-Rutledge, A.C.; Codreanu, S.G.; Sherrod, S.D.; McLean, J.A. Untargeted Metabolomics Strategies-Challenges and Emerging Directions. *J. Am. Soc. Mass Spectrom.* **2016**, *27*, 1897–1905. [CrossRef]
136. Carroll, A.J.; Zhang, P.; Whitehead, L.; Kaines, S.; Tcherkez, G.; Badger, M.R. PhenoMeter: A Metabolome Database Search Tool Using Statistical Similarity Matching of Metabolic Phenotypes for High-Confidence Detection of Functional Links. *Front. Bioeng. Biotechnol.* **2015**, *3*, 106. [CrossRef]
137. Longnecker, K.; Futrelle, J.; Coburn, E.; Kido Soule, M.C.; Kujawinski, E.B. Environmental metabolomics: Databases and tools for data analysis. *Mar. Chem.* **2015**, *177*, 366–373. [CrossRef]
138. Kumar, A.; Pathak, R.K.; Gupta, S.M.; Gaur, V.S.; Pandey, D. Systems Biology for Smart Crops and Agricultural Innovation: Filling the Gaps between Genotype and Phenotype for Complex Traits Linked with Robust Agricultural Productivity and Sustainability. *OMICS* **2015**, *19*, 581–601. [CrossRef]
139. Shameer, K.; Naika, M.B.N.; Shafi, K.M.; Sowdhamini, R. Decoding systems biology of plant stress for sustainable agriculture development and optimized food production. *Prog. Biophys. Mol. Biol.* **2019**, *145*, 19–39. [CrossRef] [PubMed]
140. de Menezes, A.B.; Lewis, E.; O'Donovan, M.; O'Neill, B.F.; Clipson, N.; Doyle, E.M. Microbiome analysis of dairy cows fed pasture or total mixed ration diets. *FEMS Microbiol. Ecol.* **2011**, *78*, 256–265. [CrossRef] [PubMed]
141. Pitta, D.W.; Kumar, S.; Vecchiarelli, B.; Shirley, D.J.; Bittinger, K.; Baker, L.D.; Ferguson, J.D.; Thomsen, N. Temporal dynamics in the ruminal microbiome of dairy cows during the transition period. *J. Anim. Sci.* **2014**, *92*, 4014–4022. [CrossRef] [PubMed]
142. D'Alessandro, A.; Zolla, L. Meat science: From proteomics to integrated omics towards system biology. *J. Proteom.* **2013**, *78*, 558–577. [CrossRef] [PubMed]
143. Fecteau, M.E.; Pitta, D.W.; Vecchiarelli, B.; Indugu, N.; Kumar, S.; Gallagher, S.C.; Fyock, T.L.; Sweeney, R.W. Dysbiosis of the Fecal Microbiota in Cattle Infected with Mycobacterium avium subsp. paratuberculosis. *PLoS ONE* **2016**, *11*, e0160353. [CrossRef]
144. Iheshiulor, O.O.M.; Esonu, B.O.; Chuwuka, O.K.; Omede, A.A.; Okoli, I.C.; Ogbuewu, I.P. Effects of Mycotoxins in Animal Nutrition: A Review. *Asian J. Anim. Sci.* **2011**, *5*, 19–33. [CrossRef]

toxins

MDPI

Article

Evaluation of Resistance to Fescue Toxicosis in Purebred Angus Cattle Utilizing Animal Performance and Cytokine Response

Daniel H. Poole [1,*], Kyle J. Mayberry [1,2], McKayla Newsome [1], Rebecca K. Poole [1,3], Justine M Galliou [1], Piush Khanal [1], Matthew H. Poore [1] and Nick V. L. Serão [1,4]

1 Department of Animal Science, North Carolina State University, Raleigh, NC 27695, USA; kmayberry@biltmore.com (K.J.M.); manewsom@ncsu.edu (M.N.); rkpoole@exchange.tamu.edu (R.K.P.); justine.galliou@wsu.edu (J.M.G.); pkhanal2@ncsu.edu (P.K.); mhpoore@ncsu.edu (M.H.P.); serao@iastate.edu (N.V.L.S.)

2 Director of Agriculture, The Biltmore Company, Asheville, NC 28803, USA

3 Department of Animal Science, Texas A&M University, College Station, TX 77843, USA

4 Department of Animal Science, Iowa State University, Ames, IA 50011, USA

* Correspondence: dhpoole@ncsu.edu; Tel.: +1-919-515-0033

Received: 15 October 2020; Accepted: 10 December 2020; Published: 14 December 2020

Abstract: Fescue toxicosis is a multifaceted syndrome common in cattle grazing endophyte-infected tall fescue; however, varying symptomatic responses potentially imply genetic tolerance to the syndrome. It was hypothesized that a subpopulation of animals within a herd would develop tolerance to ergot alkaloid toxicity. Therefore, the goals of this study were to develop selection criteria to identify tolerant and susceptible animals within a herd based on animal performance, and then examine responsive phenotypic and cytokine profiles to fescue toxicosis. Angus cows grazed endophyte-infected tall fescue at two locations for 13 weeks starting in mid-April 2016. Forage measurements were collected to evaluate ergot alkaloid exposure during the study. A post hoc analysis of animal performance was utilized to designate cattle into either tolerant or susceptible groups, and weekly physiological measurements and blood samples were collected to evaluate responses to chronic exposure to endophyte-infected tall fescue. Findings from this study support the proposed fescue toxicosis selection method formulated herein, could accurately distinguish between tolerant and susceptible animals based on the performance parameters in cattle chronically exposed to ergot alkaloids, and provides evidence to warrant additional analysis to examine the impact of ergot alkaloids on immune responsiveness in cattle experiencing fescue toxicosis.

Keywords: fescue toxicosis; genetic tolerance; cytokines; cow productivity

Key Contribution: Through accurate collection of phenotypic performance parameters in a population of cattle chronically exposed to ergot alkaloids, tolerant animals can be identified and selected to enhance productivity in an environment dominated by toxic tall fescue.

1. Introduction

Fescue toxicosis, resulting from consumption of ergot alkaloids commonly found in endophyte (*Epichloë coenophiala*)-infected tall fescue (*Lolium arundinaceum* Schreb. Darbysh), significantly impacts livestock health and production globally. Direct exposure to ergot alkaloids from tall fescue and other grass species occurs in production systems that heavily rely on grazing, such as in cow calf and stocker programs in the United States, New Zealand, and Australia; whereas importation of feedstuff has led to ergot alkaloid-induced effects in several Asian countries, including Japan and Korea [1–3]. Due to the agronomic benefits resulting from symbiosis with the endophyte, including stand persistence and drought tolerance, it is expected that endophyte-infected tall fescue will become the primary forage across more growing zones as the global climate warms. McCulley et al. [4] recently reported that warming air temperatures (+3 °C, day and night, year-round) significantly increased concentrations of ergovaline and total ergot alkaloids found in endophyte-infected tall fescue, which may exacerbate fescue toxicosis in future years. Complex plant–fungus–environment interactions exist due to variation in endophyte infection rate, alkaloid production (125 to 5000 μg/Kg), and the animal's voluntary intake [4], all of which can lead to high variation in individual animal responses. Individual variation in response to fescue toxicosis might be, in part, due to host genetics. Galliou et al. [5] tested the association between genotypes from a commercial genetic test for fescue toxicosis and performance in pregnant Angus cows. These authors found significant associations with growth, hair shedding, and calf weaning weight. Altogether, these pose a major challenge to understanding the underlying mechanisms of action responsible for the decreases in livestock productivity. While advancements have been made to better understand the mechanisms associated with fescue toxicosis symptoms, such as focusing on improving growth and reproductive performance of beef cattle grazing endophyte-infected tall fescue, a greater understanding is needed to identify markers for tolerance to fescue toxicosis, which is critical for the sustainability of the global beef industry.

Tall fescue, and more specifically the cultivar Kentucky 31, is the most abundant and economically important cool-season perennial grass in U.S. agriculture. Approximately 8.5 million head of beef cattle reside in southeast or central U.S., where fescue is the predominant forage available, most of which contains the endophyte that causes toxicosis in grazing livestock. While precise data on the rate and composition of ergot alkaloid intake in grazing cattle is limited, the symptoms of this multifaceted disease range from acute outbreaks of gangrenous ergotism to more subtle and chronic decreases in livestock productivity characterized by decreased feed intake and growth performance, compromised reproduction (pregnancy and calving rates), elevated body temperature, and reduced blood flow to the extremities, as reviewed by Strickland et al. [1]. While the gangrenous ergotism is rare and most likely due to acute exposure to very high concentrations of ergot alkaloids, combined with cold temperatures, the decreased livestock health, welfare, and productivity is expected as a result of chronic exposure to ergot alkaloids.

While there have been extensive efforts to develop an innovative and strategic pasture and grazing management system to convert toxic tall fescue to the non-toxic "Novel Endophyte Tall Fescue", many farmers have exercised some caution and hesitation to adopt the technology. This appears to be due to lack of understanding of how to renovate, expense, time, challenging terrain, and lack of feed resources to utilize while the new forage is being established. Essentially, the goal of converting all of the toxic fescue to novel endophyte tall fescue is likely impractical and highly unlikely. Therefore, there has been concurrent interest in identifying ways that producers can select animals based on their tolerance to the adverse effects of fescue toxicosis. As previously stated, animals suffering from fescue toxicosis exhibit decreased growth capability, poor reproductive performance, and have a lack of adequate heat tolerance at least in part due to their retained winter hair coat. Focusing on growth performance selection parameters and the animal's cytokine response to the toxin could be utilized to identify animals that are tolerant to the effects of fescue toxicosis. Therefore, the goals of this study were to develop selection criteria to identify tolerant and susceptible animals based on phenotypic animal performance traits, to evaluate how our criteria matches with the information from a commercial test for fescue toxicosis, and then examine responsive cytokine profiles to identify beef cattle displaying tolerance to fescue toxicosis.

2. Results

2.1. Changes in Forage Characteristics and Ergot Alkaloid Concentration

Forage fractions during period 1 (P-1) of the experiment were comprised of 72.5 and 72.9 percent fescue at the Butner Beef Cattle Field Lab (BBCFL) and Upper Piedmont Research Station (UPRS) locations, respectively. The percentage of fescue available to these animals during period 2 (P-2) decreased to 56.0 and 63.3 percent at BBCFL and UPRS, respectively (Table 1). Overall, animals that grazed pastures located at BBCFL were exposed to a lower percentage of fescue compared to UPRS. Additionally, tiller samples to determine the endophyte infection rate identified that the BBCFL pastures had a lower infection rate compared to the UPRS pastures throughout the course of the trial, 80.8 compared to 86.3 percent, respectively (Table 1). As a result of the differences in infection rate and the percent of tall fescue in the sward, ergovaline concentrations differed between these locations (Table 1). The highest concentrations of ergovaline were observed in P-1 for both locations (263.3 μg/kg for BBCFL and 436.7 μg/kg for UPRS) and decreased during P-2 (106.7 μg/kg for BBCFL and 196.7 μg/kg for UPRS) by the end of the trial (Table 1). During P-2 at UPRS, ergosine (333 μg/Kg), ergotamine (175 μg/Kg), ergocornine (53 μg/Kg), ergocriptine (150 μg/Kg), and ergocristine (48 μg/Kg) contributed to total ergot alkaloids. These alkaloids are characteristic of Claviceps purpurea [6], which can commonly infect tall fescue and other grasses. While the concentration of ergovaline decreased during P-2, these concentrations (along with the *C. purpurea* alkaloids at UPRS) were still high enough to elicit many of the negative symptoms associated with fescue toxicosis [7,8]. Lastly, the nutritive value and mineral composition of the tall fescue pastures utilized for this study are presented in Supplementary Table S1. Total digestible nutrients, crude protein (CP), neutral detergent fiber (NDF), and acid detergent fiber (ADF) were not different between the two locations and did not change over the course of the trial.

2.2. Distribution of T-Snip Genotypes Across FTSM Selected Animals

The expected proportion of T-Snip genotypes between the two FTSM groups can be see in Table 2. There is a tendency for a different distribution of T-Snip genotypes between EI-TOL and EI-SUS animals ($p = 0.089$). There was a statistical difference ($p < 0.05$) for the proportions of T-Snip genotypes 2 and 3 between EI-TOL and EI-SUS animals. EI-TOL animals had a lower proportion of genotype 2 than EI-SUS animals (0.1 + 0.07 versus 0.368 ± 0.11), whereas the opposite was found for T-Snip genotype 3 (0.6 ± 0.11 versus 0.316 ± 0.11). There were no differences in the proportion of T-Snip genotypes 1 and 4 between these animals ($p > 0.05$). These results indicated that EI-TOL animals had overall greater frequency of greater T-Snip genotypes than EI-SUS animals.

Table 1. Endophyte-infected tall fescue characteristics.

Forage Percentage	BBCFL [1]			UPRS [2]			Combined Locations		
	P-1 [3]	P-2 [4]	ES [5]	P-1 [3]	P-2 [4]	ES [5]	P-1 [3]	P-2 [4]	ES [5]
Percent fescue (%)	72.5	56.0	64.3	72.9	63.3	68.1	72.7	59.7	66.2
Tiller infection [6] (%)	84.2	77.5	80.8	82.5	90.0	86.3	83.3	83.8	83.5
Ergot alkaloid analysis [7]									
Ergosine (μg/kg)	– [8]	–	–	–	333.3 ± 322	166.7 ± 230	–	166.7 ± 230	83.3 ± 163
Ergotamine (μg/kg)	–	–	–	–	175.0 ± 202	87.5 ± 44	–	87.5 ± 142	43.8 ± 101
Ergocornine (μg/kg)	–	–	–	–	53.3 ± 61	26.7 ± 43	–	26.7 ± 43	13.3 ± 30
Ergocryptine (μg/kg)	–	–	–	–	150.0 ± 93	75.0 ± 66	–	75.0 ± 66	37.5 ± 47
Ergocristine (μg/kg)	–	–	–	–	48.3 ± 51	24.2 ± 36	–	24.2 ± 36	12.1 ± 25
Ergovaline (μg/kg)	263.3 ± 52	106.7 ± 41	185.0 ± 38	436.7 ± 83	196.7 ± 102	316.7 ± 64	350 ± 58	151.7 ± 55	250.8 ± 42
Total (μg/kg)	263.3 ± 52	106.7 ± 41	185.0 ± 38	436.7 ± 83	956.7 ± 43	696.7 ± 45	350.0 ± 58	531.7 ± 25	440.8 ± 37

[1] BBCFL: Butner Beef Cattle Field Laboratory location, Bahama, NC; [2] UPRS: Upper Piedmont Research Station location, Reidsville, NC; [3] P-1: Period 1, using values from weeks one, three, and five; [4] P-2: Period 2, using values from weeks 7, 9 and 11; [5] ES: Entire study, using values from weeks 1, 3, 5, 7, 9, and 11; [6] Phytoscreen© immunochemical endophyte test, Agrinostics Ltd. Co., Watkinsville, GA, USA; [7] HPLC analysis at the University of Missouri Veterinary Medical Diagnostic Laboratory (Columbia, MO;USA) [9]; [8] Denotes not detected via HPLC.

Table 2. Expected proportion of T-Snip genotypes [2] between tolerant (TOL) and susceptible (SUS) cows in the endophyte-infected (EI) fescue locations.

FTSM [1]	T-Snip Genotypes			
	1	2	3	4
TOL	0.150 [a] (0.08)	0.100 [a] (0.07)	0.600 [a] (0.11)	0.15 [a] (0.08)
SUS	0.263 [a] (0.10)	0.368 [b] (0.11)	0.316 [b] (0.11)	0.05 [a] (0.05)

[1] FTSM: Fescue toxicosis selection method group. There was an effect ($p = 0.089$) of FTSM on the distribution of T-Snip genotypes; [2] Values of T-Snip (AgBotanica, LCC, Columbia, MO, USA) genotypes range from 1 to 5, with expected greater tolerance to fescue toxicosis in animals with greater values; [a,b] Expected proportions within a T-Snip lacking the same superscript are statistically different from each other ($p < 0.05$).

2.3. Phenotypic Variables of Animal Performance

Animal performance is negatively impacted during ergot alkaloid exposure, thus, the focus of this study was to identify tolerant and susceptible animals within an exposed population; however, direct comparisons of performance parameters to animals not exposed to ergot alkaloids demonstrated the severity of exposure throughout the trial. Cattle grazing novel endophyte fescue pastures (EN, non-toxic) had greater body weights compared to animals grazing endophyte-infected (EI, toxic) fescue pastures ($p < 0.0001$; Figure 1). By comparison, animals deemed tolerant (EI-TOL) to ergot alkaloid exposure displayed a positive weight gain compared to animals deemed susceptible (EI-SUS) to ergot alkaloid exposure (Figure 1; Table 3; $p < 0.05$). Additionally, cattle grazing novel endophyte fescue pastures had a greater average daily gain (ADG; 0.703 kg/d) and body condition score (BCS; 6.1) compared to EI-TOL and EI-SUS animals (0.497 and −0.003 kg/d, respectively, for ADG and 5.5 and 5.6, respectively, for BCS; Table 3; $p < 0.0001$). Throughout the course of the study, EN and EI-TOL animals displayed similar hair coat scores (HCS) and were significantly less dense compared to EI-SUS animals (Table 3; $p < 0.0001$). Furthermore, EN animals were better able to shed their hair throughout the trial period compared to EI-TOL and EI-SUS animals (1.99 versus 2.36 and 2.64, respectively; Table 3; $p < 0.0001$). Although within the normal body temperature range for cattle, EN animals were cooler compared to EI-TOL and EI-SUS animals as indicated by rectal temperatures (Table 3; $p < 0.0001$).

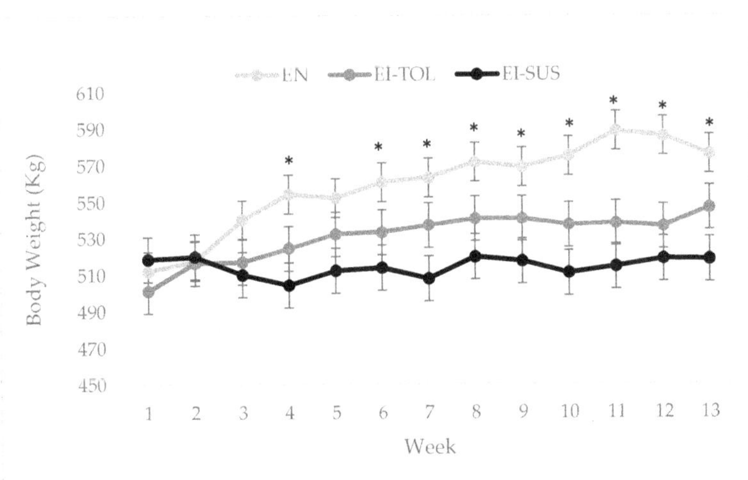

Figure 1. Body weight by treatment over time (weeks): Cows ($n = 27$) grazing on novel endophyte fescue pastures (EN, non-toxic) had greater body weights compared to grazing endophyte infected (EI, toxic) fescue pastures. Tolerant animals (EI-TOL) had greater average body weight compared to susceptible (EI-SUS) animals, determined by FTSM; ($n = 40$). The model was significant by location and treatment X location (trt, loc, trt X loc p-values: $p < 0.0001$).

Table 3. Performance parameters for Angus cows consuming either novel endophyte (EN) or endophyte-infected (EI) fescue from late April to late July 2016.

Item	Treatment [1,2]			SEM	p Value
	EN	EI-TOL	EI-SUS		
Initial BW, kg	512.1	501.4	518.6	10.6	0.5082
Final BW, kg	576.0 ^{a,*}	546.6 ^{a,b,*}	518.3 ^{b}	12.3	0.0004
ADG, kg/d	0.703 ^{a}	0.497 ^{b}	−0.003 ^{c}	0.05	<0.0001
BCS [3]	6.09 ^{a}	5.50 ^{b}	5.59 ^{b}	0.06	<0.0001
HCS [4]	1.89 ^{a,*}	1.98 ^{a,*}	2.34 ^{b}	0.04	<0.0001
HSS [5]	1.99 ^{a}	2.36 ^{b}	2.64 ^{c}	0.04	<0.0001
Rectal temp., °C	38.6 ^{a}	38.7 ^{b}	38.9 ^{c}	0.02	<0.0001

$^{a–c}$ Within the row, means without a common superscript significantly differ ($p \leq 0.05$); * $0.05 < p \leq 0.10$ determined a statistical tendency; [1] Values are reported as least square means for the experiment; [2] EN: cattle grazing novel endophyte fescue (nontoxic, n = 27); EI-TOL: cattle grazing endophyte-infected fescue deemed tolerant through FTSM (n = 20); EI-SUS cattle grazing endophyte-infected fescue deemed susceptible through FTSM (n = 20); [3] BCS = body condition score (1–9 scale); [4] HCS = hair coat score (1–5 scale); [5] HSS = hair shedding score (1–5 scale).

The average daily gain was significantly greater in EI-TOL animals during P-1 and over the entire study (ES) compared to EI-SUS animals (Table 4); however, no difference in ADG was observed during P-2 (Table 4), which could be reflective of the change in ergovaline concentrations observed in the fescue pastures between P-1 and P-2 of the trial. In addition to ADG, EI-TOL animals displayed a greater positive change in BCS during P-1, P-2, and throughout the entire study compared to the EI-SUS animals (Table 4; $p < 0.05$). Throughout the course of the study, EI-TOL animals displayed greater change in HCS and HSS in P-1 and P-2 compared to EI-SUS animals (Table 4; $p < 0.05$), with the greatest change occurring in P-1 as the temperature and humidity increased. Thus, EI-TOL animals were able to alter their hair coat density and length earlier in the year when ergot alkaloid concentrations were elevated (P-1), reducing the negative impact of fescue toxicosis on animal performance. Interestingly, no differences were observed when examining the change in HCS or HHS between EI-TOL and EI-SUS animals over the entire study (Table 4; $p > 0.05$). When examining differences in HCS and HSS between P-1 and P-2, there was an increase in hair shedding in the EI-SUS animals in P-2, which could be attributed to the decrease in ergot alkaloid concentration in the fescue pastures, in combination with the fact that elevated temperatures and humidity permitted these animals to display similar hair characteristics to that of the EI-TOL animals. However, no differences were observed in the change in body temperatures during P-1, P-2, or over the entire study (Table 4; $p > 0.05$).

Table 4. Performance parameters for Angus cattle grazing endophyte-infected fescue deemed either tolerant (EI-TOL) or susceptible (EI-SUS) through FTSM from late April to late July 2016.

Time	P-1 [1]				P-2 [2]				ES [3]			
Treatment [4,5]	EI-TOL	EI-SUS	SEM	p-Value	EI-TOL	EI-SUS	SEM	p Value	EI-TOL	EI-SUS	SEM	p-Value
ADG, kg/d	1.8 ^{a}	−0.55 ^{b}	0.13	<0.0001	0.5	0.5	0.2	0.8726	1.2 ^{a}	−0.025 ^{b}	0.1	<0.0001
ΔBCS [6]	0.488 ^{a}	0.013 ^{b}	0.09	0.0008	0.613 *	0.363 *	0.09	0.0625	1.05 ^{a}	0.375 ^{b}	0.10	<0.0001
ΔHCS [7]	−2.013 ^{a}	−0.913 ^{b}	0.174	<0.0001	−0.300 ^{a}	−1.150 ^{b}	0.207	0.0062	−2.313	−2.063	0.226	0.4398
ΔHSS [8]	−1.925 ^{a}	−0.613 ^{b}	0.198	<0.0001	−0.763 ^{a}	−1.838 ^{b}	0.202	0.0006	−2.688	−2.450	0.248	0.5027
ΔRT [9], °C	−0.250	0.105	0.176	0.1630	0.505	0.230	0.126	0.1299	0.250	0.325	0.179	0.7680

^{a,b} Within the row, means without a common superscript significantly differ ($p \leq 0.05$); * p-values $0.05 < p \leq 0.10$ determined a statistical tendency; [1] P-1: Period 1, using data collected from weeks one, three, and five; [2] P-2: Period 2, using data collected from weeks 7, 9, and 11; [3] ES: Entire study, using data collected from weeks 1, 3, 5, 7, 9, and 11; [4] Values are reported as least square means for the experiment; [5] EI-TOL: cattle grazing endophyte-infected fescue deemed tolerant through FTSM (n = 20); EI-SUS cattle grazing endophyte-infected fescue deemed susceptible through FTSM (n = 20); [6] ΔBCS = change in body condition score (1–9 scale) over the given time frame; [7] ΔHCS = change in hair coat score (1–5 scale) over the given time frame; [8] ΔHSS = change in hair shedding score (1–5 scale) over the given time frame; [9] ΔRT = change in rectal temperature (°C) over the given time frame.

2.4. Cytokine Response and Hormone Profiles in FTSM Selected Animals

While performance parameters can be directly selected, changes within the peripheral concentrations of growth factors (FGF1, FGF2, IGF1, and VEGFA); chemokines (CCL2, CCL4, CXCL9, CXCL10); cytokine receptor antagonists (IL36RN, GPRASP-1, NCAM1); anti-inflammatory (IL4 and IL13); and pro-inflammatory (IFNAR1, IFNG, IL1A, IL15, IL2, IL21 and TNF) cytokines, as well as reproductive hormones (prolactin and progesterone), in response to ergot alkaloid exposure would potentially lead to a novel marker for identifying animals resistant to fescue toxicosis.

Of the growth factors, vascular endothelial growth factor A (VEGFA) concentrations were greater in EI-TOL animals compared to EI-SUS animals (6.3 ± 1.1 versus 3.2 ± 1.1 ng/mL, respectively; Figure 2B; $p = 0.0444$). No differences were observed in VEGFA concentrations between locations or over the course of the trial (Table 5; $p > 0.05$). No differences were observed in FGF1, FGF2, or IGF1 concentrations between EI-TOL and EI-SUS animals over the course of the trial (Table 5; $p > 0.05$). Fibroblast growth factor 1, but not FGF2 or IGF1, concentrations tended to be greater in animals at UPRS compared animals at BBCFL (328.6 ± 81.6 versus 62.4 ± 81.6 ng/mL, respectively; Table 5; $p = 0.0793$).

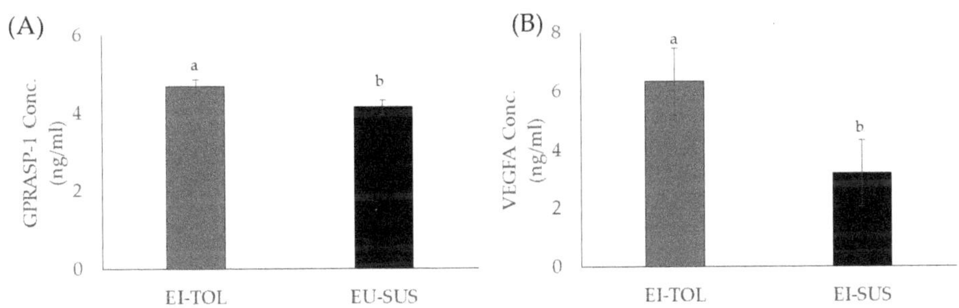

Figure 2. EI-TOL animals (n = 20) had greater concentrations of GPRASP-1 (**A**) and VEGFA (**B**) in peripheral blood compared to EI-SUS animals (n = 20) when chronically exposed to endophyte-infected tall fescue ($p < 0.05$).

Table 5. Growth factors, chemokines, and cytokine receptor antagonists.

Treatment [1,2]	EI-TOL						EI-SUS									
Location	BBCFL [3]			UPRS [4]			BBCFL [3]			UPRS [4]				p-Values		
Week	1	7	13	1	7	13	1	7	13	1	7	13	SEM	TRT	LOC	TIME
Growth factors [5]																
FGF1	196.5 *	110.3 *	77.3 *	233.8 *	145.2 *	120.0 *	98.0 *	69.3 *	52.3 *	203.8 *	163.0 *	99.8 *	60.4	0.3381	0.0793	0.0716
FGF2	9.4	28.1	46.8	27.9	16.6	16.1	34.9	26.6	19.9	25.0	1937	7.3	14.8	0.8324	0.3327	0.9849
IGF1	6.8	21.3	10.4	21.2	4.3	6.5	18.7	14.1	3.9	10.2	8.8	9.3	4.2	0.7330	0.3602	0.1441
VEGFA	572.64	429.9	441.0	1117.2	512.4	711.2	208.7	152.9	220.1	253.64	769.8	297.2	114.2	0.0444	0.1070	0.8319
Chemokines [5]																
CCL2	856.4	2817.6	269.7	2012.2	1946.2	1979.8	269.2	2124.8	277.7	1022.8	1622.4	1374.8	898.7	0.3210	0.2984	0.1452
CCL4	8.7	11.9	141.3	16.8 [a]	616.7 [b]	754.9 [b]	9.8	10.8	191.8	7.6 [a]	171.1 [a,b]	404.4 [b]	192.7	0.2831	0.0245	0.0428
CXCL9	35.7	41.6	73.2	41.8	250.9	220.1	98.2	42.2	62.8	75.9	56.3	150.1	70.2	0.4644	0.0708	0.4237
CXCL10	34.7	44.3	68.1	45.1	91.0	104.6	42.2	34.9	57.5	80.8	54.2	67.1	27.7	0.5909	0.0929	0.4521
Cytokine receptors antagonists [5]																
IL36RN	69.2	94.2	167.4	109.1	562.9	512.9	67.8	73.8	110.0	205.8	252.2	357.6	136.7	0.3590	0.0044	0.1922
GPRASP-1	4.1 [a]	6.1 [b]	4.2 [a]	4.2	4.7	4.7	3.8 [a]	5.2 [b]	3.8 [a]	4.2	4.0	3.8	0.4	0.0343	0.3268	0.0033
NCAM1	33.8	33.7	30.7	33.1	32.3	42.4	29.9	31.9	26.9	30.6	35.0	39.4	3.8	0.3428	0.0496	0.5422

[a,b] Within the row, means without a common superscript significantly differ ($p \leq 0.05$); * p-values $0.05 < p \leq 0.10$ determined a statistical tendency; [1] Values are reported as least square means for the experiment; [2] EI-TOL: cattle grazing endophyte-infected fescue deemed tolerant through FTSM ($n = 20$); EI-SUS cattle grazing endophyte-infected fescue deemed susceptible through FTSM ($n = 20$); [3] BBCFL: Butner Beef Cattle Field Laboratory location, Bahama, NC; [4] UPRS: Upper Piedmont Research Station location, Reidsville, NC; [5] Growth factors, chemokines, and cytokine receptors antagonist concentrations reported as ng/mL.

No differences were observed in chemokines, CCL2, CXCL9, or CXCL10 concentrations between EI-TOL and EI-SUS animals over the course of the trial (Table 5; $p > 0.05$). Macrophage inflammatory protein-1β (CCL4) concentrations were not difference between EI-TOL and EI-SUS animals ($p > 0.05$), however, CCL4 concentrations were greater in animals at UPRS compared to animals at BBCFL (161.0 ± 24.0 versus 100.7 ± 24.0 ng/mL, respectively; Table 5; $p = 0.0793$) and increased over the course of the trial (10.7 ± 98.9, 202.6 ± 98.9, and 373.1 ± 98.9, for weeks 1, 7, and 13, respectively; Table 5; $p = 0.0428$). Furthermore, CXCL9 and CXCL10 concentrations tended to be greater in animals at UPRS compared animals at BBCFL (132.5 ± 28.5 versus 58.9 ± 28.5 ng/mL, respectively, for CXCL9 and 73.8 ± 11.0 versus 46.9 ± 11.0 ng/mL, respectively, for CXCL10; Table 5).

G Protein-Coupled Receptor Associated Sorting Protein 1 (GPRASP-1) concentrations were greater in EI-TOL animals compared to EI-SUS animals (4.7 ± 0.2 versus 4.1 ± 0.2 ng/mL, respectively; Figure 2A; $p = 0.0343$). No differences were observed in GPRASP-1 concentrations between locations (Table 5; $p > 0.05$), however, GPRASP-1 concentrations were greatest at the end of P-1, compared to the start and end of the trial (5.0 ± 0.2 versus 4.1 ± 0.2 for weeks 1 and 13, Table 5; $p = 0.0033$). No differences were observed in IL36RN or NCAM1 concentrations between EI-TOL and EI-SUS animals over the course of the trial (Table 5; $p > 0.05$). However, IL36RN and NCAM1 concentrations were greater in animals at UPRS compared animals at BBCFL (333.4 ± 57.4 versus 97.1 ± 57.4 ng/mL, respectively, for IL36RN and 35.5 ± 1.5 versus 31.2 ± 1.5 ng/mL, respectively, for NCAM1; Table 5).

No differences were observed in anti-inflammatory cytokine, IL4, or IL13 concentrations between EI-TOL and EI-SUS animals over the course of the trial (Table 6; $p > 0.05$). However, IL4 and IL13 concentrations tended to be greater in animals at UPRS compared animals at BBCFL (218.8 ± 56.2 versus 83.3 ± 56.2 ng/mL, respectively, for IL4 and 12.5 ± 2.2 versus 6.9 ± 2.2 ng/mL, respectively, for IL13; Table 6). No differences were observed in pro-inflammatory cytokine, IFNG, IL1A, IL15, and IL2 concentrations between EI-TOL and EI-SUS animals between locations or over the course of the trial (Table 6; $p > 0.05$). While no differences were observed in IFNAR1 concentrations between EI-TOL and EI-SUS animals or between locations, IFNAR1 concentrations increased over the course of the trial (41.2 ± 62.4, 165.4 ± 62.4, and 411.9 ± 62.4, for weeks 1, 7, and 13, respectively; Table 6; $p = 0.0001$). Additionally, no differences were observed in IL21 or TNF concentrations between EI-TOL and EI-SUS animals over the course of the trial (Table 6; $p > 0.05$), but concentrations of these cytokines tended to be greater in animals at UPRS compared animals at BBCFL (135.3 ± 24.6 versus 73.7 ± 24.6 ng/mL, respectively, for IL21and 149.4 ± 29.6 versus 67.2 ± 29.6 ng/mL, respectively, for TNF; Table 6).

Table 6. Anti-inflammatory and pro-inflammatory cytokines and reproductive hormones.

Treatment [1,2]	EI-TOL						EI-SUS									
Location	BBCFL [3]			UPRS [4]			BBCFL [3]			UPRS [4]				p-Values		
Week	1	7	13	1	7	13	1	7	13	1	7	13	SEM	TRT	LOC	TIME
							Anti-inflammatory [5]									
IL4	89.7	256.8	30.1	286.1	215.8	407.7	48.9	47.8	26.4	198.2	110.1	95.2	133.9	0.1141	0.0909	0.9801
IL13	6.7	7.4	5.0	13.1	9.5	10.1	9.5	10.7	2.1	26.1	10.4	6.0	5.1	0.4710	0.0672	0.1021
							Pro-inflammatory [5]									
IFNAR1	20.0 ^{a}	44.8 ^{a}	460.2 ^{b}	71.5 ^{a}	304.5 ^{b}	518.2 ^{b}	44.4 ^{a}	52.4 ^{a}	362.7 ^{b}	29.2 ^{a}	259.9 ^{b}	306.5 ^{b}	49.5	0.3921	0.2361	0.0001
IFNG	1.7	1.8	1.8	2.5	3.2	4.5	3.9	2.0	1.2	4.3	3.5	2.9	1.5	0.6638	0.1083	0.8552
IL1A	72.0	53.3	209.0	99.2	273.7	380.6	146.0	65.6	116.6	162.4	159.3	266.7	85.1	0.5768	0.0287	0.1060
IL15	13.1	41.5	4.9	25.8	22.3	32.6	9.7	14.7	6.1	12.6	30.1	11.5	12.7	0.1847	0.2822	0.2296
IL2	107.6	218.8	255.0	100.4	191.0	87.8	216.0	110.1	143.8	30.6	43.1	60.0	93.8	0.2878	0.1113	0.9128
IL21	60.9	54.0	92.4	89.2	177.5	178.2	118.1	68.3	48.7	167.7	66.4	132.6	58.6	0.8095	0.0795	0.8667
TNF	43.0	52.0	77.3	86.3	175.5	263.5	113.9	64.2	53.1	176.4	64.2	130.4	70.5	0.7046	0.0523	0.7093
							Reproductive hormones [5]									
Prolactin	55.0 ^{a}	102.4 ^{b}	74.5 ^{a,b}	51.0 ^{a}	166.0 ^{b}	140.7 ^{a,b}	89.5 ^{a}	167.5 ^{b}	105.7 ^{a,b}	60.6 ^{a}	217.6 ^{b}	189.4 ^{a,b}	38.8	0.0787	0.0918	0.0019
Progesterone	5.0 ^{a}	2.3 ^{b}	2.9 ^{b}	2.9	1.8	3.0	3.8 ^{a}	2.5 ^{b}	3.1 ^{a,b}	2.4	2.3	2.4	0.5	0.0895	0.0018	0.0003

[a,b] Within the row, means without a common superscript significantly differ ($p \leq 0.05$); * p-values $0.05 < p \leq 0.10$ determined a statistical tendency; [1] Values are reported as least square means for the experiment; [2] EI-TOL: cattle grazing endophyte-infected fescue deemed tolerant through FTSM ($n = 20$); EI-SUS cattle grazing endophyte-infected fescue deemed susceptible through FTSM ($n = 20$); [3] BBCFL: Butner Beef Cattle Field Laboratory location, Bahama, NC; [4] UPRS: Upper Piedmont Research Station location, Reidsville, NC; [5] Anti-inflammatory and pro-inflammatory cytokines and reproductive hormone concentrations reported as ng/mL.

Of the reproductive hormones examined, prolactin (PRL) concentrations tended to be greater in EI-SUS animals compared to EI-TOL animals (138.4 ± 16.0 versus 98.3 ± 16.0 ng/mL, respectively; Table 6; $p = 0.0787$). Additionally, PRL concentrations tended to be greater in animals at UPRS compared animals at BBCFL (137.6 ± 15.8 versus 99.1 ± 15.8 ng/mL, respectively; Table 6), and was significantly less at the start of the trial compared to weeks 7 and 13 of the trial (64.0 ± 19.4 versus 163.4 ± 19.4 and 127.6 ± 19.4 for weeks 7 and 13, respectively; Table 6; $p = 0.0019$); progesterone (P4) concentrations tended to be greater in EI-TOL animals compared to EI-SUS animals (3.3 ± 0.14 versus 3.0 ± 0.14 ng/mL, respectively; Table 6; $p = 0.0895$). Additionally, P4 concentrations were greater in animals at BBCFL compared animals at UPRS (3.4 ± 1.4 versus 2.8 ± 1.4 ng/mL, respectively; Table 6) and was significantly less at week seven of the trial compared to weeks 1 and 13 of the trial (2.2 ± 0.3 versus 3.5 ± 0.3 and 2.8 ± 0.2 for weeks 1 and 13, respectively; Table 6; $p = 0.0003$).

3. Discussion

Due to the extensive use of endophyte-infected tall fescue across the globe, the goals of this study were to develop selection criteria to identify tolerant and susceptible animals based on phenotypic animal performance traits, and then examine responsive cytokine profiles to identify beef cattle displaying tolerance to fescue toxicosis. Several studies described breed differences in response to ergot alkaloids, in which breeds such as Senepol and Brahman, which are better able to handle the heat and humidity, outperformed British breeds such as Angus and Hereford [8,10–12] when exposed to toxic tall fescue. These breeds have been used in crossbreeding programs to address fescue toxicosis, but it is unclear if their apparent tolerance is to fescue or adaptation to elevated heat and humidity. However, genetic progress is obtained within the breed, and the identification of genetic variation in animals from the same breed should be exploited to allow for selection for improved response (e.g., tolerance) to fescue toxicosis. Gray et al. [13] examined Angus cattle performance in North Carolina and Mississippi on endophyte-infected fescue pasture. These researchers observed genetic variation for hair coat shedding, indicating that Angus cattle that shed their winter hair coat earlier in the year may be more heat tolerant.

At the genomic level, genes regulating prolactin production have been targeted to identify genetic markers for tolerance to fescue toxicosis. Campbell et al. [14] recently identified a single nucleotide polymorphism (SNP) within the dopamine receptor D2 gene that was associated with variation in calving rates when grazing endophyte-infected tall fescue. In addition, using part of the same data in this study, Galliou et al. [5] showed that a commercial genetic test for fescue toxicosis, T-Snip (AgBotanica, LCC, Columbia, MO, USA), is associated with growth, hair shedding, and calf weaning weight in pregnant Angus cows. With the available T-Snip data, we tested whether the distribution of T-Snip genotypes changed between EI-TOL and EI-SUS animals in our study. According to the company's instructions, T-Snip genotypes range from 1 to 5, with levels of tolerance to fescue toxicosis increasing with the value of the genotype. Although both groups of animals showed T-Snip genotypes 1 to 4, there was a greater proportion of genotype 3 in EI-TOL than in EI-SUS. Similarly, there was a greater proportion of genotype 2 in EI-SUS than in EI-TOL. Therefore, the expected genetic values based on this commercial genetic test was greater in EI-TOL than in EI-SUS. Hence, although the selection of animals in this study was fully based on their phenotypic performance, results from T-Snip genotypes further support that this selection method was able to identify animals with contrasting tolerance to fescue toxicosis. It is important to note that this commercial genetic test does not have a perfect accuracy to predict with phenotypic performance in animals during fescue toxicosis [5]. Therefore, the presence of animals with lower and greater genotype values in EI-TOL and EI-SUS groups, respectively, was expected. Considering these factors, developing a selection method based on the phenotypic performance parameter in cattle from a single breed source in controlled experiments would aid in identifying animals that are potentially tolerant or susceptible to ergot alkaloids.

A major challenge with studying this multifaceted syndrome is that endophyte production of ergot alkaloids varies with season and plant maturity, and contributes to variation in ergot alkaloid intake by grazing cattle. Total ergot alkaloids or ergovaline, the most abundant alkaloid, has been measured in fescue to describe its potential toxicity. Concentrations of ergovaline increase from 250 to 500 μg/Kg in leaf blades and from 500 to 1300 μg/Kg in leaf sheaths from April to May. Seed heads contain the greatest concentration of toxins, and reach concentrations as high as 5000 μg/Kg in June. Ergovaline concentrations decline in August and increase again during fall regrowth [9]. Total ergot alkaloid concentration displays the same seasonal changes as ergovaline [15]. The concentration of total ergot alkaloids and ergovaline declines by 81% to 85%, respectively, from December to March in stockpiled fescue (accumulated during the growing season for grazing during dormancy; [16,17]). There were differences in the percentage of fescue and endophyte infection rate between locations used in this study, which then led to differences in ergovaline concentrations between locations, as well as variation in ergovaline concentration throughout the trial. The increase in ergot alkaloid consumption during period 1 resulted in greater differences in the phenotypic performance parameter, including average daily gain (ADG), body condition scores (BCS), body weight (BW), and rectal temperatures that were evaluated in this study.

The difference in ADG and BCS in response to increased consumption of ergot alkaloids is similar to data reported Thompson et al. [18], which identified a direct negative correlation to an animal's ADG when the pasture infection rate increased. Moreover, comparison of phenotypic performance parameters of animals exposed to ergot alkaloids (EI) to animals not exposed to ergot alkaloids (EN) demonstrated the severity of exposure throughout the trial. Additionally, sufficient body condition has been shown to be one of the most important traits to cattle reproductive efficiency in a cow–calf herd, and decreases as a result of endophyte-infected tall fescue consumption [19].

Aside from animal growth performance, issues with thermotolerance have also frequently been seen in animals suffering from severe cases of fescue toxicosis [20,21]. Many of the symptoms of fescue toxicosis (ergot alkaloid exposure) are amplified during a period of heat stress, and the greatest loss in animal performance and consequently increased production losses occur during the summer months when grazing endophyte infected tall fescue [22]. It has been speculated that heat tolerant (i.e., Bos indicus) breeds of cattle have improved resistance to fescue toxicosis [10,23]. Several studies, including Poole et al. [8], have demonstrated that cattle with adaptions to manage heat stress perform better when exposed to ergot alkaloids and increased environmental temperatures. However, there is conflicting data that report no difference in growth rate [24], hormone concentration [23], or milk production [11] between Bos indicus and Bos taurus cattle grazing endophyte-infected tall fescue. Therefore, it remains unknown if the traits for heat tolerance and tolerance to ergot alkaloids are synonymous.

Correlations have been previously made in which cattle that shed their winter hair coats earlier in the fescue season suffer less from the negative effects of fescue toxicosis [13]. In this study, weekly hair shedding scores (HSS) and hair coat scores (HCS) were taken to evaluate differences based on tolerance designation week to week, as well as initial rate of shedding. Both measures were significantly different between tolerant and susceptible animals grazing the endophyte-infected tall fescue. Both HCS and HSS were similar at the start of the study, and deviation between EI-TOL and EI-SUS animal occurred at week three of the study and continued through week seven (P-1). Interestingly, HSS and HCS re-converged to similar numeric values when ergot alkaloid concentrations decreased, yet ambient temperatures and humidity remained elevated consequentially when prolactin concentrations increased. Decreases in serum prolactin have been associated with fescue toxicity, and are often used as an indicator of ergot alkaloid exposure in cattle [25]. However, increased prolactin concentrations are associated with increasing day lengths and ambient temperatures, and initiate hair shedding [26,27]. Prolactin secretion has been linked with changes in environmental temperature, such that prolactin concentrations are elevated during warmer verses cooler months [28,29], and has been shown to regulate hair shedding [30,31]. It has been speculated that fescue toxicosis-induced hypoprolactinemia prevents shedding of the winter hair coat, therefore, cattle have an elevated body temperature and increased vulnerability to heat stress [32]. In the current study, EI-TOL animals had a tendency for lower serum prolactin concentrations when compared to EI-SUS animals as determined by the FTSM, thus, it can be assumed that those animals deemed tolerant had sufficient concentrations of serum prolactin to initiate hair shedding. Based on the other physiological measurements, the cattle in this study displayed many other symptoms of fescue toxicosis, and the lack of hypoprolactinemia was unexpected. The interaction of increased THI and exposure to ergot alkaloids impacts the severity of fescue toxicosis (reviewed by one), and the relatively low inclusion rate of the ergot alkaloids in the infected pastures may have altered serum prolactin concentrations. With EI-TOL animals having lower average serum prolactin concentrations than EI-SUS animals, it is hypothesized that these cattle were genetically predisposed to eliciting a shedding response at lower thresholds of prolactin. Using the same animals from this study, Koester et al. [33] showed that EI-TOL and EI-SUS animals have distinct fecal bacterial and fungal communities, further supporting that the proposed FTSM also results in differences in the microbiome of animals under FT stress. This classification using FTSM is further supported by the differences in T-Snip genotypes between EI-TOL and EI-SUS presented in the current study. These hypotheses build from the speculation of Aiken et al. [32] that genetic predispositions exist for necessary prolactin levels to initiate shedding.

More recent studies in various livestock species have reported that immune parameters have moderate to high heritability, as well as high genetic correlation with reproductive performance (lowly heritable traits), indicating that immune parameters can be used as a genetic trait for selecting animals that are resistant to specific diseases [34]. Using the protein array approach, numerous growth factors, chemokines, cytokine receptor antagonists, and anti-inflammatory and pro-inflammatory cytokines were examined during ergot alkaloid exposure. One of the major outcomes of this analysis is that many of the factors examined were significantly different or showed a statistical tendency to be higher at UPRS compared to BBCFL (Tables 5 and 6). While cytokine differences between locations was not the primary focus of this study, these data demonstrate the immune system's responsiveness to varying ergot alkaloid exposure, and warrants further investigation with known concentrations of ergovaline to better understand this interaction. Based on these data, it is hypothesized that cattle at the UPRS location experienced a greater immune response in response to higher ergot alkaloid exposure. A similar report described a hyperactive innate immune response, which may lead to an immuno-compromised animal in stocker steers when chronically exposed to ergovaline [35].

In the current study, vascular endothelial growth factor A (VEGFA) and G Protein-Coupled Receptor Associated Sorting Protein 1 (GPRASP-1) concentrations were greater in EI-TOL animals compared to EI-SUS animals. G Protein-Coupled Receptor Associated Sorting Protein 1 has been associated with downregulation of a variety of G Protein-Coupled receptors, including the D2-dopamine receptor, through lysosomal degradation. The D2-dopamine receptor has been shown to play a direct effect in cattle prolactin secretion, and is also involved in ergot alkaloids exposure's decrease in prolactin secretion [36,37]. Prolactin is decreased when dopamine binds to the D2-dopamine receptor, and the cyclic ring structure of ergovaline closely mimics the ring structure of dopamine, allowing ergovaline to bind the D2-dopamine receptor and thus inhibit prolactin secretion [38,39]. In the current study, EI-TOL had greater GPRASP-1 concentrations, potentially causing the D2-dopamine receptor to be downregulated at a greater rate. This decrease in the prevalence of the D2-dopamine receptor decreased the opportunity for ergovaline to bind and decrease prolactin secretion, giving EI-TOL animals an advantage when grazing endophyte infected tall fescue. Furthermore, this could contribute to EI-TOL animals having greater sensitivity to prolactin through downregulation of the entire prolactin secretion pathway, resulting in lower biological thresholds needed to initiate the biological roles of the hormone, which was previously discussed.

Vascular endothelial growth factor A is a known vasodilator, and has been shown to increase micro-vascular permeability. In addition, it is known to play a particularly important role in vascular endothelial cells, where higher concentrations increase vascular permeability [40] and could provide protective effects against ergot alkaloid exposure. Vasoconstriction caused by exposure to ergot alkaloids further compounds the issues associated with fescue toxicosis, such as its ability to decrease the animal's capacity for evaporative cooling [41]. Increased VEGFA concentrations and lower rectal temperature observed in the EI-TOL animals provide evidence that the EI-TOL animals selected using the FTSM have a greater ability to avoid heat stress and the vasoconstrictions effects of fescue toxicosis compared to EI-SUS animals. Several studies have examined changes in VEGFA to heat stress, and have focused on tissue and/or cellular responsiveness as opposed to whole animal changes. Vascular endothelial growth factor concentrations in the blood of heat tolerant (Bos Indicus) cattle compared to cattle that are more sensitive to heat stress (Bos Taurus) has not been directly investigated; however, Jyotiranjan et al. [42] and Iqbal et al. [43] reported increases in VEGF gene expression in response to thermal stress in goats and Bos indicus cattle breeds, respectively. Interestingly, of the eight Bos indicus breeds examined, only two breeds (Bhagnari and Lohani) displayed increased VEGF expression, but the authors indicate that this increase in VEGF is associated with adaptation to high altitude as opposed to heat tolerances [43]. Additionally, Jones et al. [44] and Aiken et al. [45] have speculated that decreases in serum progesterone concentrations could be caused by vasoconstriction of blood flow to the ovary in cattle consuming ergot alkaloids. Poole and colleagues [7] confirmed that ergovaline exposure reduced the diameter of the ovarian artery, leading to the functional corpus luteum, thus decreasing circulating progesterone concentrations. In the current study, EI-TOL animals tended to have higher progesterone concentrations compared to EI-SUS animals, which may be linked to greater ovarian artery diameter due to greater concentrations of circulating VEGFA in these animals. Thus, the EI-TOL animal's ability to synthesize higher concentrations of VEGFA may provide the capacity to increase vessel diameter, which would result in substantial mitigation of the vasoconstriction seen with fescue toxicosis. Ultimately, further investigation is needed to evaluate these cytokines and their role in fescue toxicosis, as well as the tolerant animal's ability to mount a stronger immune response to mitigate some of the negative effects of fescue toxicosis.

Taken together, the selection strategy used in this study (i.e., FTSM) was showed to be effective in identifying groups of individuals expressing contrasting responses to ergot alkaloid exposure, i.e., fescue toxicosis [33]. Nonetheless, three items must be further discussed: first, the selection of these animals was based solely on their (adjusted) phenotypic performance. Although the T-Snip data presented in this study support that TOL animals have greater genetic tolerance to ergot alkaloid exposure than SUS animals, the accuracy of identifying individual differences from ergot alkaloid exposure using this test was not 100% [5]. Hence, there is still genetic variation in the animal's response to ergot alkaloid exposure that could be explored. One way of getting ahold of individual genetic variation is through using, for example, the expected progeny differences (EPDs) for the growth rate of these animals as part of the selection criteria. However, there are some limitations on using the current EPDs for this purpose. These EPDs are based on a nationwide genetic evaluation (American Angus Association; http://www.angus.org/), which does not take into consideration the presence of GxE, which seems to be the case for ergot alkaloid exposure and for other stress-related traits [46]. Thus, the use of EPDs for this purpose may bias the selection of animals with greater tolerance to ergot alkaloid exposure. Although no source of genetic information was used in our selection process, it was expected that our approach captured part of the genetic potential of these animals. Assuming a) unbiased estimates for the fixed effects and b) independence between fixed and random effects included in the statistical model used for selection of TOL and SUS animals (see details in Koester at al. [33]), the estimated residuals that were used for classification of TOL and SUS animals should include both genotypic effects (additive and non-additive) and the true residual effects. Thus, assuming at least a moderate narrow-sense heritability for growth rate under ergot alkaloid exposure, it is expected that the estimated residuals include a substantial contribution of additive genetic values. Hence, although the selection process used in this study did not include prior genetic information from these animals, the data presented in this study demonstrate that TOL and SUS animals differed, at some level, in their genetic potential when exposed to ergot alkaloids.

Additionally, another consideration is the potential confounding between the growth rate in the presence or absence of ergot alkaloids. In this study, it was assumed that animals expressed their growth as a function of response to ergot alkaloid exposure. In fact, the difference in results obtained between the two locations with endophyte-infected (toxic) fescue suggests that some minimum concentrations of ergot alkaloids are needed for animals to express a tolerant-related phenotype. In order to obtain accurate information on the relationship between the growth rate in the presence or absence of ergot alkaloids at the genetic level, a much larger sample size, with animals from a substantial number of sires represented across locations with varying concentrations of ergot alkaloids, would be needed to estimate the genetic correlation between the growth rate in the presence or absence of ergot alkaloids. Albeit needed, such a scenario would be difficult to achieve in a timely manner. At the phenotypic level, the use of biomarkers for ergot alkaloid exposure, such as prolactin [25] or those identified in our study (e.g., VEGFA), could provide additional information on the level of ergot alkaloid exposure on animals. This information could be used to evaluate whether a significant physiological response to ergot alkaloid exposure was observed in each animal, potentially allowing for better separation between animals showing a growth rate in the presence or absence of ergot alkaloids.

Lastly, our method was unidimensional, using only growth rate data for the selection of TOL and SUS animals. Hence, the evaluation of a fescue toxicosis selection index using multiple parameters and including other sources of information could be beneficial for more accurate selection of animals for response to ergot alkaloid exposure. Additional sources of information include the use of T-Snip, the biomarkers identified in this population (this study and Koester et al. [33]), genotypes at the dopamine receptor D2 gene locus [14], and more. Opportunities exist for the identification of additional biomarkers based on response-related traits other than growth rate, as used in this study. There is also a need for large-scale genomic studies for the identification of SNPs associated with variation in ergot alkaloid exposure related traits, which could be used to better identify animals with different genetic potential for fescue toxicosis.

4. Conclusions

Taken together, these data provide support to validate the fescue toxicosis selection method (FTSM) proposed in this study through accurate collection of phenotypic performance parameters in a population of cattle chronically exposed to ergot alkaloids, and provide evidence to warrant additional analysis on the impact of ergot alkaloids on immune responsiveness in cattle that experience fescue toxicosis. Given the increased performance and hair shedding ability, greater hormone regulation and efficiency, and stronger cytokine responses, EI-TOL animals have clear mechanisms available to provide these advantages that can and should be selected for in animals raised in a fescue-dominant environment.

5. Materials and Methods

5.1. Animal Usage

This study was conducted at two locations in the piedmont of North Carolina, which were the Butner Beef Cattle Field Laboratory (BBCFL) in Bahama, NC and the Upper Piedmont Research Station (UPRS) in Reidsville, NC. All animal procedures were approved by the North Carolina State University Institutional Animal Care and Use Committee (NCSU IACUC #13-093-A; #17-043-A). Approval for animal use from the University Institutional Animal Care and Use Committee was on 19 August 2013 (#13-093-A) and was renewed and approved on 23 March 2017 (#17-043-A).

Animal Management

Cow performance and forage data were collected from late April to late July 2016 (Figure 3) when ergot alkaloid concentrations are greatest. Purebred Angus cows (n = 148) that ranged from two to four years of age all grazed endophyte-infected tall fescue during the entire experimental period. Cattle selected for this study were confirmed pregnant at 30 days post artificial insemination via ultrasonography, and were approximately 85 days post-conception at the start of the study.

Additionally, cows selected for this study were weaned two weeks prior to the start of the study to remove the effect of lactation on animal performance. Cattle were offered ad libitum water and free choice minerals throughout the duration of the study, in addition to natural shade structures within the pastures. Additionally, a subset of cows (n = 27) were maintained under the same conditions on novel endophyte fescue pastures (non-toxic, EN) at the BBCFL to serve as a representative control group. Body weight (BW), body condition scores (BCS, as adapted from Richards et al., [47]), hair shedding score (HSS, as adapted from [13]), hair coat score (HCS, adapted from [48]), rectal temperatures, and jugular blood samples were collected weekly to evaluate the animal's physiological response to ergot alkaloid exposure. Objective scores of BCS, HSS, and HCS were all collected by two trained evaluators and composited for an average score for each animal. Blood samples were collected via jugular venipuncture using 20-gauge needles and sterile 10.0 mL vacutainer tubes that contained no additive (Becton Dickerson, Franklin Lakes, NJ, USA). All blood samples were immediately placed on ice and then transported to the laboratory for processing. Whole blood samples were centrifuged the afternoon following collection each week for 25 min at 1500× g at 4 °C, then serum was drawn from vacutainer tubes and aliquoted into a glass dram vial and a plastic micro-centrifuge tube and stored at −80 °C until progesterone, prolactin, and cytokine analyses were conducted.

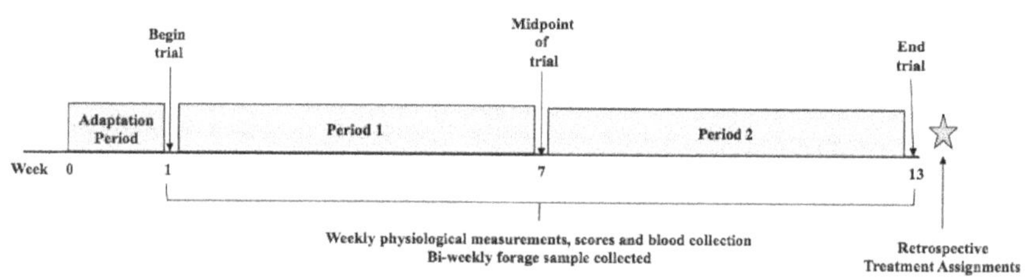

Figure 3. Experimental timeline used during cow performance phase of data collection from late-April to late-July 2016.

5.2. Fescue Toxicosis Selection Method (FTSM)

The classification of animals into tolerance (TOL) and susceptible (SUS) using the proposed Fescue Toxicosis Selection Method (FTSM) is fully described in Koester et al. [33]. In summary, growth data on each animal (Figure 3) was used to estimate the slope of the regression analysis of BW on weeks (average weekly gain; AWG, Figure 4A). This was performed based on three window periods: weeks 1 through 13 (ES), weeks 1 through 7 (P-1), and weeks 7 through 13 (P-2) to assess the effect of the increase in temperature from April to July, availability of forage (see Supplementary Table S1), and exposure of infected tall fescue (Table 1). The residuals from the analysis of AWG in a model including the fixed effects of location, parity, and initial body weight (covariate) were used to identify the window period in which the variation on the data was the largest, indicating the impact of the syndrome [46]. Results from this analysis indicated that data from P-1 resulted in the greatest residual variance [33]. Thus, the forty selected animals, those with the 10 most positive and 10 most negative residuals at each location, representing the TOL and SUS groups, respectively, from P-1 were used to evaluate performance and resistance to fescue toxicosis (Figure 4B).

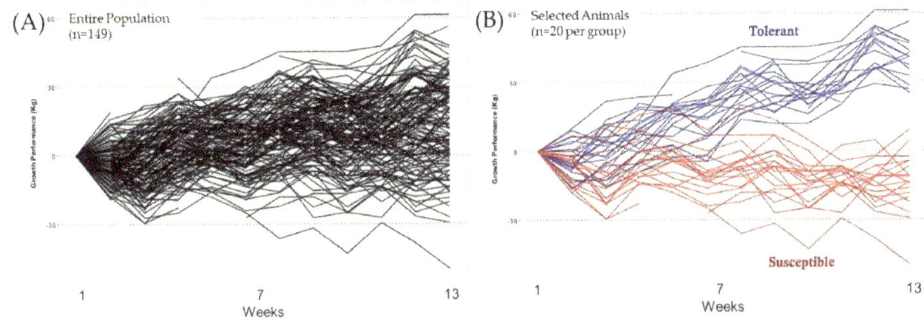

Figure 4. Graphical representation of the Fescue Toxicosis Selection Method (FTSM) to select the top 20 tolerant (EI-TOL) and 20 susceptible (EI-SUS) animals for further investigation within the entire population. (**A**) Displays the ranking of growth performance for the entire population (n = 149) observed over the 13-week period. (**B**) Displays the deviation in growth performance between the tolerant (n = 20) and susceptible (n = 20) animals over the 13-week period.

5.3. T-Snip Genotyping

A detailed description of the genotyping of animals in this study for T-Snip (AgBotanica, LLC, Columbia, MO, USA) can be found in Galliou et al. [5]. In summary, blood cards were collected on each animal and shipped to GeneSeek (Neogen Genomics, Lincoln, NE, USA) for T-Snip genotyping. There were two, one, six, and one animals from EI-TOL at BBCFL for genotypes 1 to 4, respectively. The distribution of these T-Snip genotypes was one, five, three, and one, respectively, for EI-SUS at BBCFL; one, one, six, and two, respectively, for EI-TOL UPRS; and four, two, three, and zero, respectively, for EI-SUS UPRS cows.

5.4. Forage Management

Cattle at both locations grazed endophyte-infected tall fescue (toxic) pastures throughout the 13-week study. Cattle were rotationally grazed every two weeks at each location to ensure sufficient forage utilization and availability. Composite forage samples were taken from each pasture every two weeks to evaluate the nutrient quality and percentage of forage species available. Forage samples were clipped from approximately 20 locations in each pasture and composited. Fescue tiller samples were collected in November of 2016 to evaluate the pasture endophyte infection rate of the fescue.

Nutrient quality samples were submitted within 24 h of collection for nutrient content, and then the results were averaged by experimental period (North Carolina Department of Agriculture Forage Laboratory, Raleigh, NC; USA, see Supplementary Table S1). From the same composite samples, forage was hand-separated in house by trained technicians to determine the percentage of fescue in relation to other various forage species (Table 3) making up forage dry matter. Collected fescue tiller samples were collected, rinsed, and shipped on ice to determine the pasture infection rate, and the average infection rate is reported by experimental period (Agrinostics Ltd. Co., Watkinsville, GA; USA, Table 3). Subsequent HPLC analysis of the forage samples for ergot alkaloid concentrations (MU Veterinary Medical Diagnostic Lab; [9]) demonstrated the change in ergovaline concentrations between locations and over the course of the grazing period (Table 3).

5.5. Serum Assays and Analysis

Serum progesterone concentrations (P4) were analyzed on the FTSM, EI-TOL (n = 20), and EI-SUS (n = 20) animals for both locations at weeks 1, 3, 5, 7, 9, 11 and 13, which represent a bi-weekly measure during the entirety of the collection period. Concentrations were determined by a commercially available radioimmunoassay, the Immuchem Coated Tube Progesterone I125 RIA assay (ICN Pharmaceuticals,

Inc., Costa Mesa, CA), as previously described by Lyons et al. [49]. Concentrations are reported in ng per mL, and the interassay and intraassay variation was 7.4% and 4.3%, respectively.

Serum prolactin concentrations (PRLs) were analyzed on the FTSM, EI-TOL (n = 20), and EI-SUS (n = 20) animals for both locations at weeks 1, 7, and 13, which represent the beginning, midpoint, and end of the collection period. Concentrations were determined by a commercially available Bovine Prolactin ELISA assay (MyBioSource, San Diego, CA, USA), as previously described by Poole et al. [8]. Concentrations were reported in ng per mL, and the interassay and intraassay variation was 11.7% and 4.2%, respectively.

Serum cytokine concentrations were analyzed on the FTSM, EI-TOL (n = 20), and EI-SUS (n = 20) animals for both locations at weeks 1, 7, and 13, which represent the beginning, midpoint, and end of the collection period. Cytokine concentrations were determined using a commercially available Bovine Cytokine Array (Quantibody® Cytokine Arrays; RayBiotech, Inc., Peachtree Corners, GA, USA). The Q1 array permitted detection of interferon alpha (IFNAR1), interferon gamma (IFNG), interleukin (IL)-13, IL-1A, IL-36Ra, IL-21, chemokine CXCL10, CXCL9, CCL4, and tumor necrosis factor alpha (TNF), whereas the Q2 permitted detection of fibroblast growth factor (FGF)1, FGF1, G Protein-Coupled Receptor Associated Sorting Protein (GPRASP)-1, insulin like growth factor (IGF)-1, IL-2, IL-4, IL-15, CCL2, neural cell adhesion molecule 1 (NCAM1), and vascular endothelial growth factor (VEGFA). All bovine cytokine arrays were conducted according to the manufacturer's recommendations. Following processing, all slides were returned to the manufacturer for laser scanning and data extraction. From the absorbance values provided by the manufacturer, serum concentrations were determined using a software (Q-Analyzer Software for QAB-CYT; RayBiotech, Inc., Peachtree Corners, GA, USA), and concentrations are reported in ng per mL for each individual cytokine.

5.6. Statistical Analysis

Performance and cytokine data were analyzed using the MIXED procedure of SAS 9.3 [50] with repeated measures. The individual animal was utilized as the experimental unit and the model for body condition scores, positive change in body condition scores, body weight, average daily gain, rectal temperature, hair coat score, hair shedding score, and hormone concentrations, and included Fescue Toxicosis Selection Method outcome (EI-TOL vs. EI- SUS), location (BBCFL vs. UPRS), trial period (P-1, P-2, and ES), and all respective interactions. For cytokine response data analysis, the individual animal was utilized as the experimental unit and the model for each cytokine response included treatment (EI-TOL vs. EI- SUS), location (BBCFL vs. UPRS), and time (Weeks 1, 7, and 13). The distribution of T-Snip genotypes between EI-TOL and EI-SUS was analyzed using a logistic multinomial model using the T-Snip genotype of animals as the response variable and the FTSM group (EI-TOL and EI-SUS) as a fixed effect in the model. This analysis was carried out in R [51]. Results were recorded as least square means ± SEM, where statistical significance was reported at $p \leq 0.05$ and a statistical tendency at $0.05 \leq p \leq 0.10$.

Supplementary Materials: The following are available online at http://www.mdpi.com/2072-6651/12/12/796/s1, Table S1. Nutritive Value (on a DM basis) of forage samples.

Author Contributions: Conceptualization, D.H.P., N.V.L.S and M.H.P.; methodology, D.H.P., and N.V.L.S.; software, P.K., and N.V.L.S; validation, R.K.P,; formal analysis, K.J.M., R.K.P., and D.H.P.; investigation, K.J.M., M.N., J.M.G., and R.K.P.; resources, D.H.P.; data curation, K.J.M., M.N., and R.K.P.; writing—original draft preparation, K.J.M., and D.H.P.; writing—review and editing, K.J.M., D.H.P., and M.H.P; visualization K.J.M.; supervision, D.H.P., and N.V.L.S.; project administration, D.H.P., and N.V.L.S; funding acquisition, D.H.P., and N.V.L.S. All authors have read and agreed to the published version of the manuscript.

Funding: This research was funded by the North Carolina Cattlemen's Association, North Carolina Agricultural Foundation, Inc., and Hatch Project 02420

Acknowledgments: The authors thank Harrison Dudley, Greg Shaeffer, and the rest of the Butner crew (Butner Beef Cattle Field Laboratory, Bahama, NC), as well as Joseph French and the rest of the UPRS crew (Upper Piedmont Research Station, Reidsville, NC) for their consistent help with animal care. We thank Ruby Monn, Garrett Williams, and Anna Smith for all their help with data collection and laboratory analysis.

Conflicts of Interest: The authors declare no conflict of interest.

References

1. Strickland, J.R.; Looper, M.L.; Matthews, J.C.; Rosenkrans, C.F.; Flythe, M.D., Jr.; Brown, K.R. BOARD-INVITED REVIEW: St. Anthony's fire in livestock: Causes, mechanisms, and potential solutions. *J. Anim. Sci.* **2011**, *89*, 1603–1626. [CrossRef]
2. Miyazaki, S.; Ishizaki, I.; Ishizaka, M.; Kanbara, T.; Ishiguro-Takeda, Y. Lolitrem B residue in fat tissues of cattle consuming endophyte-infected perennial ryegrass straw. *J. Vet. Diagn. Investig.* **2004**, *16*, 340–342. [CrossRef]
3. Kim, J.H.; Kim, C.W.; Ahn, G.C.; Park, E.K.; Kim, C.M.; Park, K.K. Ergovaline levels in tall fescue and its effect on performance of lactating cows. *Anim. Feed. Sci. Technol.* **2007**, *136*, 330–337. [CrossRef]
4. McCulley, R.L.; Bush, L.P.; Carlisle, A.E.; Ji, H.; Nelson, J.A. Warming reduces tall fescue abundance but stimulates toxic alkaloid concentrations in transition zone pastures of the U.S. *Front. Chem.* **2014**, *2*. [CrossRef] [PubMed]
5. Galliou, J.M.; Khanal, P.; Mayberry, K.J.; Poore, M.H.; Poole, D.H.; Serão, N.V.L. Evaluation of a commercial genetic test for fescue toxicosis in pregnant Angus beef cattle. *Transl. Anim. Sci.* **2020**, txaa181. [CrossRef]
6. Grusie, T.; Cowwan, V.; Singh, J.; McKinnon, J.; Blakley, B. Proportions of predominant Ergot alkaloids (Claviceps purpurea) detected in Western Canadian grains from 2014–2016. *World Mycotoxin J.* **2018**, *11*, 259–264. [CrossRef]
7. Poole, D.H.; Lyons, S.E.; Poole, R.K.; Poore, M.H. Ergot alkaloids induce vasoconstriction of bovine uterine and ovarian blood vessels. *J. Anim. Sci.* **2018**, *96*, 4812–4822. [CrossRef]
8. Poole, R.K.; Devine, T.L.; Mayberry, K.J.; Eisemann, J.H.; Poore, M.H.; Long, N.M.; Poole, D.H. Impact of slick hair trait on physiological and reproductive performance in beef heifers consuming ergot alkaloids from endophyte-infected tall fescue. *J. Anim. Sci.* **2019**, *97*, 1456–1467. [CrossRef]
9. Rottinghaus, G.E.; Garner, G.B.; Cornell, C.N.; Ellis, J.L. HPLC method for quantitating ergovaline in endophyte-infested tall fescue: Seasonal variation of ergovaline levels in stems with leaf sheaths, leaf blades, and seed heads. *J. Agric. Food Chem.* **1991**, *39*, 112–115. [CrossRef]
10. Browning, R., Jr. Effects of endophyte-infected tall fescue on indicators of thermal status and growth in Hereford and Senepol steers. *J. Anim. Sci.* **2004**, *82*, 634–643. [CrossRef]
11. Burke, J.M.; Coleman, S.W.; Chase, C.C., Jr.; Riley, D.G.; Looper, M.L.; Brown, M.A. Interaction of breed type and endophyte-infected tall fescue on milk production and quality in beef cattle. *J. Anim. Sci.* **2010**, *88*, 2802–2811. [CrossRef] [PubMed]
12. Cole, N.A.; Brown, M.A.; Phillips, W.A. Genetic x environment interactions on blood constituents of Angus, Brahman, and reciprocal-cross cows and calves grazing common bermudagrass of endophyte-infected tall fescue. *J. Anim. Sci.* **2001**, *79*, 1151–1161. [CrossRef] [PubMed]
13. Gray, K.; Smith, T.; Maltecca, C.; Overton, P.; Parish, J.A.; Cassady, J.P. Differences in hair coat shedding, and effects on calf weaning weight and BCS among Angus dams. *Livest. Sci.* **2011**, *140*, 68–71. [CrossRef]
14. Campbell, B.T.; Kojima, C.J.; Cooper, T.A.; Bastin, B.C.; Wojakiewicz, L.; Kallenbach, R.L.; Schrick, F.N.; Waller, J.C. A Single Nucleotide Polymorphism in the Dopamine Receptor D2 Gene May Be Informative for Resistance to Fescue Toxicosis in Angus-Based Cattle. *Anim. Biotechnol.* **2014**, *25*, 1–12. [CrossRef] [PubMed]
15. Hill, N.S.; Thompson, F.N.; Stuedemann, J.A.; Dawe, D.L.; Hiatt, E.E., 3rd. Urinary Alkaloid Excretion as a Diagnostic Tool for Fescue Toxicosis in Cattle. *J. Vet. Diagn. Investig.* **2000**, *12*, 210–217. [CrossRef]
16. Drewnoski, M.E.; Poore, M.H.; Oliphant, E.J.; Marshall, B.; Green, J.T., Jr. Agronomic Performance of Stockpiled Tall Fescue Varies with Endophyte Infection Status. Online. *Forage Grazinglands* **2007**, *5*, 1–13. [CrossRef]
17. Poore, M.H.; Drewnoski, M. Review: Utilization of stockpiled tall fescue in grazing systems for beef cattle. *Prof. Anim. Sci.* **2010**, *26*, 142–149. [CrossRef]
18. Thompson, R.W.; Fribourg, H.A.; Waller, J.C.; Sanders, W.L.; Reynolds, J.H.; Phillips, J.M.; Schmidt, S.P.; Crawford, R.J.; Allen, V.G.; Faulkner, D.B.; et al. Combined analysis of tall fescue steer grazing studies in the Eastern United States. *J. Anim. Sci.* **1993**, *71*, 1940–1946. [CrossRef]
19. Brown, M.A.; Tharel, L.M.; Brown, A.H., Jr.; Miesner, J.R.; Jackson, W.G. Reproductive Performance of Angus and Brahman Cows Grazing Common Bermudagrass or Endophyte-Infected Tall Fescue. *Prof. Anim. Sci.* **1992**, *8*, 58–65. [CrossRef]
20. Burke, J.; Bishop, C.V.; Stormshak, F. Reproductive characteristics of endophyte-infected or novel tall fescue fed ewes. *Livest. Sci.* **2006**, *104*, 103–111. [CrossRef]

21. Parish, J.A.; McCann, M.A.; Watson, R.H.; Paiva, N.N.; Hoveland, C.S.; Parks, A.H.; Upchurch, B.L.; Hill, N.S.; Bouton, J.H. Use of nonergot alkaloid-producing endophytes for alleviating tall fescue toxicosis in stocker cattle. *J. Anim. Sci.* **2003**, *81*, 2856–2868. [CrossRef] [PubMed]
22. Osborn, T.G. Effect of Consuming Fungus-Infected and Fungus-Free Tall Fescue and Ergotamine Tartrate on Certain Physiological Variables of Cattle in Environmentally-Controlled Conditions. Ph.D. Thesis, Auburn University, Auburn, AL, USA, 1988.
23. Browning, R., Jr. Physiological responses of Brahman and Hereford steers to an acute ergotamine challenge. *J. Anim. Sci.* **2000**, *78*, 124–130. [CrossRef] [PubMed]
24. Cole, N.A.; Stuedemann, J.A.; Thompson, F.N. Influence of both endophyte infestation in fescue pastures and calf genotype on subsequent feedlot performance of steers. *Prof. Anim. Sci.* **2001**, *17*, 174–182. [CrossRef]
25. Fanning, M.D.; Spitzer, J.C.; Cross, D.L.; Thompson, F.N. A preliminary study of Growth, serum prolactin and reproductive performance of beef heifers grazing Acremonium coenophialum-infected tall fescue. *Theriogenology* **1992**, *38*, 375–384. [CrossRef]
26. Smith, V.G.; Hacker, R.R.; Brown, R.G. Effect of Alterations in Ambient Temperature on Serum Prolactin Concentration in steers. *J. Anim. Sci.* **1977**, *44*, 645–649. [CrossRef] [PubMed]
27. Igono, M.O.; Johnson, H.D.; Steevens, B.J.; Hainen, W.A.; Shanklin, M.D. Effect of season on milk temperature, milk growth hormone, prolactin, and somatic cell counts of lactating cattle. *Int. J. Biometeorol.* **1988**, *32*, 194–200. [CrossRef]
28. Schams, D.; Reinhardt, V. Influence of the Season on Plasma Prolactin Level in Cattle from Birth to Maturity. *Horm. Res.* **1974**, *5*, 217–226. [CrossRef]
29. Wettemann, R.P.; Tucker, H.A. Relationship of Ambient Temperature to Serum Prolactin in Heifers. *Proc. Soc. Exp. Biol. Med.* **1974**, *146*, 908–911. [CrossRef]
30. Foitzik, K.; Krause, K.; Nixon, A.J.; Ford, C.A.; Ohnemus, U.; Pearson, A.J.; Paus, R. Prolactin and its receptor are expressed in murine hair follicle epithelium show hair cycle-dependent expression, and induce catagen. *Am. J. Pathol.* **2003**, *162*, 1611–1621. [CrossRef]
31. Nixon, A.J.; Ford, C.A.; Wildermoth, J.E.; Craven, A.J.; Ashby, M.G.; Pearson, A.J. Regulation of prolactin receptor expression in ovine skin in relation to circulating prolactin and wool follicle growth status. *J. Endocrinol.* **2002**, *172*, 605–614. [CrossRef]
32. Aiken, G.E.; Klotz, J.; Looper, M.; Tabler, S.; Schrick, F. Disrupted hair follicle activity in cattle grazing endophyte-infected tall fescue in the summer insulates core body temperatures. *Prof. Anim. Sci.* **2011**, *27*, 336–343. [CrossRef]
33. Koester, L.R.; Poole, D.H.; Serão, N.V.L.; Schmitz-Esser, S. Beef cattle that respond differently to fescue toxicosis have distinct gastrointestinal tract microbiota. *PLoS ONE* **2020**, *15*, e0229192. [CrossRef] [PubMed]
34. Serão, N.V.L.; Kemp, R.A.; Mote, B.E.; Willson, P.; Harding, J.C.S.; Bishop, S.C.; Plastow, G.S.; Dekkers, J.C.M. Genetic and genomic basis of antibody response to porcine reproductive and respiratory syndrome (PRRS) in gilts and sows. *Genet. Sel. Evol.* **2016**, *48*, 51. [CrossRef] [PubMed]
35. Poole, R.K.; Brown, A.R.; Poore, M.H.; Pickworth, C.L.; Poole, D.H. Effects of endophyte-infected tall fescue seed and protein supplementation on stocker steers: II. Adaptive and innate immune function. *J. Anim. Sci.* **2019**, *97*, 4160–4170. [CrossRef] [PubMed]
36. Freeman, M.E.; Kanyicska, B.; Lerant, A.; Nagy, G. Prolactin: Structure, Function, and Regulation of Secretion. *Physiol. Rev.* **2000**, *80*, 1523–1631. [CrossRef]
37. Lamberts, S.W.; MacLeod, R.M. Regulation of prolactin secretion at the level of the lactotroph. *Physiol. Rev.* **1990**, *70*, 279–318. [CrossRef]
38. Sibley, D.R.; Creese, I. Interactions of ergot alkaloids with anterior pituitary D-2 dopamine receptors. *Mol. Pharmacol.* **1983**, *23*, 585–593.
39. Porter, J.K.; Thompson, J.F.N. Effects of fescue toxicosis on reproduction in livestock. *J. Anim. Sci.* **1992**, *70*, 1594–1603. [CrossRef]
40. Janeway, C.A., Jr.; Travers, P.; Walport, M.; Shlomchik, M. *Immuno-Biology: The Immune System in Health and Disease*, 5th ed.; Garland Publishing: New York, NY, USA, 2001; Print.
41. Aiken, G.E.; Flythe, M.D. Vasoconstrictive responses by the carotid and auricular arteries in goats to ergot alkaloid exposure. *Front. Chem.* **2014**, *2*. [CrossRef]
42. Jyotiranjan, T.; Mohapatra, S.; Mishra, C.; Dalai, N.; Kundu, A.K. Heat tolerance in goat—A genetic update. *Pharma Innov. J.* **2017**, *6*, 237–245.

43. Iqbal, N.; Liu, X.; Yang, T.; Huang, Z.; Hanif, Q.; Asif, M.; Khan, Q.M.; Mansoor, S. Genomic variants identified from whole-genome resequencing of indicine cattle breeds from Pakistan. *PLoS ONE* **2019**, *14*, e0215065. [CrossRef] [PubMed]
44. Jones, K.; King, S.S.; Griswold, K.E.; Cazac, D.; Cross, D.L. Domperidone can ameliorate deleterious reproductive effects and reduced weight gain associated with fescue toxicosis in heifers. *J. Anim. Sci.* **2003**, *81*, 2568–2574. [CrossRef] [PubMed]
45. Aiken, G.E.; Kirch, B.H.; Strickland, J.R.; Bush, L.P.; Looper, M.L.; Schrick, F.N. Hemodynamic responses of the caudal artery to toxic tall fescue in beef heifers. *J. Anim. Sci.* **2007**, *85*, 2337–2345. [CrossRef] [PubMed]
46. Berghof, T.; Poppe, M.; Mulder, H.A. Opportunities to Improve Resilience in Animal Breeding Programs. *Front. Genet.* **2019**, *9*, 692. [CrossRef] [PubMed]
47. Richards, M.W.; Spitzer, J.C.; Warner, M.B. Effect of Varying Levels of Postpartum Nutrition and Body Condition at Calving on Subsequent Reproductive Performance in Beef Cattle. *J. Anim. Sci.* **1986**, *62*, 300–306. [CrossRef]
48. Olson, T.A.; Lucena, C.; Chase, C.C.; Hammond, A.C. Evidence of a major gene influencing hair length and heat tolerance in Bos taurus cattle. *J. Anim. Sci.* **2003**, *81*, 80–90. [CrossRef]
49. Lyons, S.E.; Shaeffer, A.D.; Drewnoski, M.E.; Poore, M.H.; Poole, D.H. Effect of protein supplementation and forage allowance on the growth and reproduction of beef heifers grazing stockpiled tall fescue. *J. Anim. Sci.* **2016**, *94*, 1677–1688. [CrossRef]
50. SAS. *SAS System for Mixed Models*; SAS Inst. Inc.: Cary, NC, USA, 1996.
51. R Core Team. R: A Language and Environment for Statistical Computing. R Foundation for Statistical Computing, Vienna, Austria. 2017. Available online: https://www.R-project.org/ (accessed on 9 August 2019).

toxins

Article

Effects of Tall Fescue Endophyte Type and Dopamine Receptor D2 Genotype on Cow-Calf Performance during Late Gestation and Early Lactation

Sarah A. Wilbanks [1], Susan Maggie Justice [1], Thomas West [1], James L. Klotz [2], John G. Andrae [3] and Susan K. Duckett [1,*]

1 Department of Animal and Veterinary Sciences, Clemson University, Clemson, SC 29634, USA; adams9@clemson.edu (S.A.W.); smj0059@auburn.edu (S.M.J.); Twest2@g.clemson.edu (T.W.)
2 USDA-ARS Forage Production Research Unit, Lexington, KY 40506, USA; James.klotz@usda.gov
3 Department of Plant and Environmental Sciences, Clemson University, Clemson, SC 29634, USA; jandrae@clemson.edu
* Correspondence: sducket@clemson.edu; Tel.: +1-(864)-656-1505

Abstract: Grazing endophyte-infected, toxic tall fescue reduces cow/calf production; therefore, this study examines alternate strategies such as use of novel endophyte fescue varieties during late gestation and early lactation or genetic selection of resistant cows. Pregnant cows (n = 75) were randomly assigned to fescue endophyte type: 1) endophyte-infected ergot alkaloid producing tall fescue (E+) or 2) novel endophyte-infected, non-toxic tall fescue (NOV) within maternal (A | A, n = 38 and G | G, n = 37) *DRD2* genotype to examine changes in cow/calf performance and milk production during late gestation and early lactation. Grazing E+ fescue pastures during late gestation reduced cow body weight gain but did not alter calf birth weight compared to NOV. Milk production and calf ADG during the first 30 day of lactation were lower for E+ than NOV. The calving rate was reduced, but not calving interval for E+ cows. The adjusted 205-day weight of calves was lower in those grazing E+ with their dams compared to NOV. There were no interactions between *DRD2* genotype and fescue endophyte type indicating that genotype was not associated with response to E+ fescue in this study. Overall, grazing NOV tall fescue pastures rather than E+ during critical stages of production improved cow gain during late gestation, calving rate, early milk production and calf growth.

Keywords: fescue toxicosis; single nucleotide polymorphism; milk production; calf growth

Key Contribution: Grazing novel, nontoxic tall fescue pastures during critical stages of cow/calf production can reduce losses associated with reduced cow weight gain during last trimester of gestation, milk production during early lactation and pre-weaning calf growth rates.

Citation: Wilbanks, S.A.; Justice, S.M.; West, T.; Klotz, J.L.; Andrae, J.G.; Duckett, S.K. Effects of Tall Fescue Endophyte Type and Dopamine Receptor D2 Genotype on Cow-Calf Performance during Late Gestation and Early Lactation. *Toxins* 2021, 13, 195. https://doi.org/10.3390/toxins13030195

Received: 11 February 2021
Accepted: 5 March 2021
Published: 9 March 2021

1. Introduction

In the southeastern United States, tall fescue [*Lolium arundinaceum* (Schreb.) Darbysh; *Schedonorus phoenix* (Scop.) Holub] is the dominant cool season, perennial forage available for cow/calf production systems [1–3]. Kallenbach [2] estimates that about 12 million beef cows graze tall fescue in this region with reproductivea and liveweight losses estimated at USD 267 per cow in 2015. Most tall fescue contains an endophyte (*Epichloë coenophialia*) that produces ergot alkaloids, a class of mycotoxins that aid the plant in establishment and drought tolerance but is the causative agent for fescue toxicosis when ingested in livestock [1–7]. Fescue toxicosis is a syndrome that includes vasoconstriction, high body temperature, low heart rate, increased blood pressure, low serum prolactin, agalactia, reduced forage intake, and poor weight gains observed in livestock grazing endophyte-infected, ergot alkaloid producing tall fescue [3,5,8]. Ergovaline is mostly responsible for vasoconstriction of vasculature in beef cattle consuming endophyte-infected tall fescue [8]. Grazing endophyte-infected tall fescue reduces cow reproductive rates [9–11], lactation and calf weaning weights [3,5,12,13]. Varieties of tall fescue containing novel endophytes are now available that do not produce ergot alkaloids [14] but have a similar agronomic response as E+ [15,16] and are able to persist under grazing conditions [17]. Research has shown that grazing novel tall fescue instead of toxic endophyte-infected (E+) improves calf birth [18] and weaning weights [18,19], and stocker cattle performance [20,21]. Roberts and Andrae [3] suggested that one solution is to renovate and replace toxic tall fescue pastures with novel endophyte varieties; however, producer adoption of the novel varieties has been slow. Others have shown that grazing novel varieties during critical stages of production can improve overall performance and profitability in beef cow/calf production systems [19]. Critical stages in cow/calf production are during the last trimester of gestation when over 75% of fetal growth occurs [22], early lactation when calf growth rates are the highest [23,24], and rebreeding to avoid the negative impact of fescue toxicosis on reproduction [9–11].

Some have suggested there may be a genetic association in the cow population that allows producers to select for cows that are tolerant to the effects of fescue toxicosis [5,25,26]. For animals ingesting ergot alkaloids, a classic response is decreased serum prolactin concentrations [10–13,18,19] and vasoconstriction [3,5,8]. Ergot alkaloids are dopamine agonists that interact with *DRD2* in a competitive manner [27] to alter serum prolactin [28] and vasoconstriction [5–8]. Campbell et al. [29] identified a SNP in dopamine receptor D2 (DRD2) and established a relationship with serum prolactin concentrations in grazing steers and cows. They identified the G | G genotype in steers as having a greater reduction in serum prolactin values than A | A when grazing E+ fescue pastures. They suggested that the A | A genotype may have resistance to fescue toxicosis. Therefore, we chose to evaluate the *DRD2* SNP in cattle to examine if there is any genetic association with response to fescue toxicosis. The objectives of this study were to evaluate: (1) the use of novel, non-ergot alkaloid producing tall fescue (NOV) compared to E+ fescue pastures during the last trimester of gestation (90 days) and early lactation (first 30 days; Figure 1) on cow/calf performance and milk production, and (2) if maternal genotype for the *DRD2* SNP interacted with the response to grazing tall fescue by endophyte type.

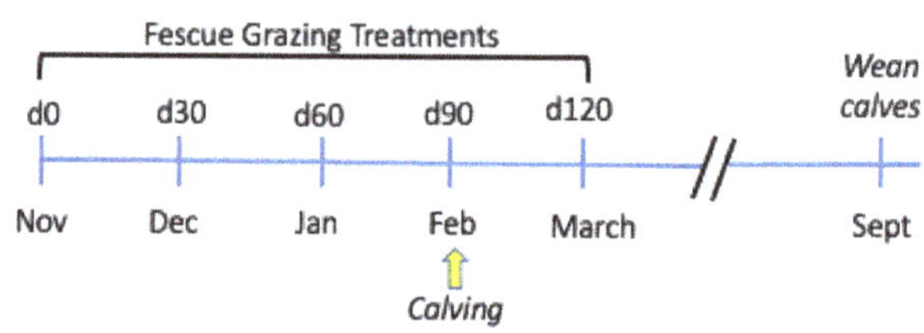

Figure 1. Timeline for the experimental study. Pregnant cows grazed NOV or E+ pastures during last trimester of gestation (90 days) and early lactation (30 days after calving).

2. Results

Total ergot alkaloid concentration (year 1) differed by fescue type ($p < 0.0001$) and month ($p = 0.022$) but the interaction was non-significant ($p = 0.40$). Pastures with E+ fescue had higher ($p < 0.001$) total ergot alkaloid levels compared to NOV during the endophyte type treatment period (Table 1). Total ergot alkaloid concentrations were higher ($p < 0.05$) in October and November than January and February (Figure 2A). Hay samples, which were fed in times when forage was limited, also had higher ($p < 0.001$) total ergot alkaloid levels for E+ than NOV (Table 1). In year 2, pasture and hay samples were examined using LC-MS/MS to quantify ergovaline and ergovalinine, which are mostly responsible for vasoconstriction [5,6]. Ergovaline and ergovalinine concentrations differed by endophyte type ($p < 0.0001$) but were non-significant by month ($p = 0.83$) or the interaction between endophyte type and month ($p = 0.16$; Figure 2B). Ergovaline and ergovalinine concentrations were greater ($p < 0.001$) in E+ pastures than NOV during the months when the endophyte type grazing experiment was conducted. Hay samples from E+ bales contained greater ($p < 0.001$) ergovaline and ergovalinine concentrations compared to NOV.

Table 1. Total ergot alkaloid or ergovaline + ergovalinine concentrations by fescue endophyte-infected type, toxic (E+) or novel endophyte infected (NOV) pasture or hay.

	NOV	E+	SEM
Total ergot alkaloids (Year 1)			
Pasture, ng/g	184 [b]	1938 [a]	93.9
Hay, ng/g	277 [b]	611 [a]	47.2
Ergovaline/ergovalinine, (Year 2)			
Pasture, ng/g	0 [b]	602 [a]	80.3
Hay, ng/g	0 [b]	477 [a]	43.8

[a,b] Means in the same row differ ($p < 0.05$) by endophyte type.

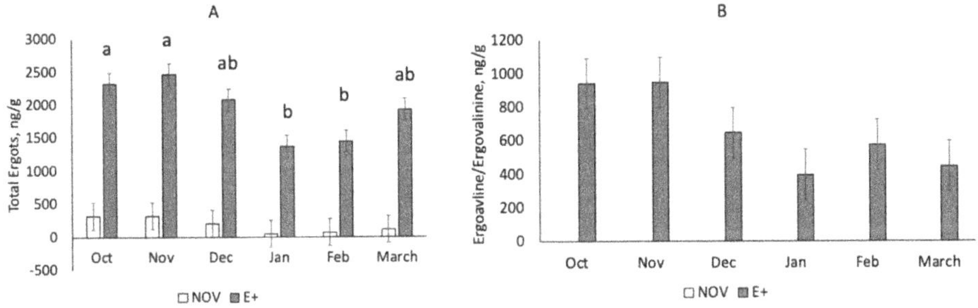

Figure 2. Total ergot alkaloid concentrations (**A**) or ergovaline/ergovalinine (**B**) for novel (NOV) or toxic (E+) endophyte-infected tall fescue pastures by month during time period when endophyte treatments were applied (year 1). Total ergot alkaloids differ ($p = 0.049$) by month but the interaction with fescue endophyte type was non-significant ($p = 0.40$). [a,b] Means with uncommon superscripts differ ($p < 0.05$). Ergovaline and ergovalinine concentrations did not differ ($p = 0.47$) by month or the interaction ($p = 0.27$) between endophyte type and month.

The Angus cow herd ($n = 227$; 3–7 years of age) at the Piedmont Research and Education Center was genotyped for *DRD2* SNP [27]. The cow herd had a distribution of 24% A | A, 56% A | G, and 20% G | G. For this study, we used the homozygote *DRD2* genotypes of A | A (resistant) and G | G (susceptible) to assess if there is any related genetic resistance to fescue toxicosis during late gestation and early lactation. Serum prolactin concentrations were measured during time on endophyte type grazing treatment prior to calving. The two-way interaction between endophyte type and time on fescue was significant ($p = 0.0088$; Figure 3). All other two-way and three-way interactions were non-significant ($p > 0.38$). Serum prolactin concentrations were similar at the start of the grazing study (day 0). Serum prolactin concentrations were lower ($p < 0.05$) for E+ compared NOV after 30, 60 and 90 days on fescue treatment. Serum prolactin concentrations increased ($p < 0.05$) during the last 30 days of the grazing study in both NOV and E+; however, concentrations were lower for E+ than NOV. Serum prolactin concentrations were lower ($p = 0.047$) for cows with G | G versus A | A *DRD2* genotypes regardless of endophyte type during the grazing study (Figure 4).

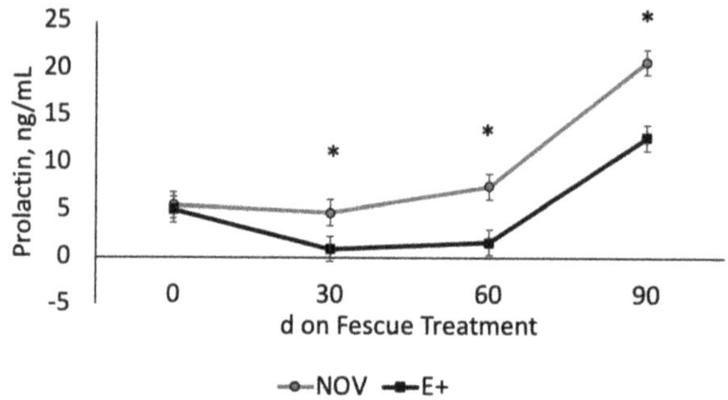

Figure 3. Serum prolactin concentrations in cows grazing endophyte-infected ergot alkaloid producing (E+) or novel endophyte-infected (NOV) during the last trimester of gestation by time on treatment. The interaction between endophyte type and time on fescue treatment was significant ($p = 0.0088$). * Denotes differences between endophyte treatments at that time point ($p < 0.05$).

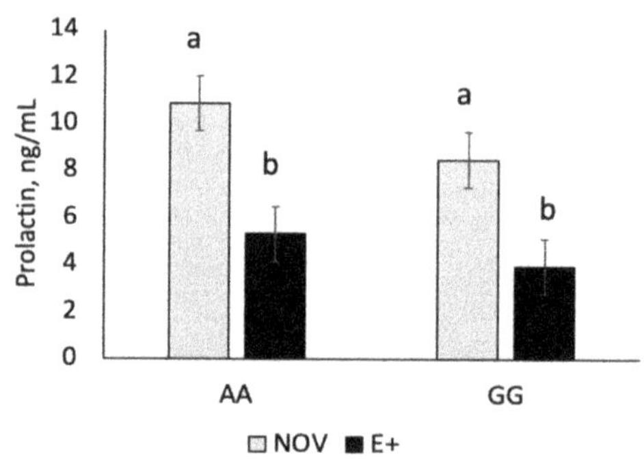

Figure 4. Maternal *DRD2* genotype for serum prolactin concentrations in cows grazing endophyte-infected ergot alkaloid producing (E+) or novel endophyte-infected (NOV) averaged over the last trimester of gestation. The interaction between endophyte type and maternal genotype, maternal genotype and time on fescue, or endophyte type, time on fescue, and maternal genotype were all non-significant ($p > 0.05$). ^{ab}Means with uncommon superscripts differ ($p < 0.05$).

Serum non-esterified fatty acids (NEFA) and cholesterol concentrations were also examined during the fescue endophyte grazing treatment. At d 30 of endophyte treatment, the interaction between maternal genotype and fescue endophyte treatment was significant ($p = 0.017$) (Figure 5). Cows with A | A genotype had elevated serum NEFA concentrations when grazing E+ tall fescue compared to NOV. Cows with G | G genotype had similar serum NEFA concentrations when grazing the different endophyte types. At 90 d on treatment, serum NEFA concentrations were higher ($p = 0.0002$) in cows grazing E+ compared NOV regardless of genotype. Serum NEFA concentrations at d 90 were greater ($p = 0.01$) for A | A than G | G genotypes regardless of fescue endophyte treatment. The interaction between genotype and fescue treatment was non-significant ($p = 0.47$) for serum NEFA on d 90. Serum cholesterol concentrations differed ($p < 0.05$) by fescue endophyte type but did not differ by maternal *DRD2* genotype ($p > 0.37$) or interaction ($p > 0.59$) on day 30 and d 90 of fescue endophyte treatment. Serum cholesterol concentrations were greater ($p = 0.0027$) in cows grazing NOV compared to E+ fescue pastures (Figure 6) on day 30 and 90 of fescue endophyte treatment.

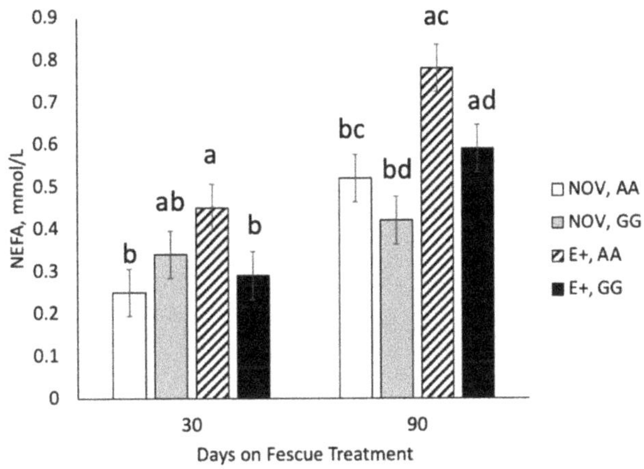

Figure 5. Serum non-esterified fatty acids (NEFA) in cows grazing endophyte-infected ergot alkaloid producing (E+) or endophyte-infected non-ergot alkaloid producing (NOV) during the last trimester of gestation. The interaction between maternal *DRD2* genotype and fescue endophyte type was significant ($p = 0.017$) on day 30 of fescue endophyte treatment. Endophyte type ($p = 0.0002$) and maternal *DRD2* genotype ($p = 0.010$) were significant but the interaction was non-significant on day 90 of fescue endophyte treatment. ^{a,b} Means with uncommon superscripts differ ($p < 0.05$) by endophyte type. ^{c,d} Means with uncommon superscripts differ ($p < 0.05$) by maternal genotype.

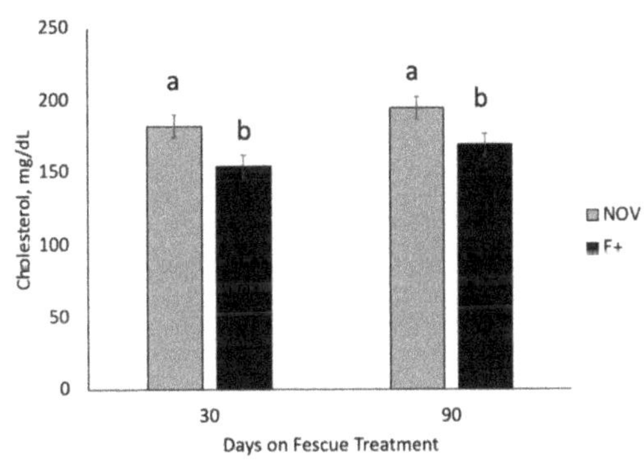

Figure 6. Serum cholesterol concentrations in cows grazing endophyte-infected ergot alkaloid producing (E+) or endophyte-infected non-ergot alkaloid producing (NOV) at 30 and 90 days on fescue endophyte treatments during the last trimester of gestation. All interactions were non-significant ($p > 0.05$) and main effects are shown in the figure for each time. ^{a,b} Means with uncommon superscripts differ ($p < 0.05$) at each time of sampling.

Cow body weight (BW) was measured over time during the fescue endophyte grazing study. Cow body weight differed by endophyte treatment ($p = 0.0019$) and time on fescue ($p < 0.001$) but the interaction was non-significant ($p = 0.45$). Cow body weight increased over time ($p = 0.001$) during late gestation. Cow BW was lower ($p = 0.0019$) in cows grazing E+ fescue compared to NOV after 30, 60, and 90 of grazing (Figure 7A). There was a trend ($p = 0.054$) for G | G cows to be heavier than A | A cows but there were no significant interactions with genotype and other variables ($p > 0.34$). Average daily gain during the last trimester of gestation differed by endophyte type ($p = 0.0012$) but maternal genotype and all interactions were non-significant ($p > 0.30$). Cows grazing E+ fescue pastures had 30% lower ($p = 0.0012$) average daily gains during the last trimester of gestation compared to NOV (Figure 7B).

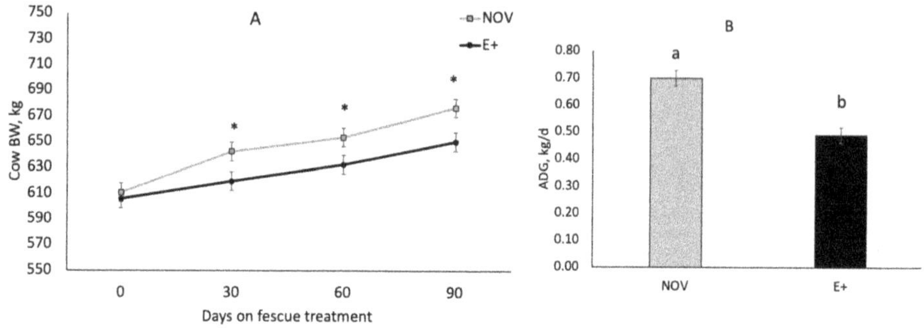

Figure 7. Changes in cow body weight (BW; **A**) and average daily gain (ADG; **B**) of cows grazing endophyte-infected ergot alkaloid producing (E+) or endophyte-infected non-ergot alkaloid producing (NOV) during last gestation. Interactions between endophyte type, year, and maternal DRD2 genotype were non-significant for body weight gain and average daily gain. * Denotes differences ($p < 0.05$) in body weight by endophyte treatment. [a,b] Means with uncommon superscripts differ ($p < 0.05$).

The interaction between endophyte type and maternal *DRD2* genotype was significant ($p = 0.012$) for cow body condition score (BCS). Cows with A|A genotype had lower ($p < 0.05$) BCS when grazing E+ fescue than NOV; however, for G|G cows there was no difference in BCS by fescue endophyte type (Figure 8A). Ultrasound measurements of rump fat thickness and rump muscle depth were obtained at the start and end of the fescue grazing experiment. The change in rump fat thickness did not differ by genotype ($p = 0.41$) or fescue endophyte type ($p = 0.71$) but the change in rump muscle depth was greater ($p = 0.036$) for cows grazing E+ fescue than NOV, regardless of genotype. The interaction between endophyte type and time on fescue was significant for hair coat score (HCS). Hair coat scores did not differ by maternal genotype ($p = 0.35$) or all interactions with maternal genotype were non-significant ($p > 0.48$). Hair coat scores increased ($p = 0.041$) over time in cows grazing E+ fescue and values were greater compared to NOV at day 60 and 90 of treatment (Figure 8B).

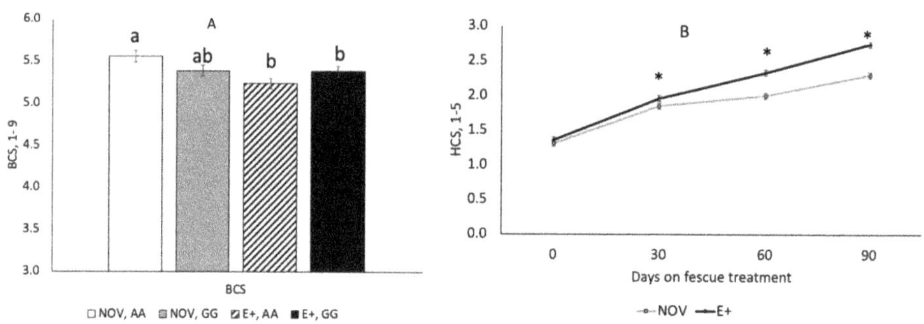

Figure 8. Body condition score (BCS; **A**) and hair coat scores (HCS; **B**) by fescue endophyte type during the last 90 days of gestation. The interaction between endophyte type and maternal genotype was significant for BCS. [a,b] Means with uncommon superscripts differ ($p < 0.05$). The interaction between endophyte type and time on fescue was significant for HCS. * Denotes significance ($p < 0.05$).

Interactions between year, endophyte type and maternal genotype were non-significant ($p > 0.37$) and the main effects are shown in Table 2. Dystocia scores did not differ ($p > 0.05$) by year, fescue endophyte treatment or maternal genotype. Calving interval from year 1 to year 2 did not differ ($p > 0.05$) by fescue endophyte type or maternal genotype. The calving rate was lower ($p = 0.012$) in cows grazing E+ than NOV. Calf birth weight nor adjusted calf birth weight did not differ ($p = 0.63$) by fescue endophyte treatment. Calf birth weight and adjusted calf birth weights were higher ($p = 0.018$) for A | A than G | G cows and by year with heavier ($p = 0.055$) calves in year 1. Calf sex did not differ ($p > 0.05$) by year, fescue endophyte type or maternal genotype. Cow milk production was estimated using the weigh-suckle-weigh procedure at 30 days after calving. Milk production was 15% lower ($p = 0.047$) in cows grazing E+ fescue compared to NOV. Milk production was higher ($p = 0.014$) in G | G than A | A cows. The interaction between fescue treatment and genotype was non-significant ($p = 0.99$).

Table 2. Main effects for year, fescue endophyte type and DRD2 genotype on adjusted calf birth weight and weaning weight of calves. All interactions were non-significant ($p > 0.05$).

	Year		Fescue Type		*DRD2* Genotype		Standard Error
	1	2	NOV	E+	AA	GG	
Cow Parameters							
Dystocia score [1]	1.09	1.07	1.09	1.08	1.07	1.09	0.37
Calving interval, d	-	-	382	371	375	378	23.2
Calving rate, %	-	-	85.7 [c]	72.5 [d]	82.5	75.7	2.90
Milk production, kg/d	11.40	10.85	12.00 [c]	10.25 [d]	10.03 [f]	12.22 [e]	4.20
Calf Parameters							
at Birth							
Birth weight, kg	39.45	36.80	38.37	37.92	39.09 [e]	37.21 [f]	4.94
Adj. birth weight, kg	40.42	37.07	38.94	38.54	39.59 [e]	37.90 [f]	4.89
Calf sex (2 = female, 3 = male)	2.59	2.55	2.59	2.56	2.61	2.53	0.50
at Weaning							
Weaning age, d	220.2	202.4	210.0	212.6	212.7	209.9	20.13
Actual weaning weight, kg	269.3 [a]	247.4 [b]	258.9	255.0	260.0	253.9	33.8
Adj. weaning weight [2], kg	277.0 [a]	247.3 [b]	264.2	260.18	264.5	259.8	33.06
Adj. 205 d wean weight [2], kg	258.6	252.5	259.4 [g]	252.7 [h]	256.6	248.5	26.79
Adj. 205 d wean ADG, kg/d	1.064	1.048	1.075 [g]	1.037 [h]	1.057	1.055	0.132

[1] Dystocia score: 1 = no difficulty, no assistance required; 2 = minor difficulty, some assistance; 3 = major difficulty, usually mechanical assistance; 4 = caesarian section or other surgery; 5 = abnormal presentation. [2] Adjusted weaning weight was adjusted by cow age and calf sex [30]. Adjusted 205-day weight was the adjusted weaning weight calculated on an equal calf age basis of 205-day. [a,b] Means in the same row with uncommon superscripts differ ($p < 0.05$) by year. [c,d] Means in the same row with uncommon superscripts differ ($p < 0.05$) fescue endophyte type. [e,f] Means in the same row with uncommon superscripts differ ($p < 0.05$) by maternal DRD2 genotype. [g,h] Means in the same row with uncommon superscripts differ ($p < 0.10$) by fescue endophyte type.

The interaction between fescue endophyte type and maternal genotype was significant ($p = 0.036$) for calf average daily gain during the first 30 days of lactation. Calf average daily gain (ADG) during the first 30 days of lactation was higher ($p = 0.036$) for calves born to A | A cows grazing NOV than G | G cows grazing NOV or E+ fescue (Figure 9). Calf weaning weight and adjusted calf weaning weights did not differ by fescue endophyte type ($p > 0.52$), genotype ($p > 0.24$) or the interaction ($p > 0.94$). Calf weaning weight and adjusted weaning weights were higher ($p = 0.021$) in year 1 compared to year 2. When weaning weight was calculated on an age adjusted basis, the adjusted 205 days weaning weights tended ($p = 0.073$) to be higher by 8.4 kg for NOV compared E+. Maternal genotype did not alter ($p > 0.47$) adjusted 205 days weaning weight and the interaction with fescue endophyte type was non-significant ($p = 0.95$ & 0.86). Adjusted 205 days ADG of the calves tended to be greater ($p = 0.072$) for NOV than E+.

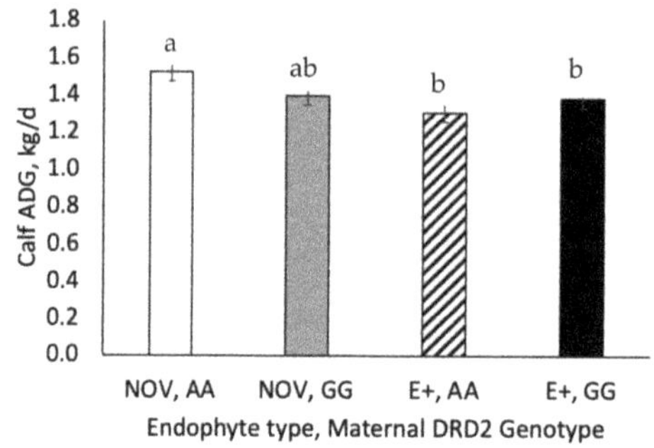

Figure 9. Calf average daily gain during the first 30 days of lactation in cows grazing different fescue endophyte type by maternal DRD2 genotype. The interaction between endophyte type and maternal DRD2 genotype was significant ($p = 0.036$) and simple effects are shown in the figure. [a,b] Means with uncommon superscripts differ ($p < 0.05$).

3. Discussion

In the current study, stockpiled tall fescue with either endophyte-infected ergot alkaloid producing (E+) or novel endophyte-infected nontoxic (NOV) pastures were utilized for grazing during the last trimester of gestation (90 days) and early lactation (30 days) in a spring-calving Angus cow herd. In year 1, total ergot alkaloid was examined during the time period when fescue endophyte type treatments were applied (October–March). Total ergot alkaloid concentrations decreased from November and December to January and February. In year 2, a LC/MS analysis was developed to determine specifically ergovaline and ergovalinine levels in forage samples because they have been shown to be the causative agents of fescue toxicosis [5,8]. The concentrations of ergovaline reported in this study are at threshold levels (300–500 ppb) where fescue toxicosis can occur [31]. Kallenbach et al. [32] reported reduced ergovaline concentrations of stockpiled tall fescue from December to March. Curtis and Kallenbach [33] showed that ergovaline concentrations declined 50% in an 84-day period with stockpiling of tall fescue during the winter months. Ergot alkaloid concentration in E+ hay was lower than in the E+ pastures. Similarly, Roberts et al. [34] found ergovaline concentrations in hay are quickly reduced within the first 3 day after mowing and decline gradually during additional months of storage. The levels of ergovaline/ergovalinine in the current study are similar to those reported by Peters et al. [12] and Kallenbach et al. [32] but the total ergot alkaloid values are higher than those published by others for cow grazing E+ fescue [18,19,33].

Pregnant cows grazed E+ or NOV fescue pastures during critical stages of production, last trimester of gestation and the first 30 days of lactation. The interaction between fescue endophyte type and maternal *DRD2* genotype was non-significant and main effects of fescue endophyte type over time and maternal *DRD2* genotype are shown in Figures 3 and 4. Before grazing treatments started (day 0, start of the last trimester of gestation), serum prolactin concentrations were similar. After 30, 60, and 90 days on fescue pastures, serum prolactin concentrations were lower for E+ than NOV. Serum prolactin levels, examined at day 90 of treatment, increased in both E+ and NOV compared to pre-treatment and day 30 and 60 values; however, concentrations were greater for NOV compared to E+. A reduction in serum prolactin concentration is commonly observed in cows grazing E+ fescue pastures [10–13,18,19]. There was a difference in prolactin concentrations among cows of different *DRD2* genotypes but the response to fescue was similar for A | A and G | G genotypes. Campbell et al. [29] identified the *DRD2* SNP suggested that A | A genotype may provide resistance to fescue toxicosis. They hypothesized that spring calving herds should have a higher prevalence of A allele. Our results showed that the A and G allele were about equally distributed in our Angus spring-calving cow herd (0.52 A allele and 0.48 G allele). The majority of our cows were heterozygous (56%) for the *DRD2* SNP with smaller percentages in the homozygous condition (24% A | A, 20% G | G). In this study, there was no interaction between maternal genotype and fescue endophyte treatment for serum prolactin concentrations indicating a lack of genetic association with prolactin values in cows grazing stockpiled E+ fescue. Others have also examined other single SNPs and found some associations with XK, Kell blood group complex subunit-related family member-4 (XKR4) and serum prolactin levels [35], prolactin gene and calving interval [36], and cytochrome P450 and lower milk production [37]. Selection for a trait like fescue resistance likely involves multiple genes and therefore the use of a single SNP marker may not be as successful [38]. A commercial test, T-Snip™ (AgBotanica, Columbia, MO, USA), is now available that provides a tolerance index for fescue toxicosis. Galliou et al. [39] found a relationship between T-Snip score and cow performance when grazing E+ tall fescue. The benefits of genetic markers for tolerance to fescue toxicosis would be advantageous for tall fescue based grazing systems, but more research is needed in this area.

Serum NEFA and cholesterol concentrations were measured at day 30 and 90 of the grazing endophyte type study to examine changes in metabolism that may be associated with fescue endophyte type or maternal genotype. Non-esterified fatty acids are released into the bloodstream upon lipolysis of adipose triglycerides and elevated levels are associated with negative energy balance in late gestation to early lactation in dairy cattle [40]. Serum NEFA values after 30 days of grazing were higher in A | A cows grazing E+ pastures but unchanged in G | G. At 90 days of treatment, NEFA concentrations were elevated in cows grazing E+ regardless of genotype. Niederecker et al. [41] also reported elevated NEFA values in cows grazing stockpiled tall fescue versus summer baled fescue hay after 56 to 99 days of feeding during late gestation. McArt et al. [40] found that cows with elevated NEFA had longer time to pregnancy, lower milk yield and increased culling rates. Cholesterol concentrations were higher for NOV than E+ at day 30 and 90 of fescue treatment. Others have shown that serum cholesterol content is lower in steers grazing E+ tall fescue pasture versus endophyte-free [42] or novel endophytes [43]. Stuedemann et al. [44] also reported lower plasma cholesterol concentrations in cows grazing E+ fescue and that cholesterol values were negatively associated with rate of N fertilization. The cows grazing high N fertilized E+ pastures had the lowest cholesterol levels and highest incidence of fat necrosis [44]. Cholesterol is the precursor for steroid hormones and higher cholesterol concentrations are associated with shorter intervals from calving to conception and successful pregnancy [45].

It is well documented that grazing of E+ forage results in production losses [1–3,5,25]. Steers grazing ergot alkaloid-producing endophyte-infected tall fescue gained 30% to 70% less on compared to cattle consuming an endophyte-free or novel endophyte tall fescue [3,20,21,46]. In this study, cows grazing E+ fescue had lighter body weights and 30% lower average daily gains during the last trimester of gestation compared to NOV. Others [12,13,18,19] have shown similar results with lower body weight gains in cows grazing E+ fescue compared to endophyte-free or novel endophyte. Reduction in BW gain for cattle consuming toxic fescue has been attributed to a number of things, but most often associated with reduced intake [9,47]. Peters et al. [12] found that forage intake was similar in the initial 28 d even though body weight loss was not. In this study, calf birth weight did not differ between fescue endophyte types even though cow body weight gain was reduced. Similarly, others have also reported no change in calf birth weight when cows grazed E+ fescue prior to calving [19,48,49]. In contrast, others have shown that calves or lambs born to dams exposed to endophyte-infected tall fescue during gestation have reduced birthweights by 5% [18], 15% [50,51] or 36% [52]. Britt et al. [50] found that exposure to ergot alkaloids during late gestation had the greatest impact on fetal development with little to no effect when fed during mid-gestation in sheep. Exposure to ergot alkaloids during late gestation reduced fetal lamb muscle and organ weights resulting in asymmetrical growth, which is indicative of intrauterine growth restriction (IUGR) [53]. These reductions in fetal growth appear related to vasoconstrictive events caused by ingestion of the ergot alkaloids, specifically ergovaline and ergovalinine [54]. In cattle, the placenta continues to increase in weight until near term [55,56] which is unlike the sheep where placental mass is established by 90 d of gestation [57]. Greenwood et al. [58] suggests that these differences in placental growth may explain why cattle appear less sensitive to under nutrition than sheep. Placental weight and birth weight are highly correlated in beef cattle [59,60]. However, in this study, cow weight gain and body weight were reduced when grazing E+ fescue pastures without any change in calf birth weight. Body condition score was lower for A | A cows grazing E+ than NOV; however, in G | G cows BCS was similar on NOV and E+ but BCS was lower than A | A cows grazing NOV. Hair coat score increased over time in both NOV and E+ but cows grazing E+ fescue had higher HCS on d 60 and 90 than NOV. Gray et al. [61] reported that cows who shed their winter coat before June 1 wean heavier calves. Real-time ultrasound measures of rump fat thickness and gluteus muscle depth at the beginning and end of the study showed that subcutaneous fat thickness was unchanged, but the change muscle depth was greater for E+ than NOV. Ultrasound rump fat and muscle depth measures are useful to monitor changes in body composition in cows grazing toxic tall fescue [62]. These results suggest the cows grazing E+ fescue during the last trimester of rapid fetal growth had reduced body weight gains that lead to mobilization of muscle tissue in order to provide nutrients for fetal growth.

This grazing study was continued through the first 30 days of lactation, which is a period of highest milk production [23,24] when the calf is almost solely dependent on the dam. Grazing E+ fescue during the last trimester and first 30 days of lactation lowered milk production by 15% compared to NOV. Milk production was also greater for G | G genotypes than A | A; however, this was consistent across endophyte treatments. Brown et al. [63] reported a 43% reduction in milk production for Angus cows grazing E+ fescue pastures compared to bermudagrass from d 61 to 200 postpartum. Peters et al. [12] found a 22% reduction in milk production of cows grazing E+ versus endophyte-free tall fescue pastures from birth to weaning. In contrast, Burke et al. [49] did not observe differences in milk production but instead found higher milk fat in cows grazing E+ fescue compared to endophyte-free. Milk production, measured at 70 days of age, did not differ in cows that grazed E+ or NOV in late gestation only [49]. Calf gains during this first 30 days period were lower for E+ than NOV. Feeding toxic endophyte-infected seed to ewes during mid and late gestation reduced milk production by 59 to 82% and reduced pre-weaning growth rates [51], which altered subsequent puberty development and post-natal growth [64]. In addition, cow calving rate was lower in cows that grazed E+ fescue during late gestation and early lactation. Calving interval from year 1 to 2 in this study did not differ among fescue treatments or genotypes. Caldwell et al. [19] also observed reduced calving rates when spring calving cows grazed E+ tall fescue compared to novel endophytes. Calf weaning weights or adjusted weaning weight for cow age and sex did not differ by endophyte type or maternal genotype. However, the adjusted 205-day weight was lower for calves whose dams grazed E+ tall fescue during late gestation and early lactation compared to NOV. Others also found that 205-day weights were lower for calves whose dams grazed E+ fescue from birth to weaning [12] and in spring-calving cows [19].

4. Conclusions

Grazing E+ fescue pastures during the last trimester of gestation and first 30 day of lactation reduced cow body weight gain but did not alter calf birth weight compared to NOV. Milk production and calf ADG during the first 30 day of lactation were lower for E+ than NOV. Calving rate was reduced but no calving interval for E+ cows. Adjusted 205-day weight of calves was lower in those grazing E+ with their dams compared to NOV. There were no interactions between *DRD2* genotype and fescue endophyte type indicating that genotype was not associated with response to E+ fescue in this study. Overall, grazing E+ tall fescue pastures during late gestation and early lactation appeared to alter the priority of nutrient utilization of the cows to adapt to reductions in gain during the time period of rapid fetal growth. Early milk production was lower for E+, which reduced early calf gains that remained to 205-day of age.

5. Materials and Methods

All animal experimental procedures were reviewed and approved by the Clemson University Institutional Animal Care and Use Committee (AUP 2017-025) on 22 May 2017. All animal experiments were conducted at the Clemson University Piedmont Research and Education Center (PREC), Pendleton, SC.

5.1. Genotyping

Hair follicles (5–10) were collected from tail switches of all cows in the Angus cow herd at Piedmont REC (n = 227; 3–7 year of age) and DNA was extracted using Quick Extract™ DNA extraction solution (Epicentre, Madison, WI, USA). The Extracted DNA was shipped to University of Tennessee for *DRD2* SNP genotyping (rs41749780) according to Campbell et al. [29]. Each cow was one of three possible genotypes: A | A, A | G or G | G. Dams with homozygous alleles (A | A and G | G) were used in this study over a two-year period.

5.2. Design

Cows were pregnancy checked by ultrasound via rectal palpation at day 95 after timed AI by an experienced technician. Pregnant cows (n = 75, 3 to 7 year of age, 608 $\pm$ 4.9 kg BW) were randomly assigned to fescue endophyte type: (1) endophyte-infected ergot alkaloid producing tall fescue (E+; Kentucky 31) or (2) novel endophyte-infected nontoxic tall fescue (NOV, Texoma MaxQ II, Pennington Seed, Madison, GA, USA) within maternal (A | A, n = 38 and G | G, n = 37) genotype during the last trimester of pregnancy (90 days) and first 30 days of lactation. Cows remained in the same fescue endophyte type treatment during year 2. Each year, there were two pasture replicates (10 ha pastures/rep; 2 rep/fescue type) per fescue endophyte type. Cows were maintained on non-fescue pastures during the first and second trimester of gestation when not on fescue endophyte treatments. For this study, cows grazed stockpiled fescue pastures with either endophyte-infected, toxic (E+) or endophyte-infected, novel (NOV) tall fescue. Pastures were fertilized 15 days prior to cows being allocated to treatment with 67.7 kg of nitrogen per ha. A strip grazing management practice was implemented through the duration of the project. Paddocks were monitored visually twice weekly and cattle were allotted additional forage when herbage availability dropped below 2000 kg/ha. When all available forage in stockpiled pastures was below 2000 kg/ha, hay of each endophyte treatment type was fed free choice in each paddock. Prior to the start of the grazing experiment each year, tiller samples were taken and sent to Agrinostics (Watkinsville, GA, USA) to determine endophyte infection levels. All pastures used in this study had >90% endophyte infection rates.

5.3. Forage Analysis

Grab samples of forage were collected during the time period when cattle were grazing the endophyte treatments (late gestation to early lactation) from each paddock when cattle were moved to a new strip of grazing and hay samples collected before feeding. All forage and hay samples were freeze dried and ground using a 1-mm screen Wiley cutting mill (Arthur H. Thomas, Philadelphia, PA, USA). Forage samples were composited on a monthly basis. In year 1, forages and hay were analyzed for total ergot alkaloid infection concentrations (Agrinostics Limited Co., Watkinsville, GA, USA). As ergovaline and its epimer ergovalinine are the most informative ergots involved in vasoconstriction [5,8], we developed an assay to measure concentration of these ergopeptines to obtain more specific information. This assay was not available in year 1 but was developed and utilized in year 2 of this experiment. In year 2, forage and hay samples were analyzed for ergovaline and ergovalinine concentrations by LC-MS/MS at the Multi-User Analytical Laboratory (MUAL) at Clemson University.

For quantification of ergovaline and ergovalinine concentrations, freeze dried forage and hay samples were extracted according to Guo et al. [65]. A tall fescue seed extract was provided by J. L. Klotz and used to establish a calibration range from 0.78 to 100 ng/mL to quantify ergovaline and ergovalinine [66]. Ergotamine tartrate (Sigma Chemical Co.,

St. Louis, MO, USA) was used as an internal standard. Samples were analyzed using a Thermo Orbitrap Fusion™ Tribrid™ Mass Spectrometer (Thermo Fisher, Waltham, MA, USA) with MS1 (ESI positive, 60,000 resolution, scan range 220–1200 m/z) and MS2 (HCD & CID on targeted mass of 25 ergot alkaloids in orbitrap at 15,000 resolution) in the Clemson University Multi-User Analytical Laboratory and Metabolomics Core. The column was Phenomenex Kinetex XB-C18 with 1.7 μm, 100A, 150 × 2.1 mm (Phenomenex, Torrance, CA, USA). Solvent A was 5 mM ammonium bicarbonate and solvent B was 90% acetonitrile with 5 mM ammonium bicarbonate. Quality control checks were performed using instrument blanks, method blanks to determine recovery, and instrument quality control to assess relative standard deviation of 12.5 ppb ergovaline standard. The percent recovery for ergovaline, ergovalinine, and ergotamine were 98, 95, and 100%, respectively. Compound detection was performed using the extracted ion chromatogram for targeted analytes within 2 ppm mass error in Skyline software. Quantification of ergovaline and ergovalinine was performed using linear external calibration curves (0.78 to 100 ng/mL. The method limit of detection for ergovaline/ergovalinine in grass and hay was 11.414 ng per g freeze-dried plant tissue (instrument LOD of 0.78 ppb).

5.4. Blood Samples

Blood was collected on cows at day 0, 30, 60, and 90 of fescue endophyte treatment in 10 mL vacutainer serum tubes (Covidien Ltd., Monoject; Dublin, Ireland) via the coccygeal vein. All serum samples were allowed to clot and centrifuged at 2000× *g* for 20 min at 4 °C to obtain serum. Serum was stored at −20°C for subsequent prolactin, non-esterified fatty acid (NEFA), and cholesterol analyses. Prolactin concentrations were measured using RIA procedures of Bernard et al. [67]. The intra- and inter-assay coefficients of variation were 5.59 and 5.75 %, respectively. Non-esterified fatty acids on day 30 and 90 were analyzed using NEFA ELISA kit (BiooScientific, Perkin-Elmer, Waltham, MA, USA) with an intra- and inter-assay coefficient of variation of 6.42% and 5.56%, respectively. Serum cholesterol was extracted in duplicate using the method of Chauveau-Duriot et al. [68] and samples had an intra- and inter-assay coefficient of variation of 5.29% and 6.00%, respectively. Cholesterol content on day 30 and 90 was analyzed by gas chromotography (Agilent 6850, Santa Clara, CA, USA) using a Restek Rxi-5ms column (15 m, 0.25 mm ID, 0.25 μm, catalog no. 13420; Restek, Bellefonte, PA, USA) according to manufacturer.

5.5. Animals

Body condition scores (BCS 1–9 with 1 = severely emaciated to 9 = very obese) [69] and hair coat scores (HCS 1–5 with 1 = slick, shiny hair coat to 5 = > 50% of body covered with old, unshed hair) [70] were collected on day 0, 30, 60, and 90 of fescue endophyte treatment. Scores for BCS and HCS were recorded by the same two technicians during the study and average for each cow. Real-time ultrasound measures of rump fat and gluteus medius muscle depth were collected on day 0 and 90 by the same technician using an Aloka 500V and 17.2 cm linear transducer. The image was captured and interpreted using BioSoft Toolbox for beef software (Biotronics, Inc., Ames, IA, USA).

At parturition, calves were processed within 12 h of birth at 800 and 1600. Data collected included date, birth weight, and sex. Calves were ear tagged, tattooed, and males were castrated. Calves were injected intramuscularly with vitamin A (500,000 IU) and D (75,000 IU; VetOne, MWI Veterinary Supply Co., Boise, ID, USA). To estimate milk production, weigh-suckle-weigh technique was utilized at 30 days post-calving [24]. Calves were separated from their dams for 3 h and then allowed to nurse the dams dry. Then, they were separated from the dam again and held for 6 h. After the separation, calves were weighed. The calves were allowed to nurse the dam and then reweighed immediately. The difference in weights between post- and pre-suckling were assumed to represent milk production in that 6 h period and was multiplied by 4 to represent a 24-h period. In April, all cows were synchronized using the fixed-time AI (TAI) 7-day Co-Synch + CIDR protocol for timed AI to a single sire (Connealy Mentor 7374, Select Sires, registration # 15832714; Spring Hill, TN, USA). At two weeks post TAI, clean up bulls were added and cows were exposed to the clean-up bulls for 70 d. Calves were vaccinated for clostridial agents (One Shot Ultra 7; Zoetis, Parsippany, NJ, USA), bovine rhinotracheitis parainfluenza 3 vaccine (TSV-2, Zoetis), and leptospirosis (Leptoferm-5, Zoetis) in May at 4 month of age. Cows were pregnancy checked by ultrasound via rectal palpation at d 95 after TAI by an experienced technician. Cows that were not pregnant in year 1 were sold. Estimated breeding dates were calculated based on rectal palpation results. Calves were weaned on September 11, 2018 and August 30, 2019 and body weight obtained. Adjusted birth and weaning weights were calculated according to the Beef Improvement Federation Guidelines [30] based on age of the dam and sex of the calf. Adjusted 205-day weights were also calculated [30].

5.6. Statistics

The univariate procedure of SAS (SAS Inst. Inc., Cary, NC, USA) was used to test all variables for normality. Serum prolactin concentrations were not normally distributed and were log-transformed for data analysis. Ergot alkaloid values were not normally distributed and were square root transformed for statistical analysis. Serum prolactin and ergot alkaloid concentrations results are presented as non-transformed for easier understanding and interpretation. For forage and hay ergot alkaloid data, paddock was the experimental unit. Data were analyzed using mixed procedure with fescue forage type, month, and two-way interaction in the model. For animal data, a 2 × 2 factorial with two-factors of fescue endophyte type (E+ vs. NOV) and maternal DRD2 genotype (A | A vs. G | G) in a split plot design. A mixed model was developed with endophyte type, DRD2 genotype and the 2-way interaction in the model as fixed effects and two different error terms applied. The error term for testing endophyte type was the random effect of paddock within endophyte type by year and the error term for testing DRD2 genotype and its interactions with endophyte type was the random effect of cow within DRD2 and endophyte type combination. The analysis was performed using the GLIMMIX procedure of SAS (SAS Inst. Inc., Cary, NC, USA). For repeated measures analyses, time and all interactions with time were included in the model. Least square means were generated and separated using a Fisher's LSD when the F-test was significant for year, fescue endophyte type, maternal genotype and or the interaction. Significance was determined at $p < 0.05$, and trends denoted when $p > 0.05$ to $p < 0.10$. When the interaction was non-significant ($p > 0.05$), main effects of endophyte type or DRD2 genotype are shown in the results. When the interaction was significant ($p < 0.05$), simple effects of endophyte type by DRD2 genotype are shown in the results.

Author Contributions: Conceptualization, S.K.D. and J.G.A.; data curation, S.A.W., S.M.J., T.W.; writing—original draft preparation, S.A.W., S.M.J., T.W.; writing—review and editing, S.K.D., J.G.A., J.L.K.; funding acquisition, S.K.D. All authors have read and agreed to the published version of the manuscript.

Funding: Technical contribution no. 6684 of the Clemson University Experiment Station. This material is based upon work support by NIFA/USDA under project number SC-1700537.

Institutional Review Board Statement: Not applicable.

Informed Consent Statement: Not applicable.

Data Availability Statement: The data presented in this study are available on request from the corresponding author.

Acknowledgments: Appreciation is expression to the Clemson University Beef Farm staff and AVS 4220 Fetal Fescue Research class for assistance with animal management and sample collection; C. Kojima and F.N. Schrick for DRD2 genotyping and prolactin assays; Clemson University Multi-User Analytical Laboratory and Metabolomics Core for ergot alkaloid analyses.

Conflicts of Interest: The authors declare no conflict of interest. The funders had no role in the design of the study; in the collection, analyses, or interpretation of data; in the writing of the manuscript, or in the decision to publish the results.

References

1. Ball, D.M.; Lacefield, G.D.; Hoveland, C.S. *The Wonder Grass—The Story of Tall Fescue in the United States*; Oregon Tall Fescue Commission: Salem, OR, USA, 2019.
2. Kallenbach, R.L.; Bill, R. Kunkle interdisciplinary beef symposium: Coping with tall fescue toxicosis: Solutions and realities. *J. Anim. Sci.* **2015**, *93*, 5487–5495. [CrossRef] [PubMed]
3. Roberts, C.; Andrae, J. Tall fescue toxicosis and management. *Crop Manag.* **2004**, *3*, 1–18. [CrossRef]
4. Young, C.A.; Charlton, N.D.; Takach, J.E.; Swoboda, G.A.; Trammell, M.A.; Huhman, D.V.; Hopkins, A.A. Characterization of EpichloÃ« coenophiala within the US: Are all tall fescue endophytes created equal? *Front. Chem.* **2014**, *2*, 95. [CrossRef]
5. Klotz, J.L. Activities and effects of ergot alkaloids on livestock physiology and production. *Toxins* **2015**, *7*, 2801–2821. [CrossRef]
6. Schardl, C.L. Introduction to the toxins special issue on ergot alkaloids. *Toxins* **2015**, *7*, 4232–4237. [CrossRef]
7. Reddy, P.; Hemsworth, J.; Guthridge, K.M.; Vinh, A.; Vassiliadis, S.; Ezernieks, V.; Spangenberg, G.C.; Rochfort, S.J. Ergot alkaloid mycotoxins: Physiological effects, metabolism and distribution of the residual toxin in mice. *Sci. Rep.* **2020**, *10*, 9714. [CrossRef]
8. Foote, A.P.; Harmon, D.L.; Brown, K.R.; Strickland, J.R.; McLeod, K.R.; Bush, L.P.; Klotz, J.L. Constriction of bovine vasculature caused by endophyte-infected tall fescue seed extract is similar to pure ergovaline. *J. Anim. Sci.* **2012**, *90*, 1603–1609. [CrossRef]
9. Gay, N.; Boling, J.A.; Dew, R.; Miksch, E.D. Effects of endophyte infected tall fescue on beef cow-calf performance. *Appl. Agric. Res.* **1988**, *3*, 182.
10. Porter, J.K.; Thompson, F.N., Jr. Effects of fescue toxicosis on reproduction in livestock. *J. Anim. Sci.* **1992**, *70*, 1594–1603. [CrossRef] [PubMed]
11. Poole, R.K.; Poole, D.H. Impact of ergot alkaloids on female reproduction in domestic livestock species. *Toxins* **2019**, *11*, 364. [CrossRef]
12. Peters, C.W.; Grigsby, K.N.; Aldrich, C.G.; Paterson, J.A.; Lipsey, R.J.; Kerley, M.S.; Garner, G.B. Performance, forage utilization, and ergovaline consumption by beef cows grazing endophyte fungus-infected tall fescue, endophyte fungus-free tall fescue, or orchardgrass pastures. *J. Anim. Sci.* **1992**, *70*, 1550–1561. [CrossRef]
13. Patterson, J.; Forcherio, C.; Larson, B.; Samford, M.; Kerley, M. The effects of fescue toxicosis on beef cattle productivity. *J. Anim. Sci.* **1995**, *73*, 889–898. [CrossRef] [PubMed]
14. Bouton, J.H.; Latch, G.C.M.; Hill, N.S.; Hoveland, C.S.; McCann, M.A.; Watson, R.H.; Parish, J.A.; Hawkins, L.L.; Thompson, F.N. Reinfection of tall fescue cultivars with non-ergot alkaloid—Producing endophytes. *Agron. J.* **2002**, *94*, 567–574. [CrossRef]
15. Drewnoski, M.E.; Poore, M.H.; Oliphant, E.J.; Marshall, B.; Green, J.T. Agronomic performance of stockpiled tall fescue varies with endophyte infection status. *Forage Grazinglands* **2007**, *5*, 1–13. [CrossRef]
16. Hopkins, A.A.; Young, C.; Panaccione, D.G.; Simpson, W.R.; Mittal, S.; Bouton, J.H. Agronomic performance and lamb health among several tall fescue novel endophyte combinations in the south-central USA. *Crop Sci.* **2010**, *50*, 1552–1561. [CrossRef]
17. Phillips, T.D.; Aiken, G.E. Novel endophyte-infected tall fescues. *Forage Grazinglands* **2009**, *7*, 1–6. [CrossRef]
18. Watson, R.H.; McCann, M.A.; Parish, J.A.; Hoveland, C.S.; Thompson, F.N.; Bouton, J.H. Productivity of cow-calf pairs grazing tall fescue pastures infected with either the wild-type endophyte or a nonergot alkaloid-producing endophyte strain, AR542. *J. Anim. Sci.* **2004**, *82*, 3388–3393. [CrossRef] [PubMed]
19. Caldwell, J.D.; Coffey, K.P.; Jennings, J.A.; Phillipp, D.; Young, A.N.; Tucker, J.D.; Hubbell III, D.S.; Hess, T.; Looper, M.L.; West, C.P.; et al. Performance by spring and fall-calving cows grazing with full, limited, or no access to toxic *Neotyphodium coenophialum*-infected tall fescue. *J. Anim. Sci.* **2013**, *91*, 465–476. [CrossRef]
20. Gunter, S.A.; Beck, P.A. Novel endophyte-infected tall fescue for growing beef cattle. *J. Anim. Sci.* **2004**, *82*, 82.
21. Parish, J.A.; McCann, M.A.; Watson, R.H.; Paiva, N.N.; Hoveland, C.S.; Parks, A.H.; Upchurch, B.L.; Hill, N.S.; Bouton, J.H. Use of nonergot alkaloid-producing endophytes for alleviating tall fescue toxicosis in stocker cattle. *J. Anim. Sci.* **2003**, *81*, 2856–2868. [CrossRef]
22. Elay, R.M.; Thatcher, W.W.; Bazer, F.W.; Wilcox, C.J.; Becker, R.B.; Head, H.H.; Adkinson, R.W. Development of the conceptus in the bovine. *J. Dairy Sci.* **1974**, *61*, 467–473. [CrossRef]
23. Boggs, D.L.; Smith, E.F.; Schalles, R.R.; Brent, B.E.; Corah, L.R.; Pruitt, R.J. Effects of milk and forage intake on calf performance. *J. Anim. Sci.* **1980**, *51*, 550–553. [CrossRef]
24. Radunz, A.E.; Fluharty, F.L.; Day, M.L.; Zerby, H.N.; Loerch, S.C. Prepartum dietary energy source fed to beef cows: I. Effects on pre- and post-partum cow performance. *J. Anim. Sci.* **2010**, *88*, 2717–2728. [CrossRef]
25. Strickland, J.R.; Looper, M.L.; Matthews, J.C.; Rosenkrans, C.F.; Flythe, M.D.; Brown, K.R. BOARD-INVITED REVIEW: St. Anthony's fire in livestock: Causes, mechanisms, and potential solutions. *J. Anim. Sci.* **2011**, *89*, 1603–1626. [CrossRef]
26. Smith, T.; Cassady, J.P.; Bill, E. Kunkle interdisciplinary beef symposium: Genetic resistance to the effects of grazing endophyte-infected tall fescue. *J. Anim. Sci.* **2015**, *93*, 5506–5511. [CrossRef]
27. Sibley, D.R.; Creese, I. Interactions of ergot alkaloids with anterior pituitary D-2 dopamine receptors. *Mol. Pharmacol.* **1983**, *23*, 585–593.

28. Strickland, J.R.; Cross, D.L.; Birrenkott, G.P.; Grimes, L.W. Effect of ergovaline, loline, and dopamine antagonists on rat pituitary cell prolactin release in vitro. *Am. J. Vet. Res.* **1994**, *55*, 716–721.
29. Campbell, B.T.; Kojima, C.J.; Cooper, T.A.; Bastin, B.C.; Wojakiewicz, L.; Kallenbach, R.L.; Schrick, F.N.; Waller, J.C. A single nucleotide polymorphism in the dopamine receptor D2 gene may be informative for resistance to fescue toxicosis in angus-based cattle. *Anim. Biotechnol.* **2013**, *25*, 1–12. [CrossRef] [PubMed]
30. BIF. Beef Improvement Guidelines Standard Adjustment Factors for Birth and Weaning Weight, 205-d Weaning Weight and Calving Difficulty Scores. Available online: http://guidelines.beefimprovement.org/index.php/Guidelines_for_Uniform_Beef_Improvement_Programs (accessed on 1 March 2018).
31. Craig, A.M.; Blythe, L.L.; Duringer, J.M. The role of the oregon state university endophyte service laboratory in diagnosing clinical cases of endophyte toxicoses. *J. Agric. Food Chem.* **2014**, *62*, 7376–7381. [CrossRef] [PubMed]
32. Kallenbach, R.L.; Bishop-Hurley, G.J.; Massie, M.D.; Rottenhaus, G.E.; West, C.P. Herbage mass, nutritive value, and ergovaline concentration of stockpiled tall fescue. *Crop Sci.* **2003**, *43*, 1001–1005. [CrossRef]
33. Curtis, L.E.; Kallenbach, R.L. Endophyte infection level of tall fescue stockpiled for winter grazing does not alter the gain of calves nursing lactating beef cows. *J. Anim. Sci.* **2007**, *85*, 2346–2353. [CrossRef]
34. Roberts, C.A.; Kallenbach, R.L.; Hill, N.S.; Rottinghaus, G.E.; Evans, T.J. Ergot alkaloid concentrations in tall fescue hay during production and storage. *Crop Sci.* **2009**, *49*, 1496–1502. [CrossRef]
35. Bastin, B.C.; Houser, A.; Bagley, C.P.; Ely, K.M.; Payton, R.R.; Saxton, A.M.; Schrick, F.N.; Waller, J.C.; Kojima, C.J. A polymorphism inXKR4is significantly associated with serum prolactin concentrations in beef cows grazing tall fescue. *Anim. Genet.* **2014**, *45*, 439–441. [CrossRef]
36. Looper, M.L.; Black, S.G.; Reiter, S.T.; Okimoto, R.; Johnson, Z.B.; Brown, M.A.; Rosenkrans, C.F., Jr. Identification of polymorphisms in the enhancer region of the bovine prolactin gene and association with profitability traits of beef cattle. *Prof. Anim. Sci.* **2010**, *26*, 103–108. [CrossRef]
37. Sales, M.A.; Larson, M.J.; Reiter, S.T.; Brown, A.H., Jr.; Brown, M.A.; Looper, M.L.; Coffey, K.P.; Rosenkrans, C.F., Jr. Effects of bovine cytochrome P450 single-nucleotide polymorphism, forage type, and body condition on production traits in cattle. *J. Anim. Physiol. Anim. Nutr.* **2011**, *96*, 545–553. [CrossRef] [PubMed]
38. Seidel, G.E., Jr. Brief introduction to whole-genome selection in cattle using single nucleotide polymorphisms. *Reprod. Fertil. Dev.* **2010**, *22*, 138–144. [CrossRef]
39. Galliou, J.M.; Khanal, P.; Mayberry, K.; Poore, M.H.; Poole, D.H.; Serão, N.V.L. Evaluation of a commercial genetic test for fescue toxicosis in pregnant Angus beef cattle. *Transl. Anim. Sci.* **2020**, *4*, txaa181. [CrossRef]
40. McArt, J.A.A.; Nydam, D.V.; Oetzel, G.R.; Overton, T.R.; Ospina, P.A. Elevated non-esterified fatty acids and β-hydroxylbutyrate and their association with transition dairy cow performance. *Vet. J.* **2013**, *198*, 560–570. [CrossRef] [PubMed]
41. Niederecker, K.N.; Larson, J.M.; Kallenbach, R.L.; Meyer, A.M. Effects of feeding stockpiled tall fescue versus summer-baled tall fescue-based hay to late gestation beef cows: I. Cow performance, maternal metabolic status, and fetal growth. *J. Anim. Sci.* **2018**, *96*, 4618–4632. [CrossRef]
42. Oliver, J.W.; Schultze, A.E.; Rohrbach, B.W.; Fribourg, H.A.; Ingle, T.; Waller, J.C. Alterations in hemograms and serum biochemical analytes of steers after prolonged consumption of endophyte-infected tall fescue. *J. Anim. Sci.* **2000**, *78*, 1029–1035. [CrossRef]
43. Nihsen, M.E.; Piper, E.L.; West, C.P.; Crawford, R.J.; Denard, T.M.; Johnson, Z.B.; Roberts, C.A.; Spiers, D.A.; Rosenkrans, C.F., Jr. Growth rates and physiology of steers grazing tall fescue inoculated with novel endophytes. *J. Anim. Sci.* **2004**, *82*, 878–883. [CrossRef]
44. Stuedemann, J.A.; Rumsey, T.S.; Bond, J.; Wilkinson, S.R.; Bush, L.P.; Williams, D.J.; Caudle, A.B. Association of blood cholesterol with occurrence of fat necrosis in cows and tall fescue summer toxicosis in steers. *Am. J. Vet. Res.* **1985**, *46*, 1990–1995. [PubMed]
45. Westwood, C.; Lean, I.; Garvin, J. Factors influencing fertility of holstein dairy cows: A multivariate description. *J. Dairy Sci.* **2002**, *85*, 3225–3237. [CrossRef]
46. Parish, J.A.; Best, T.F.; Boland, H.T.; Young, C.A. Effects of selected endophyte and tall fescue cultivar combinations on steer grazing performance, indicators of fescue toxicosis, feedlot performance, and carcass traits. *J. Anim. Sci.* **2013**, *91*, 342–355. [CrossRef] [PubMed]
47. Bond, J.; Powell, J.B.; Weinland, B.T. Behavior of steers grazing several varieties of tall fescue during summer conditions. *Agron. J.* **1984**, *76*, 707–709. [CrossRef]
48. Shoup, L.M.; Kloth, A.C.; Wilson, T.B.; González-Peña, D.; Ireland, F.A.; Rodriguez-Zas, S.; Felix, T.L.; Shike, D.W. Prepartum supplement level and age at weaning: I. Effects on pre- and postpartum beef cow performance and calf performance through weaning. *J. Anim. Sci.* **2015**, *93*, 4926–4935. [CrossRef] [PubMed]
49. Burke, J.M.; Coleman, S.W.; Chase, C.C.; Riley, D.G.; Looper, M.L.; Brown, M.A. Interaction of breed type and endophyte-infected tall fescue on milk production and quality in beef cattle. *J. Anim. Sci.* **2010**, *88*, 2802–2811. [CrossRef]
50. Britt, J.L.; Greene, M.A.; Bridges, W.C.; Klotz, J.L.; Aiken, G.E.; Andrae, J.G.; Pratt, S.L.; Long, N.M.; Schrick, F.N.; Strickland, J.R.; et al. Ergot alkaloid exposure during gestation alters. I. Maternal characteristics and placental development of pregnant ewes. *J. Anim. Sci.* **2019**, *97*, 1874–1890. [CrossRef] [PubMed]
51. Britt, J.L.; Greene, M.A.; Wilbanks, S.A.; Bertrand, J.K.; Klotz, J.L.; Bridges, J.W.; Aiken, G.; Andrae, J.G.; Duckett, S.K. Feeding tall fescue seed reduces ewe milk production, lamb birth weight and pre-weaning growth rate. *Animals* **2020**, *10*, 2291. [CrossRef] [PubMed]

52. Duckett, S.K.; Andrae, J.G.; Pratt, S.L. Exposure to ergot alkaloids during gestation reduces fetal growth in sheep. *Front. Chem.* **2014**, *2*, 68. [CrossRef] [PubMed]
53. Greene, M.; Britt, J.L.; Powell, R.R.; Feltus, F.A.; Bridges, W.C.; Bruce, T.; Klotz, J.L.; Miller, M.F.; Duckett, S.K. Ergot alkaloid exposure during gestation alters: 3. Fetal growth, muscle fiber development, and miRNA transcriptome. *J. Anim. Sci.* **2019**, *97*, 3153–3168. [CrossRef] [PubMed]
54. Klotz, J.L.; Britt, J.L.; Miller, M.F.; Snider, M.; E Aiken, G.; Long, N.M.; Pratt, S.L.; Andrae, J.G.; Duckett, S.K. Ergot alkaloid exposure during gestation alters: II. Uterine and umbilical artery vasoactivity. *J. Anim. Sci.* **2019**, *97*, 1891–1902. [CrossRef]
55. Prior, R.L.; Laster, D.B. Development of the bovine fetus. *J. Anim. Sci.* **1979**, *48*, 1546–1553. [CrossRef]
56. Ferrell, C.J. Placental Regulation of Fetal Growth. In *Animal Growth Regulation*; Campion, D.R., Hausman, G.J., Martin, R.J., Eds.; Springer: Boston, MA, USA, 1989; pp. 1–19.
57. Redmer, D.; Wallace, J.; Reynolds, L. Effect of nutrient intake during pregnancy on fetal and placental growth and vascular development. *Domest. Anim. Endocrinol.* **2004**, *27*, 199–217. [CrossRef]
58. Greenwood, P.L.; Café, L.M.; Hearnshaw, H.; Hennessy, D.W. Consequences of nutrition and growth retardation early in life for growth and composition of cattle and eating quality of beef. *Recent Adv. Anim. Nutr. Aust.* **2005**, *15*, 183–195.
59. Echternkamp, S. Relationship between placental development and calf birth weight in beef cattle. *Anim. Reprod. Sci.* **1993**, *32*, 1–13. [CrossRef]
60. Anthony, R.V.; Bellows, R.A.; Short, R.E.; Staigmiller, R.B.; Kaltenbach, C.C.; Dunn, T.G. Foetal growth of beef calves. II. Effects of sire on prenatal development of the calf and related placental characteristics. *J. Anim. Sci.* **1986**, *62*, 1375–1387. [CrossRef]
61. Gray, K.; Smith, T.; Maltecca, C.; Overton, P.; Parish, J.; Cassady, J. Differences in hair coat shedding, and effects on calf weaning weight and BCS among Angus dams. *Livest. Sci.* **2011**, *140*, 68–71. [CrossRef]
62. Looper, M.L.; Reiter, S.T.; Williamson, B.C.; Sales, M.A.; Hallford, D.M.; Rosenkrans, C.F., Jr. Effects of body condition on measures of intramuscular and rump fat, endocrine factors, and calving rate of beef cows grazing common bermudagrass or endophyte-infected tall fescue. *J. Anim. Sci.* **2010**, *88*, 4133–4141. [CrossRef]
63. Brown, M.A.; Tharel, L.M.; Brown, A.H., Jr.; Miesner, J.R.; Jackson, W.G. Reproductive performance of Angus and Brahman cows grazing common bermudagrass or endophyte-infected tall fescue. *Prof. Anim. Sci.* **1992**, *8*, 58–65. [CrossRef]
64. Greene, M.; Klotz, J.L.; Goodman, J.P.; May, J.B.; Harlow, B.; Baldwin, W.S.; Strickland, J.R.; Britt, J.L.; Schrick, F.N.; Duckett, S.K. Evaluation of oral citrulline administration as a mitigation strategy for fescue toxicosis in sheep. *Transl. Anim. Sci.* **2020**, *4*, txaa197. [CrossRef] [PubMed]
65. Guo, Q.; Shao, B.; Du, Z.; Zhang, J. Simultaneous determination of 25 ergot alkaloids in cereal samples by ultraperformance liquid chromatography-tandem mass spectrometry. *J. Agric. Food Chem.* **2016**, *64*, 7033–7039. [CrossRef] [PubMed]
66. Ji, H.; Fannin, F.; Klotz, J.; Bush, L. Tall fescue seed extraction and partial purification of ergot alkaloids. *Front. Chem.* **2014**, *2*, 110. [CrossRef]
67. Bernard, J.; Chestnut, A.; Erickson, B.; Kelly, F. Effects of prepartum consumption of endophyte-infested tall fescue on serum prolactin and subsequent milk production of Holstein cows. *J. Dairy Sci.* **1993**, *76*, 1928–1933. [CrossRef]
68. Chauveau-Duriot, B.; Doreau, M.; Nozière, P.; Graulet, B. Simultaneous quantification of carotenoids, retinol, and tocopherols in forages, bovine plasma, and milk: Validation of a novel UPLC method. *Anal. Bioanal. Chem.* **2010**, *397*, 777–790. [CrossRef]
69. Wagner, J.J.; Lusby, K.S.; Oltjen, J.W.; Rakestraw, J.; Wettemann, R.P.; Walters, L.E. Carcass composition in mature hereford cows: Estimation and effect on daily metabolizable energy requirement during winter. *J. Anim. Sci.* **1988**, *66*, 603–612. [CrossRef]
70. Saker, K.E.; Allen, V.G.; Fontenot, J.P.; Bagley, C.P.; Ivy, R.L.; Evans, R.R.; Wester, D.B. Tasco-Forage: II. Monocyte immune cell response and performance of beef steers grazing tall fescue treated with a seaweed extract. *J. Anim. Sci.* **2001**, *79*, 1022–1031. [CrossRef] [PubMed]

toxins

MDPI

Article

Rumen and Serum Metabolomes in Response to Endophyte-Infected Tall Fescue Seed and Isoflavone Supplementation in Beef Steers

Taylor B. Ault-Seay [1], Emily A. Melchior-Tiffany [1,†], Brooke A. Clemmons [1,‡], Juan F. Cordero [1], Gary E. Bates [2], Michael D. Flythe [3], James L. Klotz [3], Huihua Ji [4], Jack P. Goodman [5], Kyle J. McLean [1] and Phillip R. Myer [1,*]

1 Department of Animal Science, University of Tennessee, Knoxville, TN 37996, USA; tault1@vols.utk.edu (T.B.A.-S.); eam@nmsu.edu (E.A.M.-T.); Brooke.Clemmons@tamuc.edu (B.A.C.); Juan.Cordero@utk.edu (J.F.C.); kmclea10@utk.edu (K.J.M.)

2 Department of Plant Sciences, University of Tennessee, Knoxville, TN 37996, USA; gbates@utk.edu

3 USDA-ARS, Forage-Animal Production Research Unit, Lexington, KY 40546, USA; michael.flythe@ars.usda.gov (M.D.F.); James.Klotz@ars.usda.gov (J.L.K.)

4 Kentucky Tobacco Research and Development Center, University of Kentucky, Lexington, KY 40546, USA; hji4@email.uky.edu

5 Department of Plant and Soil Sciences, University of Kentucky, Lexington, KY 40546, USA; jpgood2@email.uky.edu

* Correspondence: pmyer@utk.edu; Tel.: +1-865-974-3184

† Current Address: Department of Animal and Range Sciences, New Mexico State University, Las Cruces, NM 88003, USA.

‡ Current Address: Department of Agriculture, Texas A&M University-Commerce, Commerce, TX 75428, USA.

Received: 4 November 2020; Accepted: 24 November 2020; Published: 26 November 2020

Abstract: Fescue toxicosis impacts beef cattle production via reductions in weight gain and muscle development. Isoflavone supplementation has displayed potential for mitigating these effects. The objective of the current study was to evaluate isoflavone supplementation with fescue seed consumption on rumen and serum metabolomes. Angus steers (n = 36) were allocated randomly in a 2 × 2 factorial arrangement of treatments including endophyte-infected (E+) or endophyte-free (E−) tall fescue seed, with (P+) or without (P−) isoflavones. Steers were provided a basal diet with fescue seed for 21 days, while isoflavones were orally administered daily. Following the trial, blood and rumen fluid were collected for metabolite analysis. Metabolites were extracted and then analyzed by UPLC-MS. The MAVEN program was implemented to identify metabolites for MetaboAnalyst 4.0 and SAS 9.4 statistical analysis. Seven differentially abundant metabolites were identified in serum by isoflavone treatment, and eleven metabolites in the rumen due to seed type ($p < 0.05$). Pathways affected by treatments were related to amino acid and nucleic acid metabolism in both rumen fluid and serum ($p < 0.05$). Therefore, metabolism was altered by fescue seed in the rumen; however, isoflavones altered metabolism systemically to potentially mitigate detrimental effects of seed and improve animal performance.

Keywords: beef cattle; endophyte; ergot alkaloid; fescue toxicosis; isoflavone; metabolites

Key Contribution: As fescue toxicosis causes multiple symptoms that negatively impact beef cattle performance, isoflavone consumption may reduce these effects. The current study found tall fescue seed type to mainly impact the rumen metabolome, while isoflavone supplementation affected the host metabolome in the serum, potentially improving animal growth and development during fescue toxicosis.

1. Introduction

Tall fescue is the major forage used to feed cattle in pasture-based systems of the southeast and covers approximately 14 million hectares across the United States [1]. The advantage of tall fescue is hardiness of the plants attributed to the presence of a fungal endophyte (*Epichloë coenophialum*, formerly known as *Neotyphodium coenophialum* and *Acremonium coenophialum*) living in a mutualistic relationship with the plant [2]. However, the endophyte produces ergot alkaloids that are toxic to animals that consume them for an extended period of time [3]. Ergot alkaloids are able to bind biogenic amine receptors on blood vessels, resulting in vasoconstriction throughout the body [4–6]. This results in a condition known as fescue toxicosis, which is commonly observed by the animal's inability to thermoregulate [7], poor reproductive performance [1], and reduced average daily gain [8], significantly reducing overall animal performance. Therefore, researchers are tasked with identifying management methods and therapeutics to alleviate these consequences to cattle producers.

Pasture management methods have been evaluated for reducing the impact of fescue toxicosis in cattle. Inter-seeding of legumes, such as red clover, to mitigate the effects of fescue toxicosis has proved beneficial in cattle grazing endophyte-infected tall fescue [9]. Recent research has found phytoestrogenic compounds, known as isoflavones, present in red clover may be responsible for reducing the effects of fescue toxicosis. Isoflavones act as an agonist on the β-adrenergic receptors present on blood vessels to promote vasodilation [10], reversing the effects of ergot alkaloid induced vasoconstriction. Additionally, isoflavones act as a natural antibiotic selective against hyper-ammonia-producing bacteria (HAB) and some cellulolytic and amylolytic bacteria [11,12]. The reduction of ammonia levels as a result of less HAB in the rumen allows more amino acids to be absorbed and used by the ruminant, while altered celluloytic and amylolytic bacteria can influence the production of volatile fatty acids for energy. Therefore, the increase in blood flow and altered rumen fermentation may improve nutrient delivery and utilization for host metabolic processes contributing to animal growth.

The objective of the present study is to evaluate the effect of isoflavone supplementation with tall fescue seed consumption on beef steer's rumen and serum metabolomes. Ruminal and circulating metabolites may provide insights into altered bacterial and host metabolic functions that improve steer performance on endophyte infected tall fescue with the administration of isoflavones.

2. Results

2.1. Global Rumen Fluid and Serum Metabolome Comparison

An orthogonal partial least squares discriminant analysis (O-PLS-DA) was used to depict the relationship between the global rumen and serum metabolomes, which illustrated distinct separation between the two metabolomes (Figure 1). A heatmap was also used to visualize the top 25 rumen fluid and serum metabolites by individual steer (Figure 2). The heatmap supports that there is very little similarity between the overall ruminal and circulating metabolites.

2.2. Rumen Fluid Metabolome

All identified rumen fluid metabolites are presented in Supplementary File 1 with means and standard errors of the mean by treatment combination group. To visualize the effect between steers of the E+P+ and E−P− groups on the rumen fluid metabolome, a partial least squares discriminant analysis (PLS-DA) was created and a significant distinct overlap among seed type was noted (Figure 3). Correlation analyses were performed to analyze the correlation of individual rumen metabolites with the treatment combination groups, and variable importance in the projection (VIP) scores were generated to determine the metabolites that contributed to variation in rumen fluid metabolomes among treatment combination groups. Xylose was negatively correlated with the treatments ($r = -0.57$) and had one of the greatest impacts on metabolome differences among all treatments ($p = 0.01$) (Figure 4, Table 1). Individual metabolite and metabolic pathway analyses were not significantly impacted by the interaction of seed type and isoflavone treatments ($p > 0.05$).

The rumen metabolome was analyzed by the main effects of seed type and isoflavone treatment. In order to visualize the difference in rumen fluid metabolomes, an O-PLS-DA was generated for the main effects of seed type (Figure 5A) and isoflavone treatment (Figure 5B). For seed type and isoflavone treatment, partial separation was observed between endophyte-infected and endophyte-free seed (Figure 5A). Additionally, partial separation was observed between steers receiving isoflavones and those that did not receive isoflavones (Figure 5B). Correlation analysis indicated hypoxanthine was negatively correlated ($r = -0.56$) and determined by VIP analysis to have a significant impact on the rumen fluid metabolome differences between endophyte-infected and endophyte-free seed treatments ($p = 0.01$; Figure 6A; Table 1). For isoflavone treatments, trehalose/sucrose was positively correlated ($r = 0.06$), but had no impact on the rumen fluid metabolome differences (Figure 6B). Metabolites that differed by seed type are presented in Table 1. Eleven metabolites differed significantly as a result of endophyte-infected versus endophyte-free treatments ($p < 0.05$, Table 1). No individual metabolite differences were observed in the rumen fluid as a result of isoflavone treatment. Metabolic pathways that differed significantly by seed type or isoflavone treatment are presented in Table 2. Twenty metabolic pathways were affected by seed type, but only two pathways were affected by isoflavones ($p < 0.05$).

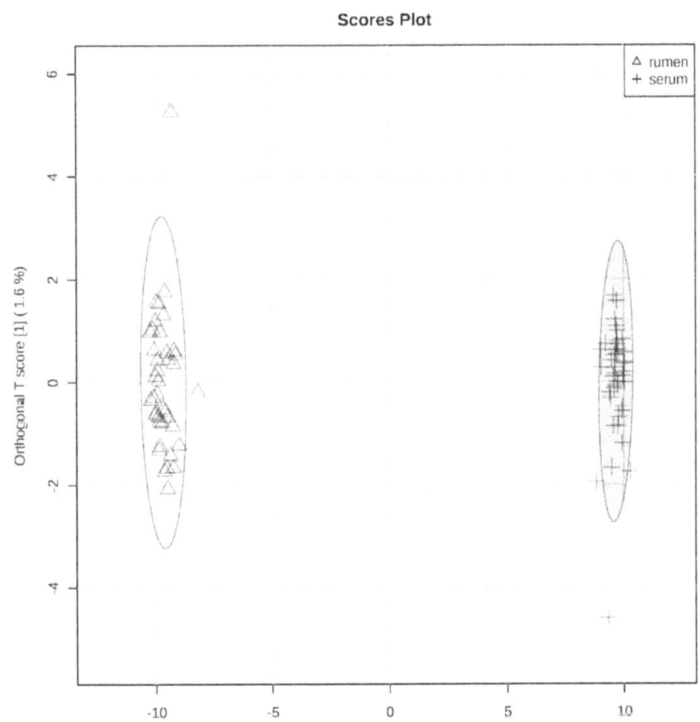

Figure 1. Orthogonal partial least squares discriminant analysis (O-PLS-DA) visualizing separation of rumen fluid (triangle) and serum (plus-sign) metabolomes. Ellipse represents a 95% confidence interval.

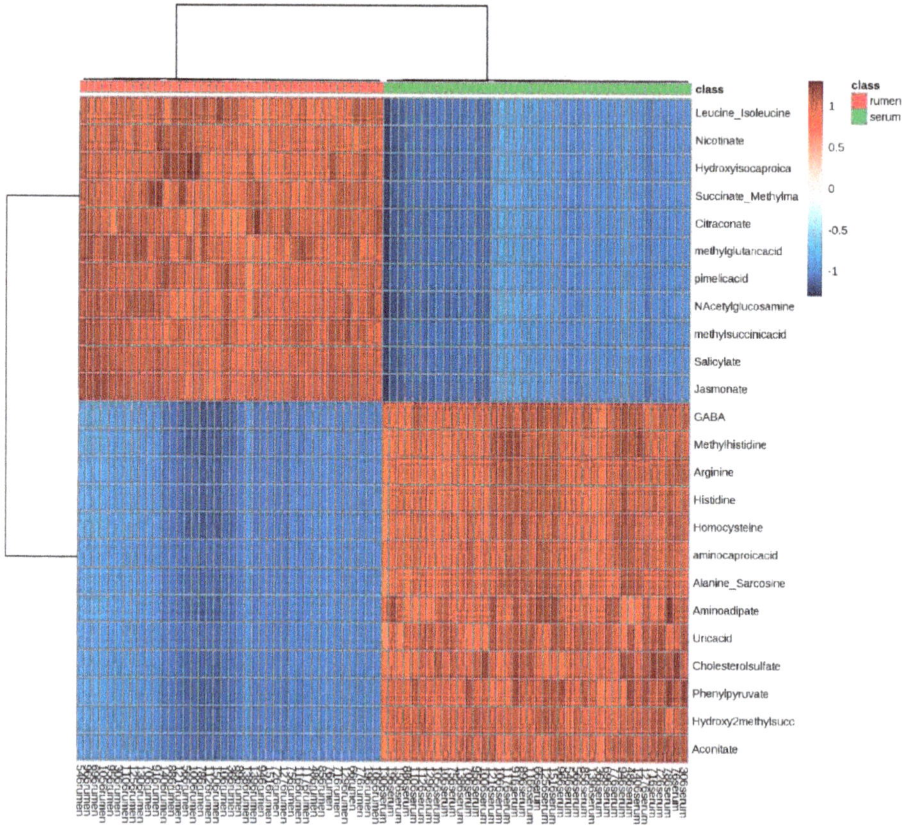

Figure 2. Heatmap of top 25 metabolites of rumen fluid and serum metabolomes by individual steers. Rumen fluid is represented by the red square at the top of the heatmap and serum metabolites are represented by the green squares.

Table 1. Rumen fluid metabolites that significantly differed by seed type.

Metabolite	Seed Type †		*p* Value €
	E+	E−	
Dihydroxybenzoate	$5.90 \times 10^7 \pm 5.35 \times 10^{6\,B}$	$8.43 \times 10^7 \pm 6.26 \times 10^{6\,A}$	0.05
Adenine *	$2.30 \times 10^7 \pm 1.13 \times 10^{7\,B}$	$6.84 \times 10^7 \pm 1.32 \times 10^{7\,A}$	0.02
CMP *	$9.17 \times 10^5 \pm 7.64 \times 10^{5\,B}$	$3.17 \times 10^6 \pm 8.95 \times 10^{5\,A}$	0.04
Deoxyuridine *	$8.04 \times 10^5 \pm 2.71 \times 10^{5\,B}$	$1.74 \times 10^6 \pm 3.18 \times 10^{5\,A}$	0.02
Glutamate *	$7.18 \times 10^7 \pm 2.17 \times 10^{7\,B}$	$1.57 \times 10^8 \pm 2.54 \times 10^{7\,A}$	0.05
Guanosine *	$3.00 \times 10^5 \pm 1.44 \times 10^{5\,B}$	$8.63 \times 10^5 \pm 1.69 \times 10^{5\,A}$	0.05
Homoserine/threonine	$1.02 \times 10^7 \pm 8.90 \times 10^{5\,B}$	$6.65 \times 10^6 \pm 7.60 \times 10^{5\,A}$	0.05
Hypoxanthine *	$4.40 \times 10^7 \pm 1.66 \times 10^{7\,B}$	$1.17 \times 10^8 \pm 1.94 \times 10^{7\,A}$	0.01
Uracil *	$5.76 \times 10^7 \pm 1.19 \times 10^{7\,B}$	$1.08 \times 10^8 \pm 1.39 \times 10^{7\,A}$	0.02
Xanthine *	$1.79 \times 10^8 \pm 4.34 \times 10^{7\,B}$	$3.48 \times 10^8 \pm 5.09 \times 10^{7\,A}$	0.01
Xylose *	$3.63 \times 10^6 \pm 1.05 \times 10^{6\,B}$	$8.69 \times 10^6 \pm 1.23 \times 10^{6\,A}$	0.01

* Analysis based on ranked data; † values are measured as mean ± SEM of area under the peak; € significance determined at $p \leq 0.05$ based on FDR-corrected *p*-values; [AB] within-row represent groupings based on Fisher's LSD.

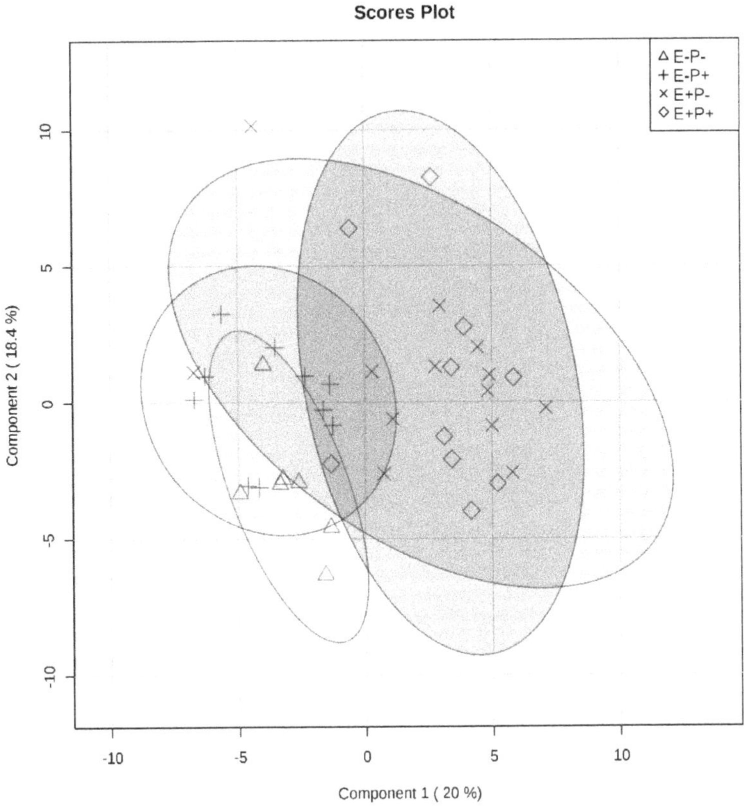

Figure 3. Partial least squares discriminant analysis (PLS-DA) visualizing differences in rumen fluid metabolomes between endophyte-free seed without isoflavones (triangle), endophyte-free with isoflavones (plus-sign), endophyte-infected without isoflavones (multiplication-sign), and endophyte-infected with isoflavones (diamond) treatment groups. Ellipse represents a 95% confidence interval.

Table 2. Rumen fluid metabolic pathways impacted by seed type or isoflavone treatments.

Pathway	FDR	Impact	p Value
Seed Type			
Purine metabolism	2.73×10^{-4}	0.338	6.89×10^{-6}
Arginine and proline metabolism	2.73×10^{-4}	0.075	1.07×10^{-5}
Pentose and glucuronate interconversions	3.42×10^{-4}	0	2.01×10^{-5}
Beta-Alanine metabolism	3.69×10^{-4}	0	2.89×10^{-5}
Pyrimidine metabolism	3.99×10^{-4}	0.494	5.92×10^{-5}
Pantothenate and CoA biosynthesis	3.99×10^{-4}	0.229	6.7×10^{-5}
Aminoacyl-tRNA biosynthesis	3.99×10^{-4}	0.2	7.81×10^{-5}
Tyrosine metabolism	3.99×10^{-4}	0	7.83×10^{-5}
Novobiocin biosynthesis	3.99×10^{-4}	0	7.83×10^{-5}
Thiamine metabolism	3.99×10^{-4}	0	7.83×10^{-5}
Phenylalanine metabolism	7.94×10^{-4}	0.001	1.71×10^{-4}

Table 2. *Cont.*

Pathway	FDR	Impact	*p* Value
Phenylalanine, tyrosine, and tryptophan biosynthesis	0.001	4.60×10^{-4}	2.78×10^{-4}
Carbapenem biosynthesis	0.001	0	3.25×10^{-4}
Butanoate metabolism	0.001	0	3.25×10^{-4}
Porphyrin and chlorophyll metabolism	0.001	0	3.25×10^{-4}
Pentose phosphate pathway	0.003	0.07	0.001
Amino sugar and nucleotide sugar metabolism	0.004	0.109	0.001
Glutathione metabolism	0.004	0.014	0.002
D-Glutamine and D-glutamate metabolism	0.004	0.172	0.002
Nitrogen metabolism	0.004	0	0.002
Isoflavone Treatment			
Methane metabolism	0.84824	0.154	0.032
Sulfur metabolism	0.84824	0	0.033

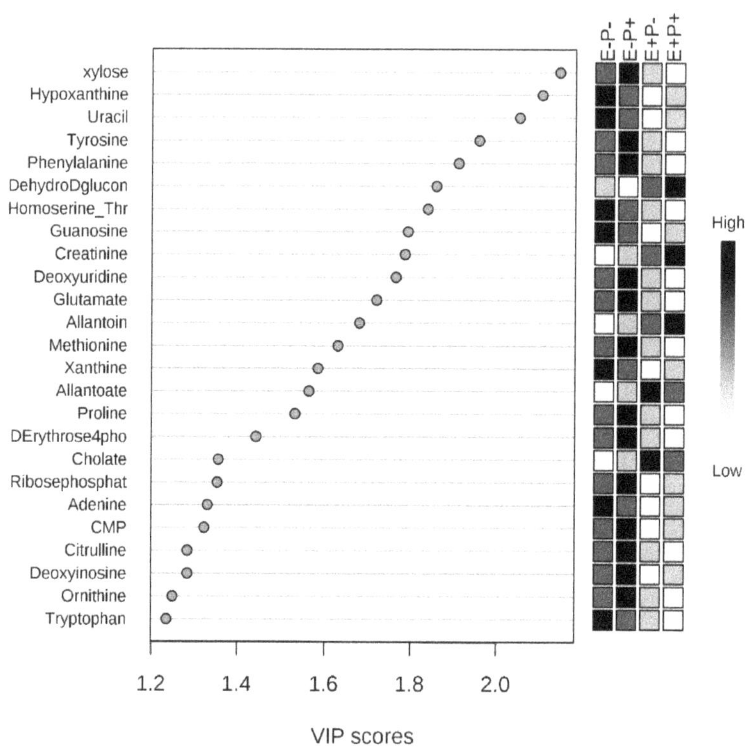

Figure 4. Variable importance in the projection (VIP) plot indicates xylose to have the greatest influence on the differences in rumen fluid metabolomes between all treatment groups.

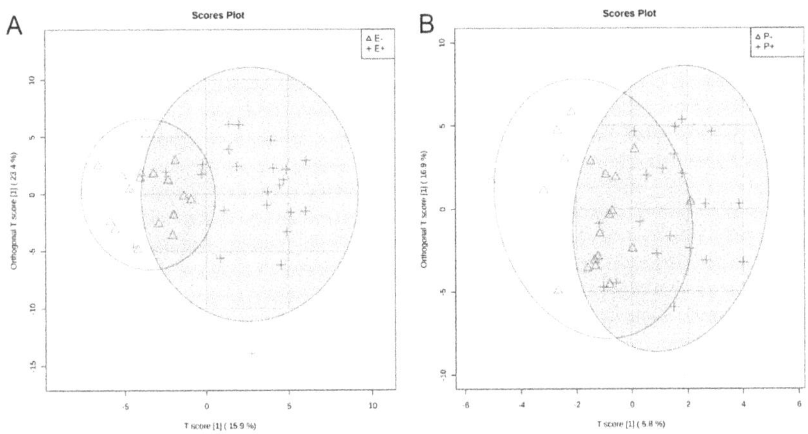

Figure 5. Orthogonal partial least squares discriminant analyses (O-PLS-DA) visualizing differences in rumen fluid metabolomes by seed type (**A**) and isoflavone (**B**) treatments. For seed type (**A**), endophyte-free (E−) steers are represented by a triangle and endophyte-infected (E+) steers by a plus-sign. For isoflavone treatments (**B**), steers receiving isoflavones (P+) are represented by a plus-sign and without isoflavones (P−) by a triangle. Ellipse represents a 95% confidence interval.

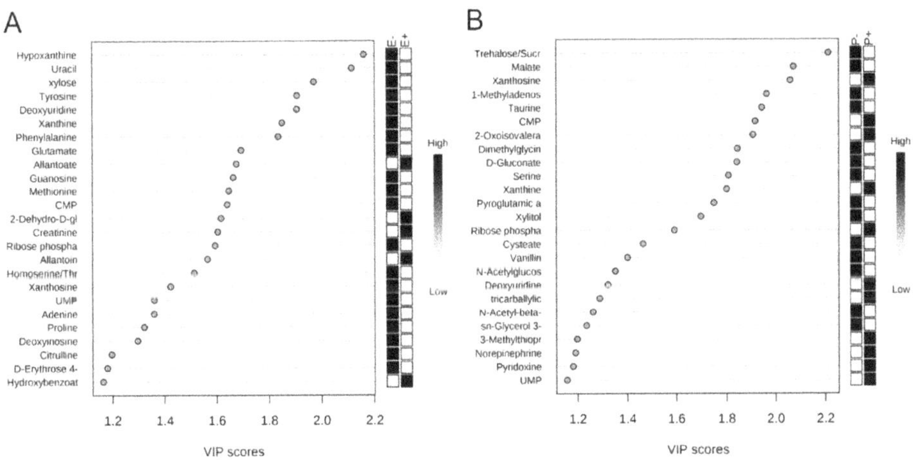

Figure 6. Variable importance in the projection (VIP) plots indicate hypoxanthine to have the greatest influence on the differences in rumen fluid metabolomes between endophyte-free (E−) and endophyte-infected (E+) seed treatment groups (**A**), and trehalose sucrose to have the greatest influence between isoflavone treated (P+) and control (P−) groups (**B**).

2.3. Serum Metabolome

All identified serum metabolites are presented in Supplementary File 2 with means and standard errors of the mean by treatment combination group. The serum metabolome was first analyzed by treatment combination group, isoflavone × seed type. The PLS-DA analysis indicated significant overlap among groups, with partial separation between the E+P+ and E−P− groups (Figure 7). Correlation analyses were performed to determine the correlation of individual serum metabolites with the treatment combination groups and VIP were generated to determine the metabolites that contributed to variation in serum metabolomes among treatment combination groups. Pantothenate was negatively correlated with interaction of seed type × isoflavone treatments (r = −0.29) and had the largest impact on metabolome differences, although not significant (p = 0.07; Figure 8).

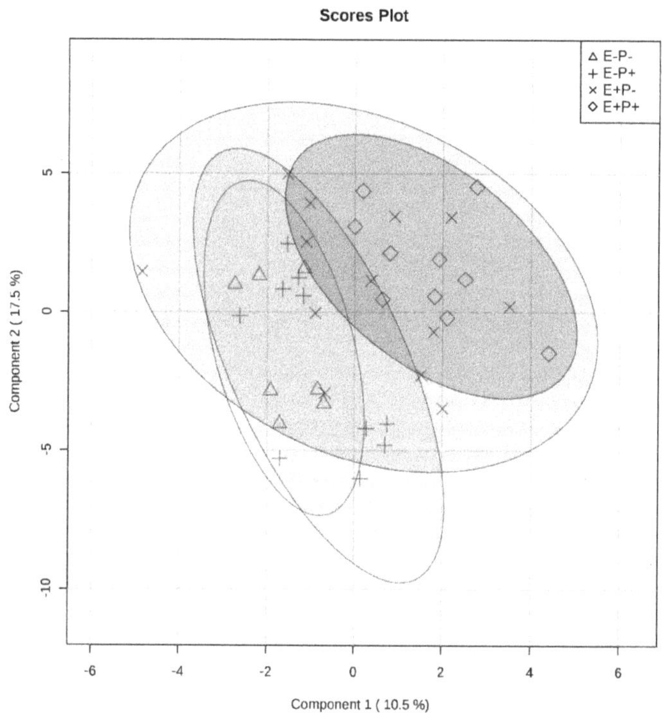

Figure 7. Partial least squares discriminant analysis (PLS-DA) visualizing differences in serum metabolomes between all treatment groups: endophyte-free seed without isoflavones (triangle), endophyte-free with isoflavones (plus-sign), endophyte-infected without isoflavones (multiplication-sign), and endophyte-infected with isoflavones (diamond). Ellipse represents a 95% confidence interval.

Similar to rumen fluid, no individual serum metabolites or metabolic pathways were affected by the interaction of seed type and isoflavone treatment ($p > 0.05$). The serum metabolome was then analyzed by the main effects of seed type or isoflavone treatment. In order to visualize the difference in serum metabolomes, O-PLS-DA analyses were generated for seed type (Figure 9A) and isoflavone treatment (Figure 9B). For seed type, partial separation was observed between E+ and E− seed groups (Figure 9A). However, complete separation of serum metabolomes was illustrated between steers receiving isoflavones and those that did not receive isoflavones (Figure 9B). Correlation analysis indicated AMP was negatively correlated with seed treatment (r = −0.35) and determined by VIP analysis to have the greatest impact on serum metabolome differences between E+ and E− steers (p = 0.03; Figure 10A). Between isoflavone treatment groups, citrulline was positively correlated (r = 0.47) and had the greatest impact on serum metabolome differences (p = 0.003; Figure 10B). Seven metabolites differed significantly as a result of isoflavone treatment ($p < 0.05$, Table 3), while no metabolites differed as a result of seed type ($p > 0.05$). Thirteen metabolic pathways differed ($p < 0.05$) as a result of seed type including glyoxylate and dicarboxylate metabolism; arginine biosynthesis; and alanine, aspartate, and glutamate metabolism ($p < 0.01$; Table 4). For isoflavone treatments, eight metabolic pathways were affected ($p < 0.05$), including pyrimidine metabolism and arginine and proline metabolism ($p < 0.01$; Table 4).

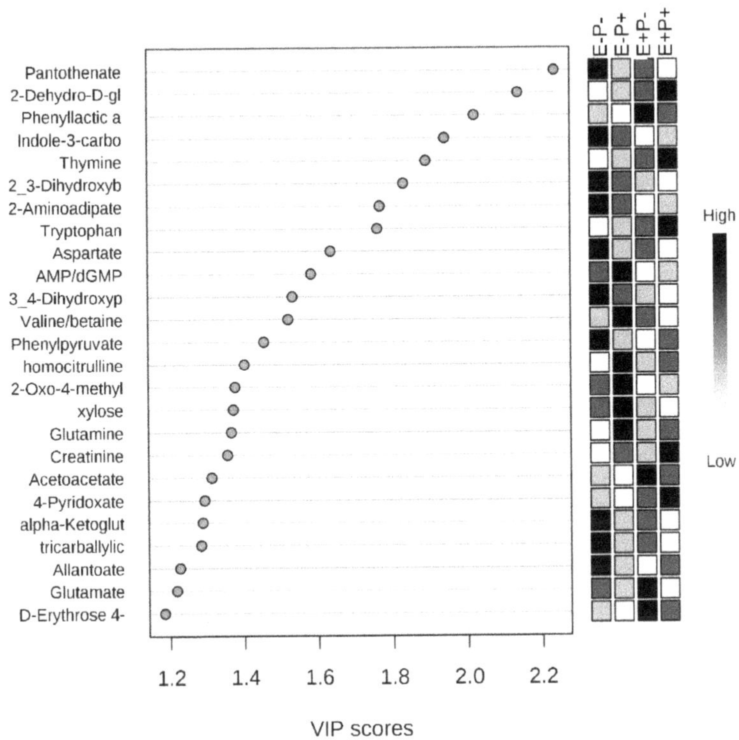

Figure 8. Variable importance in the projection (VIP) plot indicates pantothenate to have the greatest influence on the differences in serum metabolomes between all treatment groups.

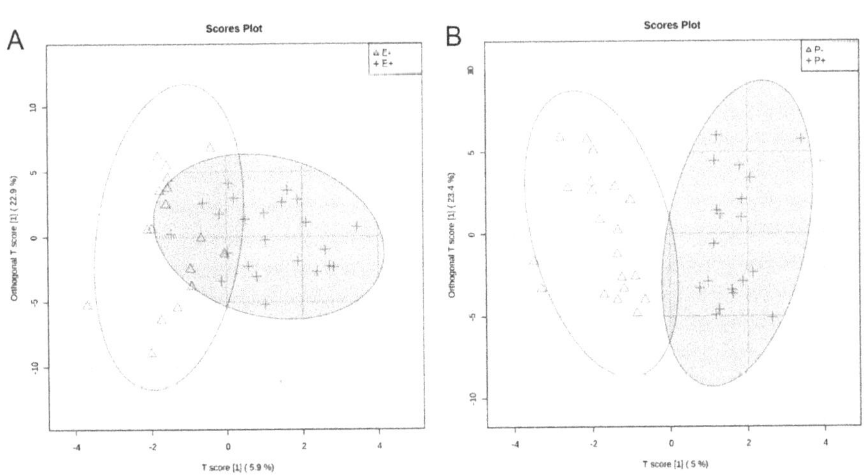

Figure 9. Orthogonal partial least squares discriminant analyses (O-PLS-DA) visualizing differences in serum metabolomes by seed type (**A**) and isoflavone (**B**) treatments. For seed type (**A**), endophyte-free (E−) steers are represented by a triangle and endophyte-infected (E+) steers by a plus-sign. For isoflavone treatments (**B**), steers receiving isoflavones (P+) are represented by a plus-sign and without isoflavones (P−) by a triangle. Ellipse represents a 95% confidence interval.

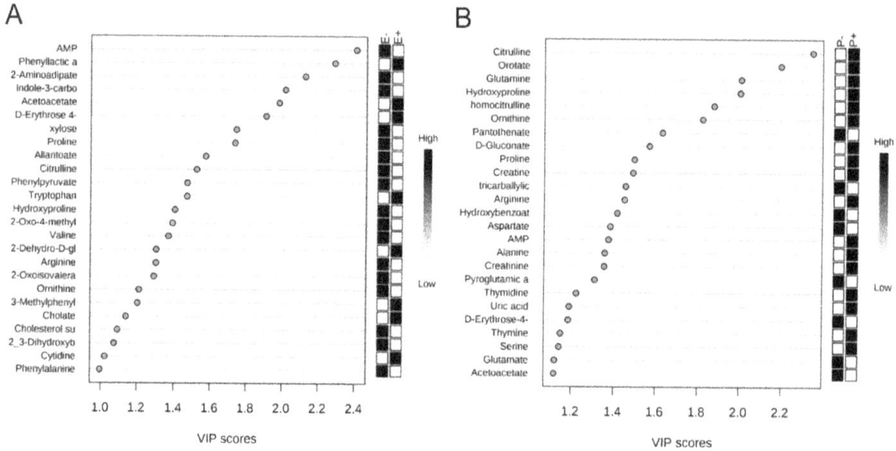

Figure 10. Variable importance in the projection (VIP) plot indicates AMP to have the greatest influence on the differences in serum metabolomes between endophyte-free (E−) and endophyte-infected (E+) seed treatment groups (**A**), and citrulline to have the greatest influence between isoflavone treated (P+) and control (P−) groups (**B**).

Table 3. Individual serum metabolites that significantly differed by isoflavone treatment.

Metabolite	Isoflavone Treatment †		p Value €
	P+	P−	
Histidine *	$8.50 \times 10^6 \pm 8.94 \times 10^5$	$1.05 \times 10^7 \pm 9.37 \times 10^5$	0.05
Cytidine *	$1.51 \times 10^6 \pm 4.89 \times 10^{5\,B}$	$2.67 \times 10^7 \pm 5.12 \times 10^{5\,A}$	0.01
Pantothenate	$6.64 \times 10^6 \pm 1.84 \times 10^{6\,B}$	$1.51 \times 10^7 \pm 1.93 \times 10^{6\,A}$	0.01
Homocysteine	$1.47 \times 10^6 \pm 1.28 \times 10^{5\,B}$	$2.02 \times 10^6 \pm 1.35 \times 10^{5\,A}$	0.02
Allantoin	$1.94 \times 10^8 \pm 1.14 \times 10^{7\,B}$	$2.37 \times 10^8 \pm 1.19 \times 10^{7\,A}$	0.03
GABA	$9.68 \times 10^5 \pm 1.40 \times 10^{5\,B}$	$1.41 \times 10^6 \pm 1.44 \times 10^{5\,A}$	0.05
Methylhistidine	$8.35 \times 10^5 \pm 6.25 \times 10^4$	$1.04 \times 10^6 \pm 6.51 \times 10^4$	0.05

* Analysis based on ranked data; † values are measured as mean ± SEM of area under the peak; € significance determined at $p \leq 0.05$ based on FDR-corrected p-values; AB within-row represent groupings based on Fisher's LSD.

Table 4. Serum metabolic pathways affected by seed type and isoflavone treatments.

Pathway	FDR	Impact	p Value
Seed Type			
Glyoxylate and dicarboxylate metabolism	0.013	0.11	0.005
Arginine biosynthesis	0.013	0.51	0.006
Alanine, aspartate, and glutamate metabolism	0.015	0.73	0.007
Cysteine and methionine metabolism	0.051	0.14	0.024
Glycine, serine, and threonine metabolism	0.054	0.16	0.029
Ubiquinone and other terpenoid-quinone biosynthesis	0.054	0	0.029
Aminobenzoate degradation	0.054	0	0.029
Vitamin B6 metabolism	0.069	0.05	0.039
Monobactam biosynthesis	0.069	0	0.041
Lysine biosynthesis	0.069	0	0.041
Nicotinate and nicotinamide metabolism	0.069	0.06	0.042
Tryptophan metabolism	0.07	0	0.044
Cyanoamino acid metabolism	0.076	0	0.05

Table 4. *Cont.*

Pathway	FDR	Impact	p Value
Isoflavone Treatment			
Pyrimidine metabolism	0.151	0.37	0.007
Arginine and proline metabolism	0.151	0.19	0.008
D-glutamine and D-glutamate metabolism	0.151	0.17	0.013
Nitrogen metabolism	0.151	0	0.013
Arginine biosynthesis	0.151	0.51	0.015
Glutathione metabolism	0.173	0.01	0.02
Purine metabolism	0.209	0.09	0.029
Glyoxylate and dicarboxylate metabolism	0.269	0.13	0.045

3. Discussion

The overall reductions in animal performance due to fescue toxicosis are estimated to cost the cattle industry over $2 billion annually [13,14]. Therefore, it is vital to discover management methods to reduce the impact and improve the efficiency of beef production. The objective of the current study was to use untargeted metabolomics to evaluate tall fescue seed and isoflavone consumption effects on metabolic intermediates, outputs, and pathways in the rumen and serum.

The metabolomes of the rumen and circulatory environments were first compared, independent of treatment groups, which observed distinctly unique metabolomes according to principal coordinate and abundance analyses. Highly abundant metabolites in each environment were not shared or only present in low abundances between the two environments. The metabolites identified between these different body systems are likely a result of the specific physiological functions of each system in the ruminant. The microbiome is a major contributor to rumen metabolome, as it supplies over 70% of the ruminant's required nutrients [15]. These microbes are highly metabolically active in order to break down feedstuffs and release metabolites to complement host metabolism of which metabolites originate from the plants and other feedstuffs consumed [16]. Therefore, the majority of metabolites identified in the rumen are of xenobiotic origin. By evaluating the rumen metabolome, the effects of tall fescue seed and isoflavone consumption on rumen microbial metabolic processes can be inferred. The serum metabolites, however, are a result of absorbed metabolic products from the rumen and other organs. Tissues throughout the body produce intermediary metabolites from protein, carbohydrate, and lipid metabolism for energy production to perform physiological functions. These metabolites are then absorbed into the blood and can travel through the circulatory system to other tissues for further catabolic or anabolic processing. Therefore, the metabolome of the circulatory system is typically dominated by endogenous metabolites. Because of this systemic nature, metabolites in blood have been used as potential biomarkers to predict feed utilization [17] and production parameters [18], as well as evaluate responses to disease [19,20] and stress [21]. Evaluating the serum metabolome will determine the systemic metabolic response to tall fescue seed and isoflavone supplementation. Together, the effects on individual metabolites and metabolic pathways in rumen fluid and serum will determine how alterations in microbial and host metabolism contribute to symptoms of fescue toxicosis or the benefits isoflavones may contribute to mitigate these detrimental impacts.

Reductions in average daily gain and delayed development of beef cattle are a major consequence of fescue toxicosis [8,22]. The rumen microbiome is crucial for providing nutrients needed by the host for energy requirements and muscle development. As the microbiome has previously been shown to be affected by consuming endophyte infected tall fescue [12,23], the metabolites and other products produced by the rumen microorganisms may be altered, potentially contributing to reductions in growth and feed efficiency. The metabolites produced by the rumen microorganisms are a result of the richness of the rumen microbiome. Several of these metabolites released may be related to the consumption of ergot alkaloids concentrated on the endophyte infected tall fescue seed. Many of the rumen metabolites have a relationship with purine, carbohydrate, and nucleic acid metabolism, such as hypoxanthine, xylose, and uracil, respectively [24,25]; these metabolites are related to feed efficiency parameters. Clemmons et al. [26] found that these metabolites are bio-indicators of feed efficiency in cattle showing low residual feed intake. Interestingly, they are negatively correlated to seed type in the rumen fluid of the current study; it is evident that animals are being affected by the detrimental symptoms of tall fescue toxicosis, failing to gain weight, and being less feed-efficient. Additionally, we did not observe a large number of different metabolites because of the reduction of the rumen microbiome. As a normal rumen environment, the significant presence of other metabolites, which are crucial for the ergot alkaloids metabolism and production of volatile fatty acids, was expected and improves feed efficiency and rumen microbial richness.

Vasoconstriction, induced by ergot alkaloids, occurs throughout the body, resulting in multiple observed symptoms of fescue toxicosis [4,6]. Specifically, the contractility of the mesenteric vasculature surrounding the digestive tract is affected by the consumption of ergot alkaloids, potentially affecting nutrient absorption and subsequent host metabolism [5]. However, the consumption of isoflavones promotes vasodilation to increase blood flow and mitigate fescue toxicosis effects [10]. Ideally, oxygen and nutrient delivery to tissues is improved, thus benefiting host metabolism. Evaluating serum metabolites during induced fescue toxicosis and treatment with isoflavones may indicate the changes in metabolism systemically. The serum metabolome in the current study was greatly affected by isoflavone treatments with complete separation of animals' global serum metabolomes between treatment groups; no metabolites differed as a result of seed type. Citrulline was identified as having the greatest influence on serum metabolomes between isoflavone treatment groups. Citrulline is an intermediary metabolite in the urea cycle, a metabolic process crucial for providing non-protein nitrogen to the ruminant [27]. As isoflavones inhibit hyper-ammonia producing bacteria in the rumen, this reduces the amount of protein degradation, leading to decreased ammonia and nitrogen availability [11,28]. Multiple metabolic pathways related to the urea cycle such as arginine biosynthesis and metabolism, pyrimidine metabolism, and nitrogen metabolism were affected in serum metabolites by isoflavone treatments. Therefore, the effect of citrulline on the serum metabolome due to isoflavone treatment may be a result of changes in the urea cycle by improving protein availability to the ruminant for muscle development.

Pantothenate was considered to be a major contributor to differences observed in the global serum metabolome among all treatment combinations, but was significantly higher in abundance in steers receiving isoflavone treatment. Previous studies have indicated different levels of pantothenate in the serum of animals differing in feed efficiency, with more feed efficient animals having greater serum levels of pantothenate [17,26]. Pantothenate is a key intermediary metabolite for the formation of Coenzyme A, which is crucial for amino acid and lipid metabolism for ruminant muscle development [29]. Additionally, the majority of metabolic pathways affected by isoflavone treatment in the current study were related to amino acid metabolism and biosynthesis. As animals experiencing fescue toxicosis often have low average daily gains [8], the use of isoflavone supplementation may mitigate the weight gain and growth consequences of fescue toxicosis. The greater amount of available pantothenate in the serum of animals consuming isoflavones, with changes in the urea cycling of the ruminant, may improve growth and muscle development in steers affected by fescue toxicosis.

A classic symptom of fescue toxicosis is a significant reduction of prolactin secretion; this is due to the similar homology of ergot alkaloids with the neurotransmitter dopamine. Ergot alkaloids will act as an agonist by binding dopamine receptors, preventing the release of prolactin [3,30]. Tyrosine is a precursor for the generation of the neurotransmitter dopamine [31]. The current study found tyrosine metabolism to be affected by seed type in the rumen; tyrosine metabolic pathways are influenced as the signals for dopamine production are reduced during fescue toxicosis. Additionally, the study found tryptophan metabolism was affected by seed type in the serum. Tryptophan is a key amino acid in regulating protein synthesis, specifically in muscle development, of multiple species [32,33]. The effects of tryptophan on muscle development are through the IGF-1 pathway [33]. Tryptophan is a precursor to the neurotransmitter serotonin, which stimulates the production of

IGF-1 [34]. Supplementation of rumen protected tryptophan improved weight gain and feed efficiency of ruminants [35,36]. The observed impact of seed type on tryptophan metabolism in the serum may indicate reduced production of serotonin and subsequent IGF-1 signaling for muscle development. Together, the impacts of fescue toxicosis on tyrosine and tryptophan metabolism were also observed previously in the plasma of steers consuming endophyte-infected or endophyte-free seed [37,38]. Therefore, tall fescue seed consumption alters neurotransmitter development, leading to commonly observed symptoms of fescue toxicosis.

4. Conclusions

In conclusion, the rumen metabolome was largely impacted by seed type, while the serum metabolome was influenced by isoflavone supplementation. In the rumen, the impact of the seed type involved carbohydrate and nucleic acids metabolism products of the fescue seed diet inclusion. In the serum, differences in global metabolomes and individual metabolites involved in urea cycling and amino acid metabolic pathways were identified in animals receiving isoflavones and those that did not. Although the low dose of isoflavones administered to cattle indicated effects on the serum and rumen metabolome, further research is needed to determine the effects at other doses. Future applications may lead to the dietary inclusion of isoflavones to reduce the harmful effects of tall fescue toxicosis.

5. Materials and Methods

All experimental procedures involving animals were approved by the University of Tennessee Institutional Animal Care and Use Committee. The ethic approval code (IACUC) was 2540-0617 and was approved on 20 June 2017.

5.1. Experimental Design and Sample Collection

Experimental design, animal treatments, and sample collection methods have been previously described in Melchior et al. [12]. Briefly, this study used 36 purebred Angus steers of approximately eight months of age weighing 250 ± 20 kg from Ames Plantation in Grand Junction, TN. Steers were transported to the Plateau Research and Education Center (PREC) in Crossville, TN for the trial, as previously described [12]. Steers were allowed a 10 d acclimation period to the diet formulated to provide 11.57% crude protein and 76.93% total digestible nutrients (DM basis). The GrowSafe System© (GrowSafe Systems Ltd., Calgary, AB, Canada) was used to monitor feed intake. Prior to the beginning of the trial, steers were genotyped for the DRD2 receptor gene, which can influence cattle's response to fescue toxicity [12]. Using this information, the study was blocked on DRD2 genotype, implementing a randomized complete block design. A 2 × 2 factorial arrangement of treatments was utilized with two types of tall fescue including endophyte-infected (E+) and endophyte-free fescue (E−), and treatment with Promensil© (P+) or without (P−) to provide red clover isoflavones. This combination of treatments resulted in four treatment groups: (1) endophyte-infected seed without Promensil (E+P−), (2) endophyte-infected with Promensil (E+P+), (3) endophyte-free without Promensil (E−P−), and (4) endophyte-free with Promensil (E−P+). Within each genotype block, steers were randomly assigned to treatments with nine steers per treatment group. The feed trial occurred over 21 days. In order to provide a consistent amount of ergot alkaloids, endophyte-infected tall fescue seed heads were incorporated into feed to provide a minimum of 0.011 mg ergovaline plus ergovalinine × kg of body weight^{-1} (BW) per day [3]. Seed heads were ground through a 5 mm screen using a Wiley Mill (Thomas Scientific, Swedesboro, NJ, USA) and included in feed. A total of 943 mg of isoflavones was provided daily before morning feeding based on previously established dosages [10] using a 28.4 g bolus (Torpac, Inc., Fairfield, NJ, USA) to provide 24.7 g of Promensil. Melchior et al. [12] previously reported the analysis and information of the components present in Promensil, and steers' response to endophyte-infected seed and altered performance parameters. On the final day of the trial (day 21), approximately 9 mL of blood was collected from the coccygeal vein using a serum separator tube (Corvac, Sherwood Medical., St. Louis, MO, USA), and approximately 100 mL of rumen content was collected via oro-gastric lavage. Blood samples were centrifuged at 2000× *g* and 4 °C for 20 min, and serum was transferred to 2 mL microvials and stored at −80 °C until metabolite extractions. Rumen samples were centrifuged at 6000× *g* at 4 °C for 20 min. The supernatant was aspirated and filtered through a 0.22 µm syringe filter, transferred to 2 mL microvials, and stored at −80 °C until metabolite extraction.

5.2. Metabolite Extraction and Identification

Metabolites were extracted and analyzed as previously described [17] at the UTK Biological and Small Molecule Mass Spectrometry Core (BSMMSC). Briefly, 50 µL of filtered rumen fluid and 50 µL of serum from each steer were extracted using 0.1% formic acid in acetonitrile/water/methanol (2:2:1) using a previously described method [39]. Mobile phases consisted of A: 97:3 water/methanol with 11 mM tributylamine and 15 mM acetic acid and B: methanol, and a gradient consisting of the following: 0.0 min, 0% B; 2.5 min 0% B; 5.0 min, 20% B; 7.5 min, 20% B; 13 min, 55% B; 15.5 min, 95% B; 18.5 min, 95% B; 19 min, 0% B; and 25 min, 0% B; Synergy Hydro-RP column (100 × 2 mm, 2.5 µm particle size) was used to separate metabolites. The flow rate was set to a constant 200 µL/min and the column temperature was kept at 25 °C. A Dionex UltiMate 3000 UPLC system (Thermo Fisher Scientific, Waltham, MA) with an autosampler tray maintained at 4 °C was used to introduce a 10 µL sample to an Exactive Plus Orbitrap MS (Thermo Fisher Scientific, Waltham, MA, USA) using negative electrospray ionization (ESI) with a capillary temperature of 300 °C; spray voltage of 3 kV; and nitrogen sheath and sweep gas at 25 and 3 units, respectively. Data acquisition was done in negative ion mode with a full-scan covering the range of 72–1000 *m/z* at 140,000 resolution with automatic gain control of 3×10^6 ions [40]. Metabolites are annotated using exact mass of the [M−H]− (±5 pmm) ion and known retention times (±0.3 min) generated from an in-house curated database. The database was created from the analysis of authentic standards and consisted of 300 compounds across various metabolic pathways, focusing on water soluble metabolites in pathways conserved among a diverse array of organisms.

5.3. Metabolite Identification

Data were analyzed similarly to those of Clemmons et al. [17]. The Xcalibur MS software (Thermo Electron Corp., Waltham, MA) was used to produce raw files, which were then converted to mzML format using ProteoWizard [41]. The software package Metabolomic Analysis and Visualization Engine for LC–MS Data (MAVEN) [42] was used to identify peaks using converted mzML files. MAVEN identifies metabolites based on non-linear retention time correction and calculates peak areas across samples, using a preliminary mass error of ±20 ppm and a retention time window of 5 min. The UTK BSMMSC used a library of 263 retention time-accurate *m/z* pairs taken from MS1 spectra for final metabolite annotations. These are based on expansions of previous work [40] and have been replicated at the UTK BSMMSC. The eluted peak of the annotated metabolite had to be found within 2 min of the expected retention time, and the metabolite mass had to be within ±5 ppm of the expected value to be identified as a known compound. The compound area of each peak was calculated using the Quan Browser function of the Xcalibur MS Software (Thermo Electron Corp., Waltham, MA, USA).

5.4. Data Analysis

Metabolomic data were analyzed using MetaboAnalyst 4.0 [43] and SAS 9.4 (SAS Institute, Cary, NC, USA). For data analysis in MetaboAnalyst, data were first pre-processed. Metabolite data were filtered using interquartile range, normalized by median, log transformed, and auto scaled prior to analysis in MetaboAnalyst 4.0. First, the rumen and serum metabolomes were collectively compared in order to determine similarity of rumen fluid and serum metabolomes for possible overlap or comparison. The rumen and serum metabolomes were visualized using orthogonal partial least squares discriminant analysis (O-PLS-DA) and partial least squares discriminant analysis (PLS-DA) with 2000 permutations. Model fitting for the O-PLS-DA was assessed using R2Y with prediction power determined using the Q2 metric. A heatmap was generated with the top 25 metabolites to illustrate differences in serum and rumen fluid metabolomes by steer. Next, rumen fluid and serum metabolomes were analyzed separately by treatment combination (i.e., isoflavone × seed type), isoflavone, and seed type. Within each of these, data were visualized via PCA and O-PLS-DA with 2000 permutations, and metabolomes by individual steers were illustrated using heatmaps. Correlation analyses between the top 25 metabolites and treatment groups or combinations were performed for both rumen fluid and serum metabolomes, as well as variable importance in projections (VIP) of the top 25 metabolites. Finally, pathway analyses were performed to determine metabolic pathways that were significantly impacted in rumen fluid and serum by isoflavone or seed type using a global test with relative-betweenness centrality and a reference pathway of Escherichia coli K-12 MG1655 [44].

Raw data were further analyzed in SAS 9.4 (SAS Institute, Cary, NC, USA), First, data were analyzed for normality using the UNIVARIATE procedure, and were considered normal with a Shapiro–Wilk statistic of ≥0.90 and visual observation of histograms and q-q plots. Data that were normally distributed were analyzed with a mixed model analysis of variance (ANOVA) using the GLIMMIX procedure with fixed effects of seed type, isoflavone treatment, and their interaction with the random effect of genotype × isoflavone × seed type. Metabolites that did not follow a normal distribution were fixed ranked and then analyzed using a mixed model ANOVA using the GLIMMIX procedure with fixed effects of seed type, isoflavone treatment, and their interaction with random effect of genotype × isoflavone × seed type.

Supplementary Materials: The following are available online at http://www.mdpi.com/2072-6651/12/12/744/s1, Supplementary File 1: Means and standard error of the means of all metabolites in the rumen fluid between treatment groups; Supplementary File 2: Means and standard error of the means of all metabolites in the serum between treatment groups.

Author Contributions: Conceptualization, E.A.M.-T. and P.R.M.; Formal analysis, B.A.C.; Funding acquisition, P.R.M.; Investigation, T.B.A.-S. and J.F.C.; Methodology, B.A.C., G.E.B., M.D.F., J.L.K., H.J., J.P.G., K.J.M. and P.R.M.; Project administration, P.R.M.; Resources, E.A.M.-T.; Visualization, T.B.A.-S. and B.A.C.; Writing—original draft, T.B.A.-S., B.A.C., J.F.C., K.J.M. and P.R.M.; Writing—review & editing, T.B.A.-S., E.A.M.-T., B.A.C., J.F.C., J.L.K., K.J.M. and P.R.M. All authors have read and agreed to the published version of the manuscript.

Funding: This research was funded by the University of Tennessee CVM, Center for Excellence in Livestock and Human Diseases.

Acknowledgments: We acknowledge the USDA-NIFA Hatch/Multistate Project W4177-TEN00524-Enhancing the Competitiveness and Value of U.S. Beef. We thank Ames Plantation, UT Plateau Research and Education Center, Rebecca Payton, Lezek Wojakiewicz, and Gloria Gellin (USDA-ARS) for their assistance. We also thank Shawn Campagna and Hector Castro at the UTK Biological and Small Molecule Mass Spectrometry Core (BSMMSC).

Conflicts of Interest: The authors declare no conflict of interest.

References

1. Poole, R.K.; Poole, D.H. Impact of Ergot Alkaloids on Female Reproduction in Domestic Livestock Species. *Toxins* **2019**, *11*, 364. [CrossRef] [PubMed]
2. Moore, J.D.; Carlisle, A.E.; Nelson, J.A.; McCulley, R.L. Fungal endophyte infection increases tall fescue's survival, growth, and flowering in a reconstructed prairie. *Restor. Ecol.* **2019**, *27*, 1000–1007. [CrossRef]
3. Klotz, J.L. Activities and Effects of Ergot Alkaloids on Livestock Physiology and Production. *Toxins* **2015**, *7*, 2801–2821. [CrossRef] [PubMed]
4. Strickland, J.R.; Aiken, G.E.; Klotz, J.L. Ergot Alkaloid Induced Blood Vessel Dysfunction Contributes to Fescue Toxicosis. *Forage Grazinglands* **2009**, *7*, 1–7. [CrossRef]
5. Egert, A.M.; Kim, D.H.; Schrick, F.N.; Harmon, D.L.; Klotz, J.L. Dietary exposure to ergot alkaloids decreases contractility of bovine mesenteric vasculature. *J. Anim. Sci.* **2014**, *92*, 1768–1779. [CrossRef]
6. Poole, D.H.; Lyons, S.E.; Poole, R.K.; Poore, M.H. Ergot alkaloids induce vasoconstriction of bovine uterine and ovarian blood vessels. *J. Anim. Sci.* **2018**, *96*, 4812–4822. [CrossRef]
7. Al-Haidary, A.; Spiers, D.; Rottinghaus, G.E.; Garner, G.B.; Ellersieck, M.R. Thermoregulatory ability of beef heifers following intake of endophyte-infected tall fescue during controlled heat challenge. *J. Anim. Sci.* **2001**, *79*, 1780–1788. [CrossRef]

8. Stuedemann, J.A.; Hoveland, C.S. Fescue Endophyte: History and Impact on Animal Agriculture. *J. Prod. Agric.* **1988**, *1*, 39–44. [CrossRef]
9. Roberts, C.A.; Andrae, J. Tall Fescue Toxicosis and Management. *Crop Manag.* **2004**, *3*, 1–18. [CrossRef]
10. Aiken, G.E.; Flythe, M.D.; Kagan, I.A.; Ji, H.; Bush, L.P. Mitigation of Ergot Vasoconstriction by Clover Isoflavones in Goats (Capra hircus). *Front. Vet. Sci.* **2016**, *3*, 17. [CrossRef]
11. Harlow, B.E.; Flythe, M.D.; Kagan, I.A.; Aiken, G.E. Biochanin A (an Isoflavone Produced by Red Clover) Promotes Weight Gain of Steers Grazed in Mixed Grass Pastures and Fed Dried-Distillers' Grains. *Crop Sci.* **2017**, *57*, 506–514. [CrossRef]
12. Melchior, E.A.; Smith, J.K.; Schneider, L.G.; Mulliniks, J.T.; Bates, G.E.; Flythe, M.D.; Klotz, J.L.; Ji, H.; Goodman, J.P.; Lee, A.R.; et al. Effects of endophyte-infected tall fescue seed and red clover isoflavones on rumen microbial populations and physiological parameters of beef cattle. *Transl. Anim. Sci.* **2018**, *3*, 315–328. [CrossRef] [PubMed]
13. Hoveland, C.S. Importance and economic significance of the Acremonium endophytes to performance of animals and grass plant. *Agric. Ecosyst. Environ.* **1993**, *44*, 3–12. [CrossRef]
14. Kallenbach, R.L. BILL E. KUNKLE INTERDISCIPLINARY BEEF SYMPOSIUM: Coping with tall fescue toxicosis: Solutions and realities. *J. Anim. Sci.* **2015**, *93*, 5487–5495. [CrossRef] [PubMed]
15. Bergman, E.N. Energy contributions of volatile fatty acids from the gastrointestinal tract in various species. *Physiol. Rev.* **1990**, *70*, 567–590. [CrossRef] [PubMed]
16. Fontanesi, L. Metabolomics and livestock genomics: Insights into a phenotyping frontier and its applications in animal breeding. *Anim. Front.* **2016**, *6*, 73–79. [CrossRef]
17. Clemmons, B.A.; Mihelic, R.I.; Beckford, R.C.; Powers, J.B.; Melchior, E.A.; McFarlane, Z.D.; Cope, E.R.; Embree, M.M.; Mulliniks, J.T.; Campagna, S.R.; et al. Serum metabolites associated with feed efficiency in black angus steers. *Metabolomics* **2017**, *13*, 147. [CrossRef]
18. Wu, X.; Sun, H.-Z.; Xue, M.; Wang, D.; Guan, L.L.; Liu, J. Serum metabolome profiling revealed potential biomarkers for milk protein yield in dairy cows. *J. Proteom.* **2018**, *184*, 54–61. [CrossRef]
19. Aich, P.; Jalal, S.; Czuba, C.; Schatte, G.; Herzog, K.; Olson, D.J.; Ross, A.R.; Potter, A.A.; Babiuk, L.A.; Griebel, P.J. Comparative Approaches to the Investigation of Responses to Stress and Viral Infection in Cattle. *OMICS A J. Integr. Biol.* **2007**, *11*, 413–434. [CrossRef]
20. Zandkarimi, F.; Vanegas, J.; Fern, X.; Maier, C.S.; Bobe, G. Metabotypes with elevated protein and lipid catabolism and inflammation precede clinical mastitis in prepartal transition dairy cows. *J. Dairy Sci.* **2018**, *101*, 5531–5548. [CrossRef]
21. Liao, Y.; Hu, R.; Wang, Z.; Peng, Q.; Dong, X.; Zhang, X.; Zou, H.; Pu, Q.; Xue, B.; Wang, L. Metabolomics Profiling of Serum and Urine in Three Beef Cattle Breeds Revealed Different Levels of Tolerance to Heat Stress. *J. Agric. Food Chem.* **2018**, *66*, 6926–6935. [CrossRef] [PubMed]
22. Crawford, R.J.; Forwood, J.R.; Belyea, R.L.; Garner, G.B. Relationship between Level of Endophyte Infection and Cattle Gains on Tall Fescue. *J. Prod. Agric.* **1989**, *2*, 147–151. [CrossRef]
23. Mote, R.S.; Hill, N.S.; Skarlupka, J.H.; Turner, Z.B.; Sanders, Z.P.; Jones, D.P.; Suen, G.; Filipov, N.M. Response of Beef Cattle Fecal Microbiota to Grazing on Toxic Tall Fescue. *Appl. Environ. Microbiol.* **2019**, *85*. [CrossRef] [PubMed]
24. Saleem, F.; Bouatra, S.; Guo, A.C.; Psychogios, N.; Mandal, R.; Dunn, S.M.; Ametaj, B.N.; Wishart, D.S. The Bovine Ruminal Fluid Metabolome. *Metabolomics* **2013**, *9*, 360–378. [CrossRef]
25. Artegoitia, V.M.; Foote, A.P.; Lewis, R.M.; Freetly, H.C. Rumen Fluid Metabolomics Analysis Associated with Feed Efficiency on Crossbred Steers. *Sci. Rep.* **2017**, *7*, 1–14. [CrossRef]
26. Clemmons, B.A.; Martino, C.; Powers, J.B.; Campagna, S.R.; Voy, B.H.; Donohoe, D.R.; Gaffney, J.; Embree, M.M.; Myer, P.R. Rumen Bacteria and Serum Metabolites Predictive of Feed Efficiency Phenotypes in Beef Cattle. *Sci. Rep.* **2019**, *9*, 1–8. [CrossRef]
27. Alemneh, T.; Getahun, D.; Akeberegn, D.; Getabalew, M.; Zewdie, D. Urea Metabolism and Recycling in Ruminants. *Biomed. J. Sci. Tech. Res.* **2019**, *20*, 14790–14796. [CrossRef]
28. Flythe, M.D.; Kagan, I. Antimicrobial Effect of Red Clover (Trifolium pratense) Phenolic Extract on the Ruminal Hyper Ammonia-Producing Bacterium, Clostridium sticklandii. *Curr. Microbiol.* **2010**, *61*, 125–131. [CrossRef]
29. Leonardi, R.; Zhang, Y.-M.; Rock, C.O.; Jackowski, S. Coenzyme A: Back in action. *Prog. Lipid Res.* **2005**, *44*, 125–153. [CrossRef]

30. Sibley, D.R.; Creese, I. Interactions of ergot alkaloids with anterior pituitary D-2 dopamine receptors. *Mol. Pharmacol.* **1983**, *23*, 585–593.
31. Fernstrom, J.D. Dietary Precursors and Brain Neurotransmitter Formation. *Annu. Rev. Med.* **1981**, *32*, 413–425. [CrossRef]
32. Lin, F.D.; Smith, T.K.; Bayley, H.S. A Role for Tryptophan in Regulation of Protein Synthesis in Porcine Muscle. *J. Nutr.* **1988**, *118*, 445–449. [CrossRef] [PubMed]
33. Dukes, A.; Davis, C.; El Refaey, M.; Upadhyay, S.; Mork, S.; Arounleut, P.; Johnson, M.H.; Hill, W.D.; Isales, C.M.; Hamrick, M.W. The aromatic amino acid tryptophan stimulates skeletal muscle IGF1/p70s6k/mTor signaling in vivo and the expression of myogenic genes in vitro. *Nutrition* **2015**, *31*, 1018–1024. [CrossRef] [PubMed]
34. Musumeci, G.; Trovato, F.; Avola, R.; Imbesi, R.; Castrogiovanni, P. Serotonin/growth hormone/insulin-like growth factors axis on pre- and post-natal development: A contemporary review. *OA Anat.* **2013**, *1*, 12. [CrossRef]
35. Ma, H.; Cheng, J.; Zhu, X.; Jia, Z. 2011. Effects of rumen-protected tryptophan on performance, nutrient utilization and plasma tryptophan in cashmere goats. *Afr. J. Biotechnol.* **2011**, *10*, 5806–5811.
36. Lee, J.S.; Priatno, W.; Ghassemi Nejad, J.; Peng, D.Q.; Park, J.S.; Moon, J.O.; Lee, H.G. Effect of dietary rumen-protected L-Tryptophan supplementation on growth performance, blood hematological and biochemical profiles, and gene expression in Korean native steers under cold environment. *Animals* **2019**, *9*, 1036. [CrossRef]
37. Mote, R.S.; Hill, N.S.; Uppal, K.; Tran, V.T.; Jones, D.P.; Filipov, N.M. Metabolomics of fescue toxicosis in grazing beef steers. *Food Chem. Toxicol.* **2017**, *105*, 285–299. [CrossRef]
38. Mote, R.S.; Hill, N.S.; Skarlupka, J.H.; Tran, V.T.; Walker, D.I.; Turner, Z.B.; Sanders, Z.P.; Jones, D.P.; Suen, G.; Filipov, N.M. Toxic tall fescue grazing increases susceptibility of the Angus steer fecal microbiota and plasma/urine metabolome to environmental effects. *Sci. Rep.* **2020**, *10*, 2497. [CrossRef]
39. Kamphorst, J.J.; Fan, J.; Lu, W.; White, E.; Rabinowitz, J.D. Liquid Chromatography–High Resolution Mass Spectrometry Analysis of Fatty Acid Metabolism. *Anal. Chem.* **2011**, *83*, 9114–9122. [CrossRef]
40. Lu, W.; Clasquin, M.F.; Melamud, E.; Amador-Noguez, D.; Caudy, A.A.; Rabinowitz, J.D. Metabolomic Analysis via Reversed-Phase Ion-Pairing Liquid Chromatography Coupled to a Stand Alone Orbitrap Mass Spectrometer. *Anal. Chem.* **2010**, *82*, 3212–3221. [CrossRef]
41. Chambers, M.C.; MacLean, B.; Burke, R.D.; Amodei, D.; Ruderman, D.L.; Neumann, S.; Gatto, L.; Fischer, B.; Pratt, B.; Egertson, J.D.; et al. A cross-platform toolkit for mass spectrometry and proteomics. *Nat. Biotechnol.* **2012**, *30*, 918–920. [CrossRef] [PubMed]
42. Clasquin, M.F.; Melamud, E.; Rabinowitz, J.D. LC-MS Data Processing with MAVEN: A Metabolomic Analysis and Visualization Engine. *Curr. Protoc. Bioinform.* **2012**, *37*. [CrossRef]
43. Chong, J.; Soufan, O.; Li, C.; Caraus, I.; Li, S.; Bourque, G.; Wishart, D.S.; Xia, J. MetaboAnalyst 4.0: Towards more transparent and integrative metabolomics analysis. *Nucleic Acids Res.* **2018**, *46*, W486–W494. [CrossRef] [PubMed]
44. Kanehisa, M.; Goto, S.; Sato, Y.; Kawashima, M.; Furumichi, M.; Tanabe, M. Data, information, knowledge and principle: Back to metabolism in KEGG. *Nucleic Acids Res.* **2013**, *42*, D199–D205. [CrossRef]

toxins

MDPI

Article

Vasoactive Effects of Acute Ergot Exposure in Sheep

Rossalin Yonpiam [1], Jair Gobbet [1], Ashok Jadhav [2], Kaushik Desai [2], Barry Blakley [3] and Ahmad Al-Dissi [1,*]

1 Department of Veterinary Pathology, Western College of Veterinary Medicine; University of Saskatchewan, Saskatoon, SK S7N 5B4, Canada; rossalin.yonpiam@usask.ca (R.Y.); jag397@mail.usask.ca (J.G.)
2 Department of Pharmacology, College of Medicine University of Saskatchewan, Saskatoon, SK S7N 5E5, Canada; ashok.jadhav@cnl.ca (A.J.); k.desai@usask.ca (K.D.)
3 Department of Veterinary Biomedical Sciences, Western College of Veterinary Medicine; University of Saskatchewan, Saskatoon, SK S7N 5B4, Canada; barry.blakley@usask.ca
* Correspondence: ahmad.aldissi@usask.ca; Tel.: +1-306-966-7643; Fax: +1-306-966-7439

Abstract: Ergotism is a common and increasing problem in Saskatchewan's livestock. Chronic exposure to low concentrations of ergot alkaloids is known to cause severe arterial vasoconstriction and gangrene through the activation of adrenergic and serotonergic receptors on vascular smooth muscles. The acute vascular effects of a single oral dose with high-level exposure to ergot alkaloids remain unknown and are examined in this study. This study had two main objectives; the first was to evaluate the role of α_1-adrenergic receptors in mediating the acute vasocontractile response after single-dose exposure in sheep. The second was to examine whether terazosin (TE) could abolish the vascular contractile effects of ergot alkaloids. Twelve adult female sheep were randomly placed into control and exposure groups ($n = 6$/group). Ergot sclerotia were collected and finely ground. The concentrations of six ergot alkaloids (ergocornine, ergocristine, ergocryptine, ergometrine, ergosine, and ergotamine) were determined using HPLC/MS at Prairie Diagnostic Services Inc., (Saskatoon, SK, Canada). Each ewe within the treatment group received a single oral treatment of ground ergot sclerotia at a dose of 600 μg/kg BW (total ergot) while each ewe in the control group received water. Animals were euthanized 12 h after the treatment, and the pedal artery (dorsal metatarsal III artery) from the left hind limb from each animal was carefully dissected and mounted in an isolated tissue bath. The vascular contractile response to phenylephrine (PE) (α_1-adrenergic agonist) was compared between the two groups before and after TE (α_1-adrenergic antagonist) treatment. Acute exposure to ergot alkaloids resulted in a 38% increase in vascular sensitivity to PE compared to control (Ctl $EC_{50} = 1.74 \times 10^{-6}$ M; Exp $EC_{50} = 1.079 \times 10^{-6}$ M, $p = 0.046$). TE treatment resulted in a significant dose-dependent increase in EC_{50} in both exposure and control groups ($p < 0.05$ for all treatments). Surprisingly, TE effect was significantly more pronounced in the ergot exposed group compared to the control group at two of the three concentrations of TE (TE 30 nM, $p = 0.36$; TE 100 nM, $p < 0.001$; TE 300 nM, $p < 0.001$). Similar to chronic exposure, acute exposure to ergot alkaloids results in increased vascular sensitivity to PE. TE is a more potent dose-dependent antagonist for the PE contractile response in sheep exposed to ergot compared to the control group. This study may indicate that the dry gangrene seen in sheep, and likely other species, might be related to the activation of α_1-adrenergic receptor. This effect may be reversed using TE, especially at early stages of the disease before cell death occurs. This study may also indicate that acute-single dose exposure scenario may be useful in the study of vascular effects of ergot alkaloids.

Keywords: acute ergot exposure; ergot toxicity; sheep; vasoconstriction; adrenergic receptors

Key Contribution: This study demonstrated that, similar to chronic exposure, an acute single-dose oral exposure of sheep to ergot alkaloids results in increased pedal artery sensitivity to phenylephrine.

Citation: Yonpiam, R.; Gobbet, J.; Jadhav, A.; Desai, K.; Blakley, B.; Al-Dissi, A. Vasoactive Effects of Acute Ergot Exposure in Sheep. *Toxins* **2021**, *13*, 291. https://doi.org/10.3390/toxins13040291

Received: 14 February 2021
Accepted: 2 April 2021
Published: 20 April 2021

1. Introduction

Ergot poisoning remains an economically important disease affecting a variety of animal species including cattle, sheep, horses, and goats with estimated annual losses of more than a billion dollars within the US [1]. Ergot poisoning is caused by the prolonged consumption of ergot alkaloids which are naturally occurring mycotoxins produced by fungi infecting crops such as triticale, cereals, and grains such as barley, wheat, and durum [2–4]. The most widely encountered species of ergot alkaloid producing fungi in Western Canada and Europe are in the family of *Clavicipitaceae* [5–7]. This fungal family includes the external spore-producing fungi (*Claviceps* spp.) and endophytic fungi (*Epichloë* spp.). The major species causing agricultural problems in Western Canada is *Claviceps purpura* [5,6]. The active ingredients of ergot alkaloids are confined and concentrated within the sclerotia which are external fungal bodies [8]. Clinical signs of lameness, hoof loss, and dry gangrene of the lower limbs, tail, ear tips, and teats are commonly seen in chronic ergotism and are related to the effect of ergot alkaloids on the vasculature causing severe vasoconstriction [9–11]. A complete discussion of the impact of ergot alkaloids on various organs or systems is beyond the scope of this paper. Table 1 provides a brief summary of the main effects reported in the literature (reviewed by Strickland J et al., 2011).

Table 1. Summary of the main reported effects of ergot alkaloids on different organ or systems in animals (reviewed by Strickland J et al., 2011).

System/Organ Affected	Reported Effect
Gastrointestinal System	- Decreased blood flow to the duodenum and colon - Inhibition of cyclical contractions in the rumen and reticulum - Decreased abomasal motility - Reduced digestibility
Pancreas	- Stimulated insulin secretion - Decreased plasma insulin - Increased activity of protease, trypsin, amylase, and lipase
Neural & Neuroendocrine System	- Endocrine and hypothalamic disruption - Variations in blood concentrations of epinephrine and norepinephrine - Increased nervousness and excitability - Depressed endogenous catecholamine activity - Increased dopamine release - Mixed effects on adrenergic, dopaminergic, and serotoninergic receptors.
Reproductive & Mammary Gland	- Decreased reproductive performance - Decrease pregnancy rate - Decreased prolactin - Decreased milk production - Reduced progesterone - Reduced follicle stimulating and luteinizing hormones - Increased uterine contraction - Decreased number of mature ovarian follicle - Erratic estrous cycles - Decreased dry matter take and nutrition - Decreased neonatal weight & weight gain - reduced sperm motility
Cardiovascular	- Decreased heart rate - Increased vascular tone - Gangrene

The precise in vivo vasoactive mechanisms of ergot alkaloids have not been determined. However, in vitro tissue bath studies, where normal dissected and isolated arterial rings were exposed to purified ergot alkaloids, have previously shown that the adrenergic and serotonergic receptors on vascular smooth muscles are activated. This is also supported by the fact that the chemical structure, i.e., the ergoline ring, of ergot alkaloids resembles that of physiologic neurotransmitters such as dopamine, norepinephrine, epinephrine, and serotonin, which are known to be vasoactive [4].

It is important to note that despite the rapid metabolism and excretion of ergot alkaloids which occurs within several hours after exposure [12,13], the clinical vascular manifestations of ergot alkaloids are always seen after the prolonged (several weeks to months) consumption of ergot-contaminated plants. While these clinical vascular manifestations could be explained by the repeated exposure to ergot alkaloids, these effects often remain long after the ergot-contaminated feed is removed! Recent evidence suggests that ergot alkaloids may bioaccumulate within the vasculature [14,15]. It is also possible that other unknown vasoactive mechanisms may be involved in mediating these effects.

It is unknown whether acute ergot exposure affects vascular contractility in a similar manner to chronic exposure. If similar effects are found, then acute exposure scenarios may be useful to study the mechanisms of vascular alteration by ergot alkaloids. Many studies have focused on finding an antagonist to counteract the clinical effects of ergotism. Elucidating the mechanisms of vascular contractile response induced by ergot alkaloids may prove useful to identify treatment options for the vascular-related clinical manifestations of ergot poisoning.

This study aimed to examine the role of adrenergic receptors in mediating the vascular effects of ergot alkaloids after an in vivo acute exposure scenario to these alkaloids. Vascular sensitivity to phenylephrine (PE), an α1-adrenergic agonist, was compared between ergot exposed and control groups before and after terazosin (TE), α_1–adrenergic antagonist, treatment.

We hypothesized that an acute single dose oral exposure to ergot alkaloids results in increased vascular sensitivity (decreased EC50) to PE in the pedal artery; an effect that is mediated through the activation of α_1-adrenergic receptors. We also hypothesized that the acute vascular effects of oral exposure to ergot alkaloids can be reversed via TE (the α_1-adrenergic antagonist).

2. Results

All animals remained healthy after treatment and did not exhibit any clinical signs during the 12 h period between the administration of ergot alkaloids and euthanasia. No gross or histological changes were seen in either group in the lung, liver, kidneys, heart, spleen, intestines, fat, and pedal arteries. The concentration of ergot alkaloids used to formulate the single oral dose is shown in Table 2.

Table 2. The concentration of six ergot alkaloids determined within ground sclerotia using HPLC/MS *. The total concentration of these alkaloids was used to formulate a single oral dose (600 μg/kg BW) which was administered to each sheep using a stomach tube.

Alkaloid	Concentration (ppb) Dry Weight	Oral Dose (μg/kg BW)
Ergocornine	216,500	26.4
Ergocristine	3,653,000	445.9
Ergocryptine	540,100	65.9
Ergometrine	78,850	9.6
Ergosine	89,570	10.9
Ergotamine	338,300	41.3
Total	4,915,900	600

* The detection limit for each alkaloid was 1.25 ppb. HPLC/MS, high performance liquid chromatography and mass spectrometry; μg/kg BW, microgram per kilogram body weight; ppb, part per billion.

2.1. *Phenylephrine Dose Response Curve Compared between Ergot Exposure and Control Groups*

In the control group, the PE contractile response was first observed at 1×10^{-7} M concentration, and the maximum contractile response recorded at the highest PE concentration

(1 × 10^{-4} M) was 22.8 g. The contractile response in the exposure group was first observed at 0.5 × 10^{-7} M while the highest PE concentration yielded a maximum contraction of 18.0 g. Ergot exposure resulted in a significant decrease in EC_{50} compared to the control group (p = 0.0462). Comparisons of EC_{50} between the two groups are presented in Figure 1. Details of EC_{50} are for all groups are presented in Table 3.

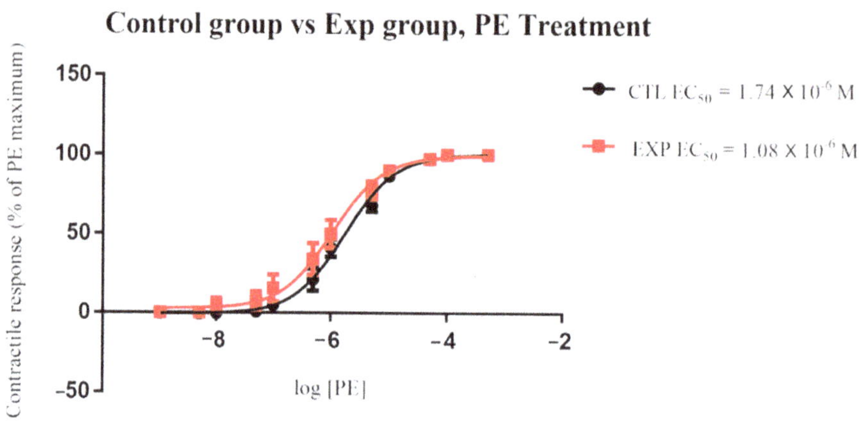

Figure 1. Mean arterial contractile responses to increasing concentration of PE compared between control and ergot exposed group. The pedal artery was collected 12 h after single oral exposure to 600 µg/kg BW (total ergot) or after placebo water treatment (n = 6/group). Contractile response data were presented as percentage means ± *SEM* of the maximum contractile effect induced by 1 × 10^{-4} M PE treatment. Ergot exposure resulted in a significant decrease in EC_{50} compared to the control group ($p < 0.05$). EC_{50}, the concentration of phenylephrine producing 50% of the maximum contractile response; PE, phenylephrine; BW, body weight; CTL, control group; EXP, exposure group; M, Molar.

Table 3. Phenylephrine (PE) EC_{50} compared between ergot exposed and control sheep (n = 6/group) before and after terazosin treatment in dissected pedal arteries using an arterial tissue bath. Ergot exposed sheep received a single oral dose of 600 µg/kg BW total ergot dissolved in a water based on the levels of six ergot alkaloids determined previously. Control sheep received a water placebo treatment. The effect of terazosin was determined using three increasing concentrations of terazosin: 30, 100, and 300 nM. For each treatment type, a sigmoidal dose-response curve was plotted using nonlinear regression which was used to calculate EC_{50}. Statistical differences in EC_{50} among the different treatment types were calculated by the extra sum-of-squares F-test. A p-value less than 0.05 was considered significant.

Tissue Bath Treatment Type	Control EC_{50} and 95% CI	Ergot Exposed EC_{50} and 95% CI	p-Value
PE	1.74 × 10^{-6} (1.39 × 10^{-6}–2.18 × 10^{-6}) [a,d]	1.08 × 10^{-6} (7.4 × 10^{-7}– 1.57 × 10^{-6}) [a,g]	$p < 0.05$
PE + TE (30 nM)	6.11 × 10^{-6} (4.78 × 10^{-6}–7.8 × 10^{-6}) [d,e]	7.74 × 10^{-6} (4.63 × 10^{-6}–1.3 × 10^{-5}) [g,h]	$p = 0.37$
PE + TE (100 nM)	9.6 × 10^{-6} (7.69 × 10^{-6}–1.2 × 10^{-5}) [b,e,f]	2.57 × 10^{-5} (1.7 × 10^{-5}–3.9 × 10^{-5}) [b,h,i]	$p < 0.0001$
PE + TE (300 nM)	1.77 × 10^{-5} (1.34 × 10^{-5}–2.33 × 10^{-5}) [c,f]	6.73 × 10^{-5} (4.24 × 10^{-5}–1.07 × 10^{-4}) [c,i]	$p < 0.0001$

[a–i] letters with the same superscript are significantly different. EC_{50}, the concentration of phenylephrine producing 50% of the maximum contractile response; µg/kg BW, microgram per kilogram body weight; PE, phenylephrine; TE, Terazosin; CI, confidence interval; nM, nanomolar.

2.2. Effect of Terazosin Treatment on Phenylephrine Dose Response Curve

In the control group, TE treatment resulted in a significant and dose-dependent increase in EC_{50} ($p < 0.0001$ for all concentrations; 30, 100, and 300 nM). Similarly, EC_{50} significantly increased in a dose-dependent manner in the exposure group after terazosin treatment ($p < 0.0001$ for all concentration; 30, 100, and 300 nM) (Figures 2 and 3). The blocking effect of TE was greater in the exposure group when compared to the control group when given at 100 nM and 300 nM ($p < 0.0001$). A similar trend of increasing EC_{50} in the exposure group compared to the control group after the 30 nM TE treatment was seen, but the difference was not statistically significant ($p = 0.076$) (Figures 4–6). (See Table 3 for details).

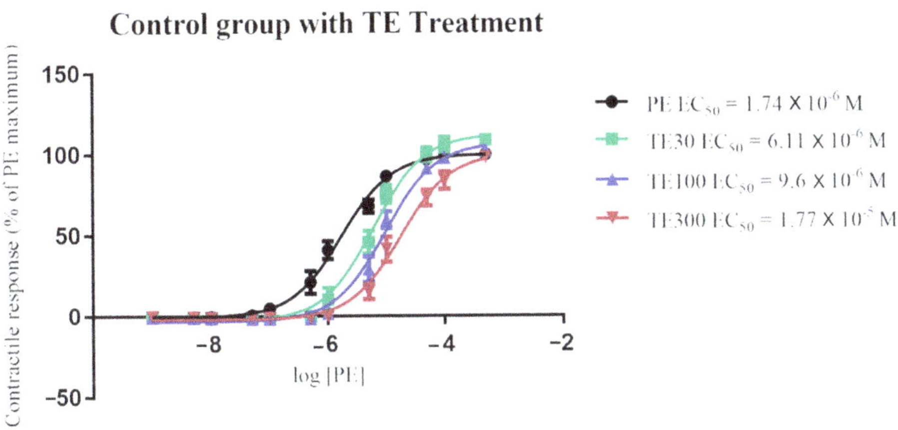

Figure 2. Mean arterial contractile responses to increasing concentration of PE in control animals compared before and after 30, 100, or 300 nM TE treatment ($n = 6$). Contractile response data were presented as percentage means ± *SEM* of the maximum contractile effect induced by 1×10^{-4} M PE treatment. TE treatment resulted in a significant dose-dependent increase in EC_{50} compared to PE alone ($p < 0.0001$). EC_{50}, the concentration of phenylephrine producing 50% of the maximum contractile response; PE, phenylephrine; TE, terazosin; M, Molar; nM, nanomolar.

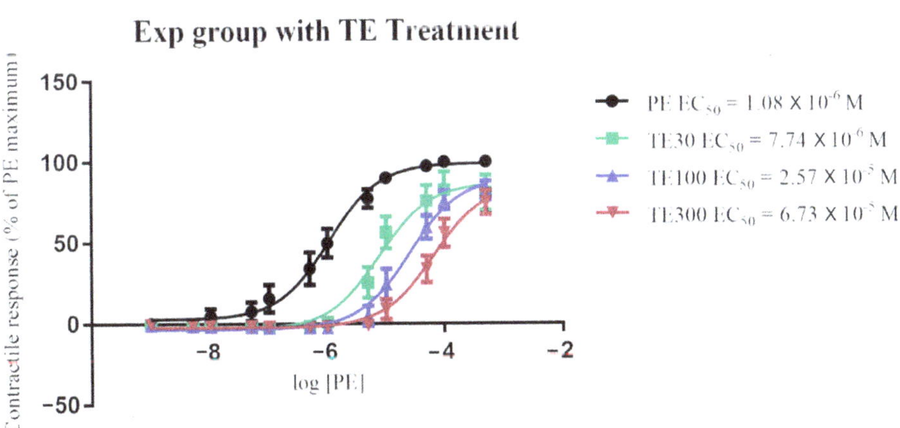

Figure 3. Mean arterial contractile responses to increasing concentration of PE in ergot exposed animals compared before and after 30, 100, or 300 nM TE treatment ($n = 6$). Contractile response data were presented as percentage means ± *SEM* of the maximum contractile effect induced by 1×10^{-4} M PE treatment. TE treatment resulted in a significant dose-dependent increase in EC_{50} compared to PE alone ($p < 0.0001$). EC_{50}, the concentration of phenylephrine producing 50% of the maximum contractile response; PE, phenylephrine; TE, terazosin; Exp, exposure group; M, Molar; nM, nanomolar.

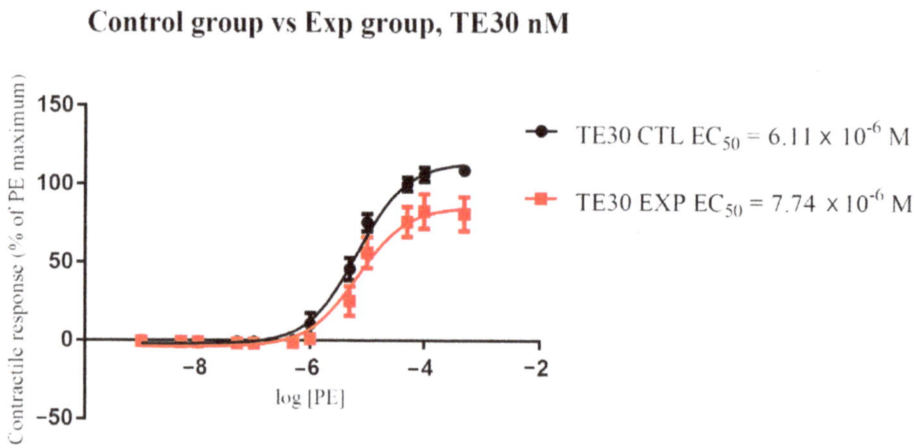

Figure 4. Mean arterial contractile responses to increasing concentration of PE compared between control and ergot exposed groups after TE treatment at 30 nM. EC_{50} was not significantly different between the two groups (p = 0.37). Contractile response data were presented as percentage means ± *SEM* of the maximum contractile effect induced by 1×10^{-4} M PE treatment. EC_{50}, the concentration of phenylephrine producing 50% of the maximum contractile response; PE, phenylephrine; TE, terazosin; CTL, control group; EXP, exposure group; M, Molar; nM, nanomolar.

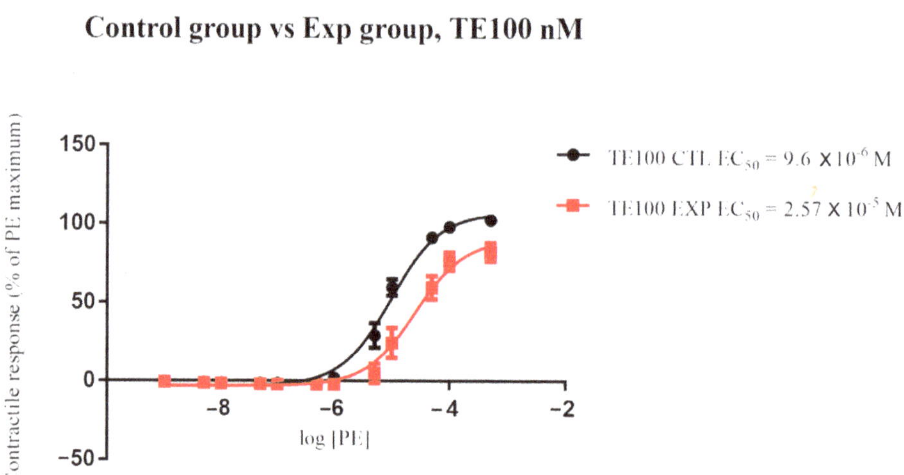

Figure 5. Mean arterial contractile responses to increasing concentration of PE compared between control and exposure after TE treatment at 100 nM. EC_{50} was significantly different between the two groups ($p < 0.0001$). Contractile response data were presented as percentage means ± *SEM* of the maximum contractile effect induced by 1×10^{-4} M PE treatment. EC_{50}, the concentration of phenylephrine producing 50% of the maximum contractile response; PE, phenylephrine; TE, terazosin; CTL, control group; EXP, exposure group; M, Molar; nM, nanomolar.

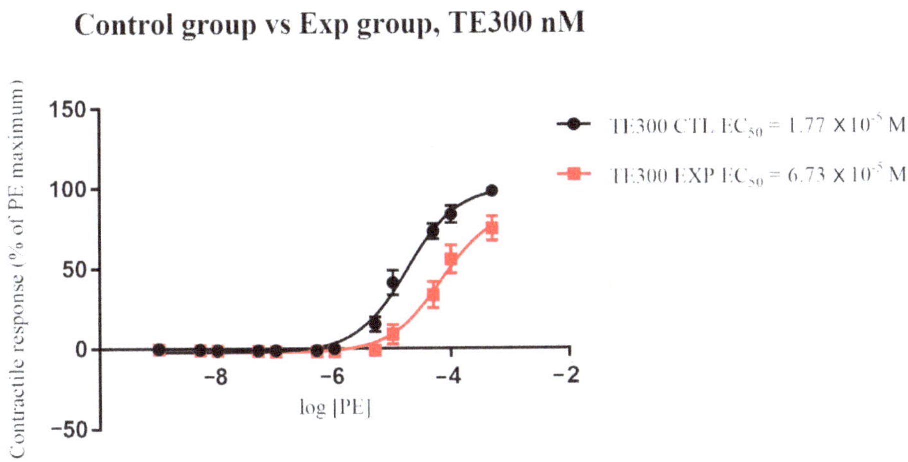

Figure 6. Mean arterial contractile responses to increasing concentration of PE compared between control and exposure after TE treatment at 300 nM. EC_{50} was significantly different between the two groups ($p < 0.0001$). Contractile response data were presented as percentage means ± *SEM* of the maximum contractile effect induced by 1×10^{-4} M PE treatment. EC_{50}, the concentration of phenylephrine producing 50% of the maximum contractile response; PE, phenylephrine; TE, terazosin; CTL, control group; EXP, exposure group; M, Molar; nM, nanomolar.

3. Discussion

The biological effects of ergot alkaloids on livestock are known to be diverse. This diversity is not only related to differences in alkaloid concentration and specific alkaloid content in different plants, but also due to their ability to affect multiple biological processes [4,7,16,17]. The ergoline ring system, which is a structure common to ergot alkaloids, is similar to the ring structure of epinephrine, dopamine, and serotonin thus allowing ergot alkaloids to mimic their function. In the vasculature, ergot alkaloids bind with a variety of serotonergic and adrenergic receptors to modify vascular tone [11]. The effects of these alkaloids are known to be diverse with differing potencies among different animal species; however, within a species, the effects are dependent on the animal's general health, body condition, reproductive status and previous exposure [4,18].

Ergotism in livestock is known to cause dry gangrene due to severe vasoconstriction within peripheral vasculature. Ergotism occurs after the prolonged ingestion of ergot alkaloids. Therefore, previous studies have focused on examining the mechanisms of vasoconstriction following chronic exposure scenarios [19–21]. Thus, the vascular effects following acute exposure remain unknown. In this study, we wanted to investigate the role of α_1-adrenergic receptor activation on vascular contractile response following a single acute high-dose of oral exposure scenario to ergot alkaloids using sheep as a model. Similar to other livestock species, sheep are chronically affected by ergotism and develop dry gangrene after prolonged exposure. We chose to examine the pedal artery due to its peripheral location on the ovine limb.

Ideally, pure individual ergot alkaloids should be used in prolonged feeding trials to precisely examine their vascular effects and the mechanism of these effects. However, because pure ergot alkaloids are very expensive, previous studies often used ergot or endophyte-infected tall fescue. It is often difficult to accurately estimate the individual dose in these studies as the concentration of alkaloids within feed is subject to significant variability due to feed storage conditions and uneven distribution. Alternatively, pure individual alkaloids are often used on dissected arteries to examine their vascular contractile effects in vitro using arterial tissue bath systems. In order to achieve a more defined dosing protocol, we used ground sclerotia in which the concentration of six different ergot alkaloids was determined [16,17]. The dose was adjusted in every animal depending on the body weight to receive a dose of 600 μg/kg BW of total ergot alkaloid content. To the author's knowledge, no studies have been performed to examine the vascular effects of acute single-dose oral exposure to ergot alkaloids. All previous studies reported vascular effects after repeated low-dose exposure. It is likely that a single low-dose oral exposure will not have a detectable impact on the vasculature. Therefore, to increase the odds of detecting vascular contractile effects after single-dose oral exposure, this study opted to use a high-dose of ergot alkaloids. While the repeated exposure to such a high level is unlikely to occur due to feed refusal, pelleted feed submitted to our diagnostic laboratory from livestock producers for ergot testing occasionally contained similarly high levels.

It is known that the degree of vasoconstriction induced by ergot alkaloids is alkaloid dependent. For example, the vasoconstrictive effect elicited by ergocryptine is 100 times less potent as compared to ergotamine, whereas ergocristine and ergocornine are only 10 times less potent. Ergovaline, the predominant alkaloid in tall fescue grass, is thought to have a similar potency to ergotamine [11,22]. The potency of these alkaloids varies depending on their relative binding affinity to α-adrenergic and serotonergic receptors and their ability to specifically activate them. Most studies examining the vascular effects of ergot alkaloids have focused on studying the serotonergic receptors [19,23–25]. However, very few studies examined the activation of α-adrenergic receptors by different alkaloids. For example, the contractile response in the lateral saphenous vein of cattle grazing tall fescue was significantly enhanced compared to control animals by BHT-920, an α_2-adrenergic agonist, but not by PE (α_1-adrenergic agonist) [26]. In addition, Schöning et al. reported that ergovaline stimulated α_1-adrenergic receptors but with low efficacy in rat thoracic aorta [22]. In vivo studies focusing on heart rate and blood pressure changes after exposure to ergot alkaloids also indicate α-adrenergic receptor activation. Bradycardia induced by ergotamine in anesthetized rats was reduced by yohimbine, an α_2-adrenergic antagonist. In addition, ergotamine treatment reduced the tachycardia induced by electrical stimulation of the spinal cord, and the reduction was similarly blocked after yohimbine treatment [27]. Similarly, in rats, ergotamine has been shown to act as an agonist on α_2-adrenergic receptors and an antagonist on α_1-adrenergic receptor [28]. In our study, a significant increase in vascular sensitivity to PE was found in ergot exposed sheep compared to control animals, which might suggest that α_1-adrenergic receptors mediate that response.

Similar to what we expected, TE decreased the vascular sensitivity to PE in ergot exposed and control sheep due to its antagonistic effects on the α_1-adrenergic receptor. However, surprisingly, the potency of TE as an α_1-adrenergic receptor antagonist was significantly enhanced in ergot exposed sheep compared to controls. It has been recently shown that previous exposure to high concentrations of ergot alkaloids may decrease vascular contractility making the vasculature less susceptible to the effects of ergot alkaloids. Klotz et al. examined the contractile response to ergovaline in cattle chronically grazing high and low-endophyte-infected tall fescue [23]. This study demonstrated that the maximum contractile response was significantly higher in steers consuming low-endophyte-infected tall fescue. This is contrary to other studies, which found that the increase in vascular contractile response to ergot alkaloids is dose-dependent [24,29]. It is, thus, possible that the vascular contractile effect of ergot alkaloids is dose-dependent but may become less effective at very high doses. It is possible that the high dose of ergot alkaloids we used resulted in a relatively low contractile response, and also enhanced the blocking effect of TE resulting in a reduced contractile response compared to control tissues. The increased blocking sensitivity of TE in exposed animals may also indicate that the full impact of ergot exposure was not realized due to the short exposure duration.

Alternatively, it is also possible that the effect of the blocker was enhanced in the ergot exposed group due to the unique mixture of alkaloids in the diet. Interestingly, it has been shown that the presence of ergocristine, ergocornine, and ergocryptine together produces adrenergic blockade [30–32]. Additionally, Roquebert and Demichel reported that ergocristine acts as an α_1-adrenergic blocker in rat tail artery [28,33,34]. Ergocristine had the highest concentration in the diet used in this study and may have acted as an antagonist. The enhanced blocking effect of TE in ergot exposed animals may indicate that this blocker may be useful in counteracting the vascular effects of ergot alkaloid exposure.

Several studies have shown that ergot alkaloids interact with serotonin receptors in chronic exposure scenario [19,24,25]. However, Kalkman et al. reported that in rats injected intravenously with ergometrine, the vasoconstrictor response was related to the activation of α_1- and α_2-adrenergic receptors, but not serotonergic receptors [35]. It would be interesting to examine the role of serotonergic receptors in mediating arterial contraction after acute exposure to ergot alkaloids. High-level exposure in ruminants can result in nervous signs such as hyperexcitability, hypermetria, and tremors [11,36]. The dose we used was relatively high but was well tolerated by all animals, with none showing clinical signs of illness.

Currently, ergot toxicity is thought to be only related to the prolonged consumption of ergot alkaloids, and it is presumed that a short-term exposure will have no significant clinical effects. However, we show for the first time that even a single oral dose of ergot alkaloids causes a significant increase in sensitivity in arteries supplying distal extremities. This finding is of significance to the livestock industry and regulators, as it may indicate that in cold weather conditions, short-term exposure to ergot alkaloids may result in a significant decrease in blood supply to the extremities, making animals prone to gangrene. It would be interesting to examine whether a similar but lower level exposure scenario would result in a similar response in livestock. In addition, it is also important to examine the effects of a short-term exposure on other systems as it is now presumed that the effects are only seen after chronic exposure. If similar negative effects are seen in other systems, it may indicate the need to lower the allowable limits of ergot alkaloids within feed to reflect the true nature of the negative impact of this disease.

The finding that a blocking effect of TE was more potent in ergot exposed animals may indicate that this drug could be used to treat animals who have been recently exposed to ergot alkaloids. If it is proved to be useful, this drug may significantly reduce the economic impact of ergotism to the livestock industry. It would be interesting to examine whether TE has any impact on other systems affected by this disease.

We recently showed that the S-epimers of ergot alkaloids are vasoactive causing vasoconstriction of bovine dorsal metatarsal arteries in vitro [37]. It is, therefore, possible that the effects seen in this study are related to the combined activity of the R and S epimers and not just the R-epimers.

In summary, this study found that acute high-level exposure to ergot alkaloids results in increased vascular sensitivity to PE and increased blocking effect of TE. Additional studies are immensely needed to examine the role of adrenergic and, serotonergic receptors in other vascular beds in vivo and in vitro after acute exposure.

4. Materials and Methods

4.1. Animals

All protocols were approved by the Animal Care and Ethics Committee at the University of Saskatchewan (Animal Use Protocol # 20150047, approval date: 31 August 2016). Before the experiment, all animals were weighed and clinically examined with body temperature and heart rate recorded. A blood sample was also collected from each animal, and a complete blood count was performed evaluating red and white blood cell counts as well as platelets count and total plasma protein to ensure that all animals were healthy.

4.2. Tissue Collection & Stock Solutions

Twelve healthy adult ewes were randomly assigned into treatment or control groups (n = 6/group). Animals were allowed to acclimatize for fourteen days and were fed alfalfa hay and water ad libitum. Ergot alkaloids containing sclerotia were collected, finely ground and the concentrations of six alkaloids (ergocornine, ergocristine, ergocryptine, ergometrine, ergosine, and ergotamine) were determined using HPLC/MS at Prairie Diagnostic Services Inc. (PDS), Saskatoon, SK, Canada (16). Each ewe within the treatment group received a single dose of ground ergot sclerotia at a dose of 600 μg/kg BW (total ergot) dissolved in 50 mL of water via a stomach tube. The concentrations of ergot alkaloids within sclerotia are recorded in Table 1. The control group received water placebo. Twelve hours after treatment, animals were euthanized using a captive bolt and a necropsy was performed. A 15 cm segment of the pedal artery (dorsal metatarsal artery III) was carefully dissected and collected from each animal, soaked in a diluted heparin solution (10 Unit/1 mL) for 2 min and transferred into a container containing modified Krebs–Henseleit buffer solution [in mM: 118 NaCl, 4.7 KCl, 1.2 $MgSO_4$, 1.2 KH_2PO_4, 22.0 $NaHCO_3$, 5.0 glucose and 2.5 $CaCl_2$; (Sigma-Aldrich Canada Ltd. Oakville, ON, Canada) (pH 7.4 gassed with 95% O_2, 5% CO_2 at 37 °C)] on ice until transport to the laboratory. Immediately upon arrival to the lab, adipose and connective tissue were carefully removed from each arterial segment, which was later sliced into four 3 to 5 mm cross sections. Each arterial section was suspended between the bases of two triangular-shaped wires within an isolated 10 mL tissue bath (Chengdu equipment manufacturing, China) containing modified Krebs–Henseleit buffer solution maintained at the above conditions. Arterial rings were allowed to equilibrate for 1 h under a resting tension of 2 g with the bath solution changed every 15 min. Each day, a fresh stock solution was prepared for phenylephrine (PE) (Sigma-Aldrich Canada Ltd. Oakville, ON, Canada) at a concentration of 1 M, followed by a 10-fold serial dilution to prepare the remaining working solutions. 10 μL were added from each dilution to the 10 mL incubation buffer to obtain the desired final concentration (1×10^{-9}–1×10^{-4} M). Similarly, a fresh initial 1 M stock solution of terazosin (Sigma-Aldrich Canada Ltd. Oakville, ON, Canada) was prepared from which a 10 μM and 100 μM dilutions were made. The desired final concentrations of 30, 100, and 300 nM were prepared by adding 30 μL from the first stock or 10 and 30 μL from the second stock as appropriate. Arterial rings were treated with PE (1×10^{-4} M) (Sigma-Aldrich Canada Ltd. Oakville, ON, Canada) to initiate contraction and to confirm tissue viability and responsiveness. The tissues were later washed with incubation buffer until resting tension was achieved.

4.3. Contractile Response

Three vascular rings from the pedal artery of each animal were used to assess the PE contractile response before and after the incubation of each ring with a different concentration of TE. Initially, a cumulative concentration-dependent contraction in response to PE was obtained by adding increasing concentrations of PE (1×10^{-9} M to 1×10^{-4} M). After each PE treatment, arterial rings were allowed to achieve maximum tension which plateaued for 2 min before the next concentration was added. After the last PE treatment, arterial rings were allowed to return to resting tension with buffer replacement occurring every 15 min for 1 h. This was followed by incubating each of the three rings with 30, 100 or 300 nM TE for 20 min after which the cumulative PE contractile response was repeated in each chamber as above. Following completion of the exposure, all rings were exposed to 1×10^{-4} M PE to verify their viability.

4.4. Data Collection, Analysis and Statistical Analysis

All measured isometric contractile responses were recorded in grams of tension using 'Chart' software and Powerlab equipment (AD Instruments Inc., Colorado Springs, CO, USA). For each PE treatment, the maximum tension in grams achieved before the 2 min plateau period was recorded and corrected for a baseline. To minimize variation due to arterial size, each contractile response from an individual ring was normalized to its maximum contractile tension induced by 1×10^{-4} M PE treatment.

Contractile response data were presented as percentage means ± *SEM* of the maximum contractile effect induced by 1×10^{-4} M PE treatment. For each treatment type, a sigmoidal dose-response curve was plotted using nonlinear regression with variable slope utilizing GraphPad Prism 7 (GraphPad Software Inc., La Jolla, CA, USA), which was later used to calculate potency presented as the concentration producing 50% of the maximum response (EC_{50}). Results were presented as the log of the EC_{50} value. Statistical differences in EC_{50} among the different dose-response curves were calculated by the extra sum-of-squares F-test where a p-value less than 0.05 was considered significant.

Author Contributions: Formal analysis, R.Y. and A.A.-D.; Funding acquisition, B.B. and A.A.-D.; Investigation, R.Y., J.G., A.J., K.D., and A.A.-D.; Methodology, A.J., K.D., B.B., and A.A.-D.; Project administration, R.Y., B.B., and A.A.-D.; Resources, K.D. and A.A.-D.; Supervision, A.A.-D.; Writing–original draft, R.Y.; Writing–review & editing, K.D., B.B., and A.A.-D. All authors have read and agreed to the published version of the manuscript.

Funding: This research and the APC were funded by Agriculture Development Fund of Saskatchewan, grant number 20140186.

Institutional Review Board Statement: This study was approved by the University of Saskatchewan Animal Research Ethics Board. Animal Use Protocol # 20150047. Approval date: 31 August 2016.

Informed Consent Statement: Informed consent was not needed because all studied subjects were animals.

Data Availability Statement: Not applicable.

Acknowledgments: Funding for this project was kindly provided by a reseach grant from the Saskatchewan Agriculture Development Fund recevived by AN. Al-Dissi as the primary investigator.

Conflicts of Interest: The authors declare no conflict of interest.

References

1. Hoveland, C.S. Importance and economic significance of the Acremonium endophytes to performance of animals and grass plant. *Agric. Ecosyst. Environ.* **1993**, *44*, 3–12. [CrossRef]
2. Canty, M.J.; Fogarty, U.; Sheridan, M.K.; Ensley, S.M.; E Schrunk, D.; More, S.J. Ergot alkaloid intoxication in perennial ryegrass (Lolium perenne): An emerging animal health concern in Ireland? *Ir. Veter J.* **2014**, *67*, 21. [CrossRef] [PubMed]
3. Haarmann, T.; Rolke, Y.; Giesbert, S.; Tudzynski, P. Ergot: From witchcraft to biotechnology. *Mol. Plant Pathol.* **2009**, *10*, 563–577. [CrossRef] [PubMed]
4. Strickland, J.R.; Looper, M.L.; Matthews, J.C.; Rosenkrans, C.F., Jr.; Flythe, M.D.; Brown, K.R. Board-invited review: St. Anthony's Fire in livestock: Causes, mechanisms, and potential solutions. *J. Anim. Sci.* **2011**, *89*, 1603–1626. [CrossRef] [PubMed]
5. Mills, J.T. Mycotoxins and toxigenic fungi on cereal grains in western Canada. *Can. J. Physiol. Pharmacol.* **1990**, *68*, 982–986. [CrossRef]
6. Menzies, J.G.; Turkington, T.K. An overview of the ergot (Claviceps purpurea) issue in western Canada: Challenges and solutions. *Can. J. Plant Pathol.* **2014**, *37*, 40–51. [CrossRef]
7. European Food Safety Authority. Scientific Opinion on Ergot alkaloids in food and feed. *EFSA J.* **2012**, *10*, 2798–2956.
8. Schardl, C.L.; Panaccione, D.G.; Tudzynski, P. Ergot alkaloids–biology and molecular biology. *Alkaloids Chem. Biol.* **2006**, *63*, 45–86. [PubMed]
9. Strickland, J.; Aiken, G.; Spiers, D.; Fletcher, L.; Oliver, J. Physiological basis of fescue toxicosis. *Tall Fescue Twenty-First Century* **2009**, *53*, 203–227.
10. Strickland, J.R.; Aiken, G.E.; Klotz, J.L. Ergot Alkaloid Induced Blood Vessel Dysfunction Contributes to Fescue Toxicosis. *Forage Grazinglands* **2009**, *7*, 1–7. [CrossRef]
11. Craig, A.M.; Klotz, J.L.; Duringer, J.M. Cases of ergotism in livestock and associated ergot alkaloid concentrations in feed. *Front. Chem.* **2015**, *3*, 8–14. [CrossRef]
12. Stuedemann, J.A.; Hill, N.S.; Thompson, F.N.; Fayrer-Hosken, R.A.; Hay, W.P.; Dawe, D.L.; Seman, D.H.; Martin, S.A. Urinary and biliary excretion of ergot alkaloids from steers that grazed endophyte-infected tall fescue. *J. Anim. Sci.* **1998**, *76*, 2146–2154. [CrossRef]
13. Hill, N.; Thompson, F.; Stuedemann, J.; Dawe, D.; Hiatt, E., III. Urinary alkaloid excretion as a diagnostic tool for fescue toxicosis in cattle. *J. Vet. Diagn. Investig.* **2000**, *12*, 210–217. [CrossRef] [PubMed]
14. Klotz, J.L.; Kirch, B.H.; Aiken, G.E.; Bush, L.P.; Strickland, J.R. Bioaccumulation of ergovaline in bovine lateral saphenous veins in vitro1,2. *J. Anim. Sci.* **2009**, *87*, 2437–2447. [CrossRef]
15. Realini, C.E.; Duckett, S.K.; Hill, N.S.; Hoveland, C.S.; Lyon, B.G.; Sackmann, J.R.; Gillis, M.H. Effect of endophyte type on carcass traits, meat quality, and fatty acid composition of beef cattle grazing tall fescue. *J. Anim. Sci.* **2005**, *83*, 430–439. [CrossRef] [PubMed]
16. Krska, R.; Stubbings, G.; Macarthur, R.; Crews, C. Simultaneous determination of six major ergot alkaloids and their epimers in cereals and foodstuffs by LC–MS–MS. *Anal. Bioanal. Chem.* **2008**, *391*, 563. [CrossRef] [PubMed]

17. European Food Safety Authority. Opinion of the Scientific Panel on contaminants in the food chain [CONTAM] related to ergot as undesirable substance in animal feed. *EFSA J.* **2005**, *3*, 225–252. [CrossRef]
18. Strickland, J.R.; Brown, K.R.; Aiken, G.E.; Klotz, J.L.; Flythe, M.D. *Ergot Alkaloids: Toxicokinetics and Vascular Effects*; Ardmore: Samuel Roberts Noble Foundation: Ardmore, PA, USA, 2012.
19. Klotz, J.L.; Aiken, G.E.; Bussard, J.R.; Foote, A.P.; Harmon, D.L.; Goff, B.M.; Schrick, F.N.; Strickland, J.R. Vasoactivity and Vasoconstriction Changes in Cattle Related to Time off Toxic Endophyte-Infected Tall Fescue. *Toxins* **2016**, *8*, 271. [CrossRef]
20. Foote, A.P.; Harmon, D.L.; Brown, K.R.; Strickland, J.R.; McLeod, K.R.; Bush, L.P.; Klotz, J.L. Constriction of bovine vasculature caused by endophyte-infected tall fescue seed extract is similar to pure ergovaline1,2. *J. Anim. Sci.* **2012**, *90*, 1603–1609. [CrossRef]
21. Egert, A.M.; Kim, D.H.; Schrick, F.N.; Harmon, D.L.; Klotz, J.L. Dietary exposure to ergot alkaloids decreases contractility of bovine mesenteric vasculature1,2. *J. Anim. Sci.* **2014**, *92*, 1768–1779. [CrossRef]
22. Schöning, C.; Flieger, M.; Pertz, H.H. Complex interaction of ergovaline with 5-HT2A, 5-HT1B/1D, and alpha1 receptors in isolated arteries of rat and guinea pig. *J. Anim. Sci.* **2001**, *79*, 2202–2209. [CrossRef]
23. Klotz, J.L.; Brown, K.R.; Xue, Y.; Matthews, J.C.; Boling, J.A.; Burris, W.R.; Bush, L.P.; Strickland, J.R. Alterations in serotonin receptor-induced contractility of bovine lateral saphenous vein in cattle grazing endophyte-infected tall fescue12. *J. Anim. Sci.* **2012**, *90*, 682–693. [CrossRef]
24. Oliver, J.W.; Abney, L.K.; Strickland, J.R.; Linnabary, R.D. Vasoconstriction in bovine vasculature induced by the tall fescue alkaloid lysergamide. *J. Anim. Sci.* **1993**, *71*, 2708–2713. [CrossRef] [PubMed]
25. Dyer, D.C. Evidence that ergovaline acts on serotonin receptors. *Life Sci.* **1993**, *53*, PL223–PL228. [CrossRef]
26. Oliver, J.W.; Strickland, J.R.; Waller, J.C.; Fribourg, H.A.; Linnabary, R.D.; Abney, L.K. Endophytic fungal toxin effect on adrenergic receptors in lateral saphenous veins (cranial branch) of cattle grazing tall fescue. *J. Anim. Sci.* **1998**, *76*, 2853–2856. [CrossRef]
27. Roquebert, J.; Grenie, B. Heart rate lowering effects of dihydroergotamine in rats. *Arch. Int. Pharmacodyn. Ther.* **1987**, *290*, 25–35.
28. Roquebert, J.; Grenie, B. Alpha 2-adrenergic agonist and alpha 1-adrenergic antagonist activity of ergotamine and dihydroergotamine in rats. *Arch. Int. Pharmacodyn. Ther.* **1986**, *284*, 30–37. [PubMed]
29. MacLennan, S.J.; Martin, G.R. Actions of non-peptide ergot alkaloids at 5-HT1-like and 5-HT2 receptors mediating vascular smooth muscle contraction. *Naunyn-Schmiedeberg's Arch. Pharmacol.* **1990**, *342*, 120–129. [CrossRef]
30. El-Etr, A.A.; Glisson, S.N. Alpha-Adrenergic Blocking Agents. *Int. Anesthesiol. Clin.* **1978**, *16*, 239–260. [CrossRef] [PubMed]
31. Tomb, J.W. Ergotoxine in Shock. *Br. Med. J.* **1941**, *1*, 764. [CrossRef]
32. Stoll, A.; Hofmann, A. Chapter 21 The Ergot Alkaloids. *Alkaloids Chem. Physiol.* **1965**, *8*, 725–783.
33. Roquebert, J.; Demichel, P. Agonist/Antagonist Activity of Ergocristine at Alpha-Adrenoceptors in the Rat. *Fundam. Clin. Pharmacol.* **1987**, *1*, 23–33. [CrossRef] [PubMed]
34. Roquebert, J.; Malek, A.; Gomond, P.; Demichel, P. Effect of dihydroergocristine on blood pressure and activity at peripheral alpha-adrenoceptors in pithed rats. *Eur. J. Pharmacol.* **1984**, *97*, 21–27. [CrossRef]
35. Kalkman, H.O.; Van Gelderen, E.M.; Timmermans, P.B.; Van Zwieten, P.A. Involvement of α1- and α2-adrenoceptors in the vasoconstriction caused by ergometrine. *Eur. J. Pharmacol.* **1982**, *78*, 107–111. [CrossRef]
36. Doherty, M.L.; Murphy, M.G. Case report: Suspected ergot poisoning in cattle. *Ir. Vet. News* **1986**, *8*, 22–23.
37. Cherewyk, J.E.; Parker, S.E.; Blakley, B.R.; Al-Dissi, A.N. Assessment of the vasoactive effects of the (S)-epimers of ergot alkaloids in vitro. *J. Anim. Sci.* **2020**, *98*, skaa203. [CrossRef] [PubMed]

toxins

Article

Influence of Prolonged Serotonin and Ergovaline Pre-Exposure on Vasoconstriction Ex Vivo †

Eriton E. L. Valente [1], David L. Harmon [2] and James L. Klotz [3,*]

1 Animal Science Department, State University of Western Parana, Marechal Cândido Rondon 85960-000, PR, Brazil; eriton.valente@unioeste.br
2 Department of Animal and Food Science, University of Kentucky, Lexington, KY 40546, USA; dharmon@uky.edu
3 Forage-Animal Production Research Unit, USDA-ARS, Lexington, KY 40546, USA
* Correspondence: Klotz james.klotz@usda.gov; Tel.: +1-859-257-1647; Fax: +1-859-257-3334
† Mention of trade name, proprietary product of specified equipment does not constitute a guarantee or warranty by the USDA and does not imply approval to the exclusion of other products that may be available.

Abstract: Ergot alkaloid mycotoxins interfere in many functions associated with serotonergic neurotransmitters. Therefore, the objective was to evaluate whether the association of serotonin (5-hydroxytryptamine, 5-HT) and ergot alkaloids during a 24 h pre-incubation could affect the vascular contractile response to ergot alkaloids. To evaluate the effects of 24 h exposure to 5-HT and ergot alkaloids (ergovaline, ERV), two assays were conducted. The first assay determined the half-maximal inhibitory concentration (IC_{50}) following the 24 h pre-exposure period, while the second assay evaluated the effect of IC_{50} concentrations of 5-HT and ERV either individually or in combination. There was an interaction between previous exposure to 5-HT and ERV. Previous exposure to 5-HT at the IC_{50} concentration of 7.57×10^{-7} M reduced the contractile response by more than 50% of control, while the exposure to ERV at IC_{50} dose of 1.57×10^{-10} M tended to decrease ($p = 0.081$) vessel contractility with a response higher than 50% of control. The 24 h previous exposure to both 5-HT and ERV did not potentiate the inhibitory response of blood vessels in comparison with incubation with each compound alone. These results suggest receptor competition between 5-HT and ERV. More studies are necessary to determine the potential of 5-HT to treat toxicosis caused by ergot alkaloids.

Keywords: blood vessel; ergovaline; myograph; serotonin

Key Contribution: The inhibitory effect of prolonged exposure to 5-HT and ergot alkaloids on vascular contractility was demonstrated. The possible competition between 5-HT and ergot alkaloids for serotonergic receptors should be explored as a potential therapy for ergot toxicosis.

Citation: Valente, E.E.L.; Harmon, D.L.; Klotz, J.L. Influence of Prolonged Serotonin and Ergovaline Pre-Exposure on Vasoconstriction Ex Vivo. *Toxins* **2022**, *14*, 9. https://doi.org/10.3390/toxins14010009

Received: 5 November 2021
Accepted: 18 December 2021
Published: 23 December 2021

Publisher's Note: MDPI stays neutral with regard to jurisdictional claims in published maps and institutional affiliations.

1. Introduction

Ergot alkaloid mycotoxins have a significant impact on livestock health and productivity globally [1]. The similarities between the tetracyclic ergoline ring, common to naturally occurring ergot alkaloids, and the ring structure of the biogenic amine neurotransmitters allows ergot alkaloids, like ergovaline to interfere in the many functions associated with these neurotransmitters [2]. After a period of feeding ergot alkaloids, some serotonergic receptors become less responsive to serotonin (5-HT) [3–5]. Drugs that are derivatives of ergot alkaloids have an association and dissociation rate to serotonin $5\text{-}HT_2$ receptors that is much slower than 5-HT [6], causing long-term stimulation. This persistent stimulation of serotonergic receptors has been shown to result in blunted signaling [7] and this alters how tissues respond to receptor-driven stimuli. This is exemplified in ergot alkaloid mycotoxicosis by a sustained vasoconstriction that limits effective blood flow to visceral and peripheral tissues [2]. Many symptoms of ergotism, like fescue toxicosis, are related to vasoconstriction, which is associated with lower blood flow [8,9] causing gangrene [10] and affecting thermal regulation [11]. Currently, there is no effective treatment for mycotoxicosis caused by ergot alkaloids in livestock.

Agonist therapy is the use of stimulant-like medications to treat stimulant addictions and has proven an effective treatment for dependence on neuromodulator drugs, such as nicotine and opioids [12]. This strategy involves the administration of medications that share neurobiological mechanisms with the undesirable drug in pursuit of neurochemical normalization [13]. The exposure to ergot alkaloids can produce neurochemical deficits in 5-HT [14]. Thus, the ideal agonist therapy would normalize 5-HT dysfunction.

Traditionally, agonist therapy uses agonists with lower potency [12]. The less persistent receptor association with 5-HT in comparison with ergot alkaloid derivatives [6] qualifies 5-HT as a potential molecule for agonist therapy for ergot alkaloid toxicosis. Hypothetically, the increase of 5-HT in the synaptic cleft could intensify competition with ergot alkaloids at the serotonergic receptors and reduce the overstimulation through a reduction in ergot alkaloid receptor association. Prolonged overstimulation caused by elevated 5-HT is much more difficult to achieve in comparison to ergot alkaloids because of the many mechanisms that control 5-HT neuron firing [7]. However, there is no evidence to show if a controlled increase in 5-HT can compete with ergot alkaloids for receptors to offset and reduce the overstimulation caused by ergot alkaloids. The potential uses of 5-HT as agonist therapy for ergotism in cattle have not been evaluated previously. Therefore, the objective was to evaluate whether the association of 5-HT and ergovaline in a 24 h pre-incubation could affect the vascular contractile response.

2. Results

2.1. Pre-Exposure to Serotonin

Bovine lateral saphenous veins were constricted as evidenced by decreased ($p < 0.05$) internal and external diameters after the 24 h pre-incubation with 5-HT at doses of 1×10^{-7} M or higher (Table 1). However, no difference ($p > 0.05$) was observed in vessel wall thickness. The pre-incubation with 5-HT produced a dose-dependent decrease in vessel contractility when exposed to increasing concentrations of 5-HT (Figure 1). Pre-incubation concentrations of 1×10^{-6} M 5-HT and higher resulted in contractile responses to 5-HT that were lower ($p < 0.05$) than control (Figure 2A). The previous exposure dose of 1×10^{-4} and 1×10^{-5} M 5-HT almost completely suppressed the contractile response of the vessel. The resultant IC_{50} determined from a 24 h previous exposure to increasing concentrations of 5-HT was 7.57×10^{-7} M (Figure 2B).

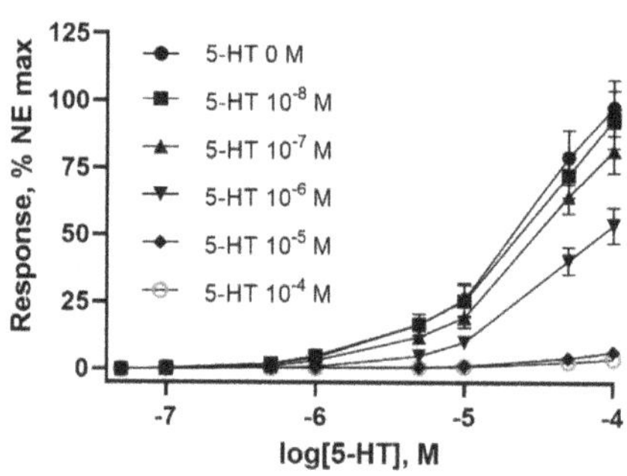

Figure 1. Effects of pre-incubation with serotonin (5-HT) at concentrations of 0.1×10^{-8}, 1×10^{-7}, 1×10^{-6}, 1×10^{-5}, 1×10^{-4} M on contractile responses in the isolated lateral saphenous vein from cattle exposed to increasing concentrations of 5-HT. Points represent the mean values and vertical bars show the SEM, $n = 12$.

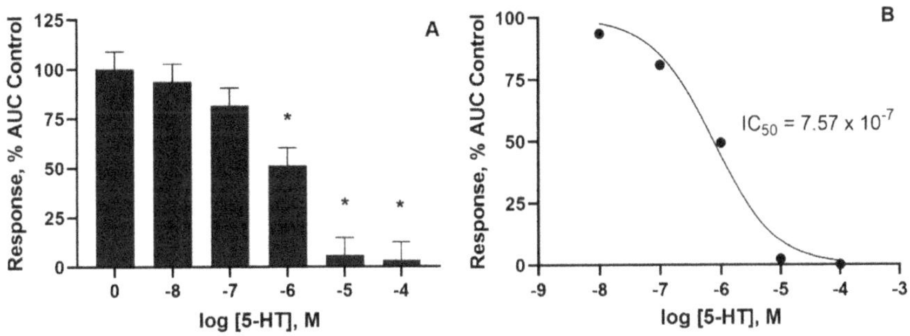

Figure 2. (**A**) Effect of pre-incubation with serotonin (5-HT) on contractile concentration–response curve represented as area under curve (AUC) relative to control. * Indicates a difference relative to control ($p < 0.05$). (**B**)The half-maximal inhibitory concentration (IC_{50}) was 7.57×10^{-7} M, $n = 12$.

Table 1. Dimensions of bovine lateral saphenous veins after a 24 h incubation with buffer containing 5-HT.

	5-HT, log [M]						SE	*p*-Value
	0	−8	−7	−6	−5	−4		
Length, mm	2.14	2.17	2.27	2.36 *	2.22	2.21	0.06	0.001
ID, mm [1]	1.41	1.32	0.89 *	0.87 *	0.89 *	0.86 *	0.19	<0.001
OD, mm [2]	2.97	2.78	2.55 *	2.33 *	2.47 *	2.47 *	0.41	<0.001
Wall, mm [3]	0.78	0.73	0.83	0.73	0.79	0.8	0.13	0.026

* Indicates a difference relative to control (0 M) by Dunnett test, $p < 0.05$. [1] Internal diameter. [2] External diameter. [3] Wall thickness.

2.2. Pre-Exposure to Ergovaline

Bovine lateral saphenous veins that were pre-incubated with ERV (in a tall fescue seed extract) at doses of 1×10^{-9} M or higher decreased ($p < 0.05$) the internal diameter while doses of 1×10^{-10} M or higher decreased the external diameter of the vessels (Table 2). Similar to 5-HT, the vessel wall thickness was not affected ($p > 0.05$) by a 24 h pre-incubation with ERV. Like pre-incubation with 5-HT, the pre-incubation with ERV produced a dose-dependent decrease in contractility in vessels incubated with increasing concentrations of ERV (Figure 3). Only the pre-incubation with 1×10^{-11} M ERV produced little effect on vessel contractility, whereas the 1×10^{-7} M pre-incubation dose almost completely suppressed the contraction of the vessel (Figure 4A). The IC_{50} for ERV from a 24 h exposure was 1.5×10^{-10} M (Figure 4B).

Table 2. Dimensions of bovine lateral saphenous veins after a 24 h incubation with buffer containing ergovaline.

	Ergovaline [1], log [M]						SE	*p*-Value
	0	−11	−10	−9	−8	−7		
Length, mm	2.32	2.34	2.27	2.3	2.52	2.21	0.07	0.019
ID, mm [2]	1.06	0.99	0.91	0.84 *	0.65 *	0.75 *	0.13	<0.001
OD, mm [3]	2.76	2.57	2.51 *	2.40 *	2.29 *	2.43 *	0.36	<0.001
Wall, mm [4]	0.85	0.79	0.80	0.78	0.82	0.84	0.12	0.463

* Indicates a difference relative to control (0 M) by Dunnett test, $p < 0.05$. [1] Tall fescue seed purified extract. [2] Internal diameter. [3] External diameter. [4] Wall thickness.

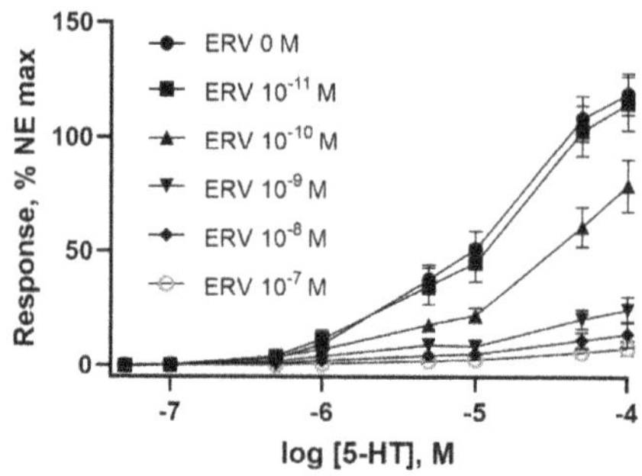

Figure 3. Effects of pre-incubation with ergovaline (ERV) at concentrations of 0, 1×10^{-11}, 1×10^{-10}, 1×10^{-9}, 1×10^{-8}, 1×10^{-7} M on the contractile responses to serotonin (5-HT) in the isolated lateral saphenous vein from cattle. Points represent the mean values and vertical bars show the SEM, $n = 12$.

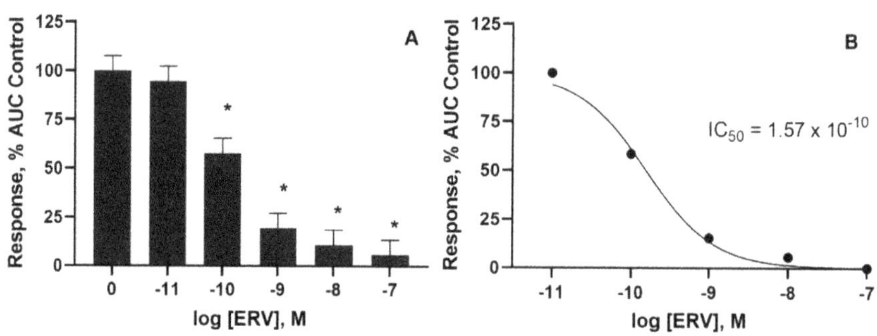

Figure 4. (**A**) Effect of pre-incubation with ergovaline on contractile concentration–response curve represented as area under curve (AUC) relative to control. * Indicates a difference relative to control $p < 0.05$. (**B**) The half-maximal inhibitory concentration (IC_{50}) was 1.57×10^{-10} M, $n = 12$.

2.3. Pre-Exposure to Combined Serotonin and Ergovaline

The final assay evaluated the effects of the IC_{50} concentrations of ERV and 5-HT alone and 5-HT + ERV in combination during a pre-incubation on vessel dimensions (Table 3) and subsequent contractile response to increasing concentrations of 5-HT (Figure 5). There were no interactions ($p > 0.05$) between treatments for blood vessel length, internal diameter, or external diameter. Bovine lateral saphenous veins pre-incubated with 5-HT at the IC_{50} concentration had greater ($p < 0.05$) length, lower ($p < 0.05$) internal and external diameters, and no difference ($p > 0.05$) in wall thickness. Conversely, those pre-incubated with ERV at the IC_{50} concentration tended to have a shorter ($p = 0.081$) length with no difference ($p > 0.05$) in internal and external diameter, or wall thickness.

There was an interaction ($p < 0.05$) between pre-incubation with 5-HT and ERV on the contractile response to increasing 5-HT, as seen in the AUC response. Previous exposure to 5-HT or 5-HT + ERV reduced ($p < 0.05$) the contractile response by an average of 60% while the pre-incubation with ERV at the IC50 concentration decreased the AUC by only 26%. Interestingly, the 24 h pre-incubation with both 5-HT (7.57×10^{-7} M) and ERV (1.57×10^{-10} M) did not potentiate the inhibitory response (Figure 5). There was no interaction ($p > 0.05$) between 5-HT and ERV on the maximum response to norepinephrine (Table 3). The pre-incubation with 5-HT decreased ($p < 0.05$) the blood vessel contractility response to norepinephrine. Conversely, the pre-incubation with ERV did not affect ($p > 0.05$) the response to norepinephrine.

Table 3. Area under curve (AUC) of contractile concentration–response curve to increasing concentrations of serotonin (5-HT) and vascular dimensions of bovine lateral saphenous veins pre-incubated with buffer containing 5-HT, ergovaline (ERV), or both 5-HT and ERV.

	Treatments [1]					p-Value [2]		
	CTRL	5-HT	ERV	5-HT + ERV	SE	5-HT	ERV	5-HT × ERV
Length, mm	2.50	2.59	2.43	2.64	0.06	0.006	0.081	0.267
ID, mm [5]	1.21	0.90	1.19	0.79	0.14	<0.001	0.44	0.571
OD, mm [6]	3.16	2.84	3.12	2.76	0.17	<0.001	0.498	0.838
Wall, mm [7]	0.976	0.969	0.964	0.986	0.05	0.785	0.924	0.565
AUC [3]	98.4	37.9	72.7	40.7	9	<0.001	0.096	0.039
NE Response [4]	98.9	49.8	92.3	51.7	6.8	<0.001	0.57	0.313

[1] CTRL= only buffer; 5-HT = buffer with serotonin at IC_{50} concentration (7.57×10^{-7} M); ERV = buffer with ergovaline at IC_{50} concentration (1.57×10^{-10} M); 5-HT + ERV = buffer with serotonin (7.57×10^{-7} M) and ergovaline (1.57×10^{-10} M). [2] Main effects of serotonin (5-HT), ergovaline (ERV) and interaction (5-HT × ERV). [3] Area relative to control = 100. [4] Contractile response to norepinephrine (NE, 1×10^{-4} M). [5] Internal diameter. [6] External diameter. [7] Wall thickness.

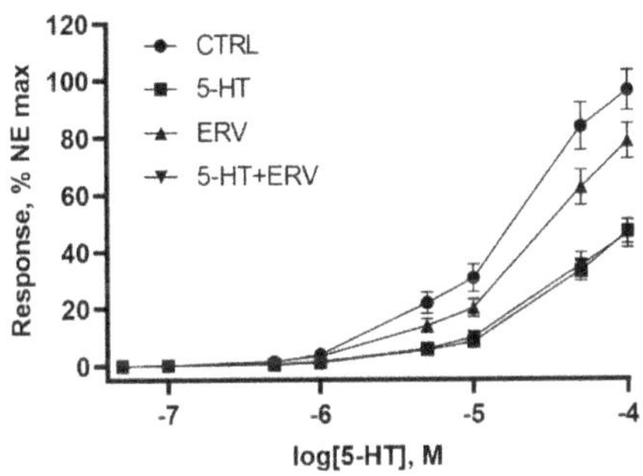

Figure 5. Effect of pre-incubation with buffer (CTRL), serotonin (5-HT), ergovaline (ERV), and 5-HT + ERV on contractile concentration–response curve to 5-HT in the isolated lateral saphenous vein from cattle, $n = 13$ to 15.

3. Discussion

This study aimed to perform the first evaluation of the use of 5-HT to mitigate the negative effects of ergot alkaloids on blood vessel contractility following a prolonged ex vivo pre-incubation. The pre-incubation with 5-HT modified the contractility of vessels exposed at physiological concentrations. The results suggest that 5-HT might compete with ergot alkaloids for serotonergic receptor binding.

The 24 h incubation with 5-HT and ERV represented normal cellular conditions in an animal. It has been established that 5-HT receptors mediate the contraction of the bovine lateral saphenous vein [15,16]. The reduction of vessel diameter occurred at 5-HT doses below physiological concentrations (1×10^{-7} M). Previous studies have found that bovine plasma 5-HT concentrations are close to 1×10^{-6} M [17,18]. Platelets contain a highly efficient transporter that controls 5-HT in the plasma, which represents the free active 5-HT [19]. The prolonged exposure of a static 5-HT concentration during the pre-incubation may have caused an overstimulation of the receptors causing a persistent contracted state.

The reduction of vessel contractility in the myograph is evidence that prolonged overstimulation can decrease the serotonergic receptor response. The concentration of 5-HT near the nerve terminal may substantially alter the activation or desensitization of serotonergic receptors [20]. Chronic infusion of 5-HT produces a residual desensitization of the receptors in many organs [21]. During the contractility evaluation in the myograph, the response to high doses of 5-HT applied for 15 min were rapidly washed out. However, the effect of prolonged exposure to 5-HT during the 24 h pre-incubation was not as readily washed out and a drastic reduction of vessel vasoactivity to 5-HT was observed.

Similar to 5-HT, pre-incubation with ERV provoked an intense reduction of the lateral saphenous vein diameter. In contrast, Trotta et al. [5] did not find a difference in the diameter of bovine ruminal or mesenteric blood vessels after a 2 h incubation with 1×10^{-8} or 1×10^{-6} M ERV. However, in the current study, the pre-incubation with ERV with lateral saphenous veins decreased the vessel diameter and the contractility at much lower concentrations. The inclusion of a control treatment ensured that the changes in contractility were caused by the ERV and not by loss of viability. The dose-dependent decrease in contractility after pre-incubation with ERV and the occurrence of this effect at a very low dose (1×10^{-10} M) was evidence that the effect of ergot alkaloids on blood vessels is mediated by both the dose and the time that the receptors are exposed. Without previous exposure, the ergopeptine ergot alkaloids normally reduce the contractility of vessels at concentrations higher than 1×10^{-7} to 1×10^{-6} M [4,22,23], which are considerably higher than the ergovaline levels expected in plasma of ruminants consuming ergot alkaloids [24]. The difference in the concentration of ergot alkaloids and the observed responses in the literature indicates that the duration of exposure can influence the ergot alkaloid effect.

Reductions in the contractile response to 5-HT in lateral saphenous veins collected from endophyte-infected tall fescue of grazing steers has been reported [25]. The reduction in the contractile response to 5-HT of vessels can be associated with a reduction of mRNA of both 5-HT receptors [26,27] and calcium channels [26]. The persistent contracted state of the vessels after ergot alkaloid administration and resistance to washout [15] can be caused by the lower receptor dissociation of the ergot alkaloids [6]. However, prolonged exposure to 5-HT also decreased the relaxation of vessels and the contractility in the organ bath, demonstrating that many mechanisms are involved in changing vessel contractility, rather than receptor kinetics alone.

The lower contractility of blood vessels exposed to ergot alkaloids can be partly explained by the reduction of 5-HT receptors [6]. However, the reduction of contractility after chronic exposure to 5-HT, which is easily washed out compared to ergovaline, indicates other mechanisms also act to modify the contractile response. One plausible mechanism is that the overstimulation of serotonergic receptors, either by chronic 5-HT or ergot alkaloid exposure, can alter calcium dynamics and consequently the smooth muscle contractility.

The administration of ergot alkaloids does not affect the maximum contractile response of vessels to KCl [4,5]. The KCl-derived stimulation bypasses G protein-coupled receptors and activates smooth muscle by the activation of voltage-operated Ca^{2+} channels [28]. Although it is not clear how ergot alkaloids act on smooth muscle to influence the contractile response, it is possible that they act on G protein-coupled receptors and in different parts of the calcium pathway, other than by voltage-operated Ca^{2+} channels.

Although prolonged 5-HT exposure mimicked the response of ergot alkaloids, the difference in concentration for a similar response was large. The IC_{50} concentration for ERV was almost 5000-fold lower than for 5-HT. The chronic infusion of 5-HT produces a residual desensitization of serotonergic receptors in smooth muscle [21]. The prolonged stimulation of 5-HT receptors leads to phosphorylation, which modifies cell surface expression, coupling profiles, and interactions with protein partners, usually leading to blunted signaling [7]. Serotonin has many mechanisms to control synaptic signaling and prevent neural malfunction. Conversely, the ergot derivatives do not have these mechanisms to control receptor stimuli and their long durations of action are not easily attributed to circulating levels [6]. The vasoconstriction caused by ergot alkaloids can provoke several modifications in cattle, including decreased blood flow and increased blood pressure [9,11,29]. These physiologic modifications can result in localized gangrene and the loss of animal productivity [2,10].

The combined incubation with 5-HT and ERV did not potentiate the inhibitory effect on contractility compared with these substances individually. The competition for receptors may have reduced the potential inhibitory activity when they were combined. These results indicate that an increase in 5-HT has the potential to mitigate vasoconstriction caused by ergot alkaloids and the refinement of optimal concentrations is warranted. The reduction of the contractile response of vascular smooth muscle to norepinephrine after prolonged exposure to 5-HT or ERV indicates that the residual effects of ergot alkaloids can be associated with triggering the G-protein coupled pathways that can be activated by serotonergic as well as adrenergic receptors. By simultaneously administering ERV and a selective agonist for the 5-HT_{2A} receptor, Trotta et al. [5] observed that the increase in the agonist concentration decreased the contraction caused by ERV in ruminal and mesenteric veins. The use of pharmacologically similar agents as a substitute for an undesired drug is considered agonist therapy [30]. Its efficacy has been proven for dependence on many substances, such as nicotine and opioids [31,32]. The kinetics of ergovaline at vascular serotonergic receptors complicates the development of treatment protocols, yet the prophylactic administration of molecules to modify the 5-HT receptors binding might be effective [33].

In contrast with many illicit drugs, ergotism acts directly on monoamine receptors instead of causing an increase of monoamines in the presynaptic terminal [2]. Thus, strategies to avoid receptor binding potentially achieve better results by reducing neuronal overstimulation and prolonged signaling. Ergot alkaloids (e.g., ergotamine) are extensively metabolized by the liver and cleared from the blood by first-pass hepatic metabolism [34,35]. Jaussaud [24] observed a half-life of 23.6 min for ERV in sheep, which suggests that strategies to avoid receptor binding can accelerate its clearance, reducing the neuronal excitation by the circulating ergot alkaloid.

The 5-HT receptors have complex and integrated functions throughout the body. Although many 5-HT receptors can cause vasoconstriction, activation of the 5HT_7 receptor can provoke a hypotensive response [36]. The stimulation of the 5-HT_7 receptor reduces the contraction induced by the 5-HT_{2A} receptor in the isolated abdominal vena cava [36]. It has been shown that the chronic administration of 5-HT can provoke a sustained fall in blood pressure in rats [37]. Therefore, an increase of 5-HT has the potential to activate receptors for relaxation responses, as well as to compete with ergot alkaloids reducing the persistent overstimulation of the receptor with contractile responses.

4. Conclusions

The increased exposure of blood vessels to 5-HT appears to compete with ergot alkaloids through receptors in the vascular smooth muscle, indicating that the manipulation of 5-HT has the potential to be used as a treatment of the vasoconstrictive effects induced by ergovaline. However, it is not clear how prolonged exposure to 5-HT and ergot alkaloids can reduce receptor sensitivity. This is further complicated by the fact that ergovaline has a 5000-fold higher potency than 5-HT. Future research should focus on the study of mechanisms related to the induction of blood vessel relaxation when under prolonged serotonergic stimulation, test doses of 5-HT or related compounds, and the evaluation of these in animal models.

5. Materials and Methods

No live animals were involved in the experiments that make up this study; thus, approval from the University of Kentucky Animal Care and Use Committee was not required.

5.1. Tissue Collection

The cranial branch of the lateral saphenous vein was collected from 14 predominantly Angus heifers (448 $\pm$ 21 kg) immediately after slaughter at the University of Kentucky Meats Laboratory using procedures according to Klotz [38]. Veins from nine heifers were used for calculation of IC_{50} (12 vessels sections per treatment) and veins from five heifers were used for testing the IC_{50} doses (15 vessels sections per treatment). As part of a separate study, the heifers were kept in the barn and maintained on a silage-based diet for 17 days prior to slaughter. Prior to that, they were managed on endophyte-infected tall fescue pastures with the same silage diet for at least 90 days. Segments (10 cm in length) of lateral saphenous vein at the cranial branch were removed and placed in a modified

Krebs–Henseleit buffer for transport (95% O_2/5% CO_2; pH = 7.4; 11.1 mM D-glucose; 1.2 mM MgSO4; 1.2 mM KH_2PO_4; 4.7 mM KCl; 118.1 mM NaCl; 3.4 mM $CaCl_2$; 24.9 mM $NaHCO_3$; Sigma Chemical Co., St. Louis, MO), and were kept on ice during transportation from the collection site to the laboratory.

Blood vessels were dissected and the surrounding adipose and connective tissues were removed. The clean vessel segments were sliced into 2 mm cross-sections using an adjustable acrylic tissue matrix (Braintree Scientific, Inc., Braintree, MA, USA). The vessel cross-sections were inspected under magnification (12.5×) for abnormalities (e.g., structural damage incurred during dissection and points of vascular branches), and abnormal sections were discarded and replaced with viable sections.

5.2. Previous Exposure of Vascular 5-HT Receptors

To simulate the previous exposure of 5-HT receptors in the blood vessels to the treatments, the vein cross-sections were incubated at 37 °C in 5% of CO_2 for 24 h in a 12-well culture plate (1 cross section/well) with 5 mL of Krebs–Henseleit buffer containing either control, serotonin, tall fescue seed extract (source of ergovaline; ERV), or a combination of the treatments. The previous exposures were conducted in two separate assays. The first assay determined the IC_{50} for 5-HT and the ergovaline-containing extract. The second assay evaluated the effect of IC_{50} doses of 5-HT and ergovaline determined in the previous assay, either individually or in combination.

5.3. IC_{50} Determination

To determine the IC_{50}, stock solutions of 5-HT (Sigma-Aldrich, St. Louis, MO) were diluted to the corresponding final working concentrations of 1×10^{-8}, 1×10^{-7}, 1×10^{-6}, 1×10^{-5} and 1×10^{-4} M. Working concentration ranges were prepared based on previous bovine vascular bioassay research using 5-HT [3]. Tall fescue seed extract was prepared and purified as described by Ji et al. [39] and Foote et al. [40]. Stock solutions of the extract were based on ERV concentrations (ergovaline + ergovalinine) and were prepared to final working concentrations of 1×10^{-11}, 1×10^{-10}, 1×10^{-9}, 1×10^{-8} and 1×10^{-7} M. Working concentration ranges were prepared based on previous bovine vascular bioassay studies using ERV [3,41]. The vein segments were incubated with Krebs–Henseleit buffer alone or with working solutions of 5-HT or ERV for 24 h before contractile response evaluation.

5.4. Associated 5-HT and ERV Pre-Exposure

The blood vessels were incubated for 24 h with: (1) exclusively Krebs–Henseleit buffer; (2) Krebs–Henseleit buffer with 5-HT at IC_{50} concentration; (3) Krebs–Henseleit buffer with ERV at IC_{50} concentration; (4) Krebs–Henseleit buffer with 5-HT and ERV both at their respective IC_{50} concentrations. Following the completion of the 24 h pre-incubation period, the contractile response to increasing concentrations of 5-HT of each vein cross-section was evaluated.

5.5. Vascular Dimensions

Following the 24 h incubation, cross-sections were removed from treatment-containing buffer and examined using a dissecting microscope (Stemi 2000-C, Carl Zeiss Inc., Oberkochen, Germany) at 12.5× magnification to measure dimensions of the vessels. Vessel cross-sections were then transferred to a treatment-free Krebs–Henseleit buffer until mounting in the myograph chambers.

5.6. Contractile Response

After the 24 h incubation period of the treatments, blood vessels were mounted onto luminal supports in individual chambers of a multimyograph (DMT 610M, Danish Myo Technology, Atlanta, GA) with 5 mL Krebs–Henseleit buffer and constant gassing (95% O_2/5% CO_2; pH = 7.4; 37 °C). For each myograph experiment, the incubation treatments were run in duplicate during the determination of 5-HT and ERV IC_{50} and run in triplicate during evaluation of the determined IC_{50} doses associated with 5-HT and ERV.

The incubation buffer had the same composition of the transport buffer plus 3×10^{-5} M desipramine (Sigma Chemical, Co.) and 1×10^{-6} M propranolol (Sigma Chemical, Co). These compounds were included to inhibit biogenic amine reuptake and non-specific binding to β-adrenergic receptors, respectively. After mounting the blood vessels on the luminal supports in the myograph, an equilibration period was conducted with constant gassing (95% O_2/5% CO_2; pH = 7.4; 37 °C) for 90 min, with buffer changes every 15 min to allow blood vessels to reach a resting tension of approximately 1 g. At completion of the 90 min equilibration period, 1×10^{-4} M norepinephrine (Sigma Chemical, Co) was added to each chamber and incubated for 15 min. This served as a reference and to confirm vessel responsiveness and viability. Myograph chambers were then emptied and refilled with incubation buffer to remove norepinephrine and allow the vessels to return to an approximate 1 g tension. Once vessels returned to a resting tension, cumulative additions of 5-HT were added to reach the final concentrations of 5×10^{-8}, 1×10^{-7}, 5×10^{-7}, 1×10^{-6}, 5×10^{-6}, 1×10^{-5}, 5×10^{-5}, and 1×10^{-4} M to build the contractile response curve. Afterwards, 1×10^{-4} M norepinephrine was again added to each chamber for evaluation of the final contractile response. Each 5-HT addition occurred in 15 min intervals in order of increasing concentration. Each cycle (addition) was a 9 min treatment incubation period, two 2.5 min buffer washes, a third buffer replacement, and followed by a 1 min recovery period leading into the next 5-HT addition.

5.7. Data Analysis

Isometric contractions in blood vessels to norepinephrine and 5-HT were digitized and recorded in grams of tension using PowerLab 16/35 and Chart software (version 8, ADInstruments, Colorado Springs, CO, USA). Baseline tensions were recorded before addition of each compound. For all contractile response data, the maximum tension (measured in g) during the 9 min incubation was recorded as the contractile response and corrected for baseline tension.

Due to variation between animals and tissues, contractile response data were normalized as a percentage of the maximum of tension induced by the reference addition of norepinephrine to compensate for differences in vessel responsiveness. Vessel contractile response data were reported as the percentage mean contractile response ± SEM of the maximum contractile response produced by the 1×10^{-4} M norepinephrine reference addition in vessels incubated for 24 h without treatment compounds. All contractile response data presented were plotted using GraphPad Prism software (San Diego, CA, USA). Sigmoidal concentration response curves to each treatment were calculated and plotted using a nonlinear regression equation:

$$y = \frac{100}{1 + 10^{(b-X)*a}}$$

where y is the contractile response, X is the concentration of the test solution, a is the Hill slope, and b is the concentration corresponding to the response midway between bottom and top (IC_{50}).

5.8. Statistical Analysis

All data were analyzed as a completely randomized design using the MIXED model of SAS with the fixed effect of previous exposure treatment and the random effect of the incubation run.

The area under curve (AUC) was calculated using a linear trapezoidal method with baseline set to zero. The AUC was transformed, setting the control value to 100. Comparison of AUC and dimension of vessels to control in previous exposure doses was conducted by Dunnett test.

The heifer from which the vessel was collected and the myograph run were included as random effects to compare associated 5-HT and ERV previous exposure. Pairwise comparisons were conducted using the LSD test. Statistical significance was considered at $p \leq 0.05$ and tendency at $0.05 < p \leq 0.10$.

Author Contributions: Conceptualization, E.E.L.V., D.L.H., J.L.K.; formal analysis, E.E.L.V., D.L.H., J.L.K.; investigation, E.E.L.V., J.L.K.; writing—original draft preparation, E.E.L.V., D.L.H., J.L.K.; writing—review and editing, E.E.L.V., D.L.H., J.L.K.; funding acquisition, J.L.K. All authors have read and agreed to the published version of the manuscript.

Funding: This work is funded by Hatch Capacity Grant Project no. KY007088 from the USDA National Institute of Food and Agriculture and Project 201807121511 from USDA/ARS and the University of Kentucky Agricultural Experiment Station.

Institutional Review Board Statement: Not applicable.

Informed Consent Statement: Not applicable.

Data Availability Statement: The data presented in this study are available on request from the corresponding author.

Acknowledgments: The authors acknowledge Adam J. Barnes of the Forage-Animal Production Research Unit (Lexington, KY, USA) for his technical assistance.

Conflicts of Interest: The authors declare no conflict of interest.

References

1. Poole, D.H.; Mayberry, K.J.; Newsome, M.; Poole, R.K.; Galliou, J.M.; Khanal, P.; Poore, M.H.; Serão, N.V.L. Evaluation of Resistance to Fescue Toxicosis in Purebred Angus Cattle Utilizing Animal Performance and Cytokine Response. *Toxins* **2020**, *12*, 796. [CrossRef]
2. Klotz, J.L. Activities and effects of ergot alkaloids on livestock physiology and production. *Toxins* **2015**, *7*, 2801–2821. [CrossRef]
3. Klotz, J.L.; Aiken, G.E.; Johnson, J.M.; Brown, K.R.; Bush, L.P.; Strickland, J.R. Antagonism of lateral saphenous vein serotonin receptors from steers grazing endophyte-free, wild-type, or novel endophyte-infected tall fescue. *J. Anim. Sci.* **2013**, *91*, 4492–4500. [CrossRef]
4. Egert, A.M.; Kim, D.H.; Schrick, F.N.; Harmon, D.L.; Klotz, J.L. Dietary exposure to ergot alkaloids decreases contractility of bovine mesenteric vasculature. *J. Anim. Sci.* **2014**, *92*, 1768–1779. [CrossRef]
5. Trotta, R.J.; Harmon, D.L.; Klotz, J.L. Interaction of ergovaline with serotonin receptor 5-HT2A in bovine ruminal and mesenteric vasculature. *J. Anim. Sci.* **2018**, *96*, 4912–4922. [CrossRef]
6. Unett, D.J.; Gatlin, J.; Anthony, T.L.; Buzard, D.J.; Chang, S.; Chen, C.; Chen, X.; Dang, H.T.M.; Frazer, J.; Le, M.K.; et al. Kinetics of 5-HT 2B Receptor Signaling: Profound Agonist-Dependent Effects on Signaling Onset and Duration. *J. Pharmacol. Exp. Ther.* **2013**, *347*, 645–659. [CrossRef]
7. Millan, M.; Marin, P.; Bockaert, J.; Mannourylacour, C. Signaling at G-protein-coupled serotonin receptors: Recent advances and future research directions. *Trends Pharmacol. Sci.* **2008**, *29*, 454–464. [CrossRef]
8. Aiken, G.E.; Kirch, B.H.; Strickland, J.R.; Bush, L.P.; Looper, M.L.; Schrick, F.N. Hemodynamic responses of the caudal artery to toxic tall fescue in beef heifers. *J. Anim. Sci.* **2007**, *85*, 2337–2345. [CrossRef]
9. Aiken, G.E.; Strickland, J.R.; Looper, M.L.; Bush, L.P.; Schrick, F.N. Hemodynamics are altered in the caudal artery of beef heifers fed different ergot alkaloid concentrations. *J. Anim. Sci.* **2009**, *87*, 2142–2150. [CrossRef]
10. Strickland, J.R.; Aiken, G.E.; Spiers, D.E.; Fletcher, L.R.; Oliver, J.W. Physiological Basis of Fescue Toxicosis. In *Tall Fescue for the Twenty-first Century. Agronomy Monograph 53*; Fribourg, H.A., Hannaway, D.B., West, C.P., Eds.; American Society Agron: Madison, WI, USA, 2015; pp. 203–227.
11. Eisemann, J.H.; Huntington, G.B.; Williamson, M.; Hanna, M.; Poore, M. Physiological responses to known intake of ergot alkaloids by steers at environmental temperatures within or greater than their thermoneutral zone. *Front. Chem.* **2014**, *2*, 96. [CrossRef]
12. Jordan, C.J.; Cao, J.; Newman, A.H.; Xi, Z.-X. Progress in agonist therapy for substance use disorders: Lessons learned from methadone and buprenorphine. *Neuropharmacology* **2019**, *158*, 107609. [CrossRef]
13. Rothman, R.B.; Baumann, M.H. Balance between Dopamine and Serotonin Release Modulates Behavioral Effects of Amphetamine-Type Drugs. *Ann. N. Y. Acad. Sci.* **2006**, *1074*, 245–260. [CrossRef]
14. Valente, E.E.L.; Klotz, J.L.; Ahn, G.; McLeod, K.R.; Herzing, H.M.; King, M.; Harmon, D.L. Ergot alkaloids reduce circulating serotonin in the bovine. *J. Anim. Sci.* **2020**, *98*, skaa362. [CrossRef]
15. Pesqueira, A.; Harmon, D.L.; Branco, A.F.; Klotz, J.L. Bovine lateral saphenous veins exposed to ergopeptine alkaloids do not relax. *J. Anim. Sci.* **2014**, *92*, 1213–1218. [CrossRef]

16. Klotz, J.L.; Kirch, B.H.; Aiken, G.E.; Bush, L.P.; Strickland, J.R. Contractile response of fescue-naïve bovine lateral saphenous veins to increasing concentrations of tall fescue alkaloids. *J. Anim. Sci.* **2010**, *88*, 408–415. [CrossRef]
17. Valente, E.E.L.; Klotz, J.L.; Harmon, D.L. 5-Hydroxytryptophan strongly stimulates serotonin synthesis in Holstein steers. *Domest. Anim. Endocrinol.* **2021**, *74*, 106560. [CrossRef]
18. Valente, E.E.L.; Damasceno, M.L.; Klotz, J.L.; Harmon, D.L. Residual effects of abomasal 5-hydroxytryptophan administration on serotonin metabolism in cattle. *Domest. Anim. Endocrinol.* **2021**, *76*, 106627. [CrossRef]
19. Mercado, C.P.; Kilic, F. Molecular Mechanisms of SERT in Platelets: Regulation of Plasma Serotonin Levels. *Mol. Interv.* **2010**, *10*, 231–241. [CrossRef]
20. Bertrand, P.P.; Bertrand, R.L. Serotonin release and uptake in the gastrointestinal tract. *Auton. Neurosci.* **2010**, *153*, 47–57. [CrossRef]
21. McLean, P.G.; Coupar, I.M.; Molenaar, P. Changes in sensitivity of 5-HT receptor mediated functional responses in the rat oesophagus, fundus and jejunum following chronic infusion with 5-hydroxytryptamine. *Naunyn- Schmiedebergs. Arch. Pharmacol.* **1996**, *354*, 513–519. [CrossRef]
22. Foote, A.P.; Harmon, D.L.; Strickland, J.R.; Bush, L.P.; Klotz, J.L. Effect of ergot alkaloids on contractility of bovine right ruminal artery and vein. *J. Anim. Sci.* **2011**, *89*, 2944–2949. [CrossRef]
23. Cherewyk, J.E.; Parker, S.E.; Blakley, B.R.; Al-Dissi, A.N. Assessment of the vasoactive effects of the (S)-epimers of ergot alkaloids in vitro. *J. Anim. Sci.* **2020**, *98*, skaa203. [CrossRef]
24. Jaussaud, P.; Durix, A.; Videmann, B.; Vigié, A.; Bony, S. Rapid analysis of ergovaline in ovine plasma using high-performance liquid chromatography with fluorimetric detection. *J. Chromatogr. A* **1998**, *815*, 147–153. [CrossRef]
25. Klotz, J.L.; Aiken, G.E.; Egert-McLean, A.M.; Schrick, F.N.; Chattopadhyay, N.; Harmon, D.L. Effects of grazing different ergovaline concentrations on vasoactivity of bovine lateral saphenous vein. *J. Anim. Sci.* **2018**, *96*, 3022–3030. [CrossRef]
26. Povlsen, G.K.; Waldsee, R.; Ahnstedt, H.; Kristiansen, K.A.; Johansen, F.F.; Edvinsson, L. In vivo experimental stroke and in vitro organ culture induce similar changes in vasoconstrictor receptors and intracellular calcium handling in rat cerebral arteries. *Exp. Brain Res.* **2012**, *219*, 507–520. [CrossRef]
27. Klotz, J.L.; Kim, D.; Foote, A.P.; Harmon, D.L. Effects of ergot alkaloid exposure on serotonin receptor mRNA in the smooth muscle of the bovine gastrointestinal tract. In Proceedings of the Joint Meeting of the ADSA, AMSA, ASAS and PSA, Kansas City, MO, USA, 20–24 July 2014; p. 890p.
28. Ratz, P.H.; Berg, K.M.; Urban, N.H.; Miner, A.S. Regulation of smooth muscle calcium sensitivity: KCl as a calcium-sensitizing stimulus. *Am. J. Physiol. -Cell Physiol.* **2005**, *288*, C769–C783. [CrossRef]
29. Cowan, V.; Grusie, T.; McKinnon, J.; Blakley, B.; Singh, J. Arterial Responses in Periparturient Beef Cows Following a 9-Week Exposure to Ergot (Claviceps purpurea) in Feed. *Front. Vet. Sci.* **2019**, *6*, 262. [CrossRef]
30. Rush, C.R.; Stoops, W.W. Agonist replacement therapy for cocaine dependence: A translational review. *Future Med. Chem.* **2012**, *4*, 245–265. [CrossRef]
31. Rice, D.; Corace, K.; Wolfe, D.; Esmaeilisaraji, L.; Michaud, A.; Grima, A.; Austin, B.; Douma, R.; Barbeau, P.; Butler, C.; et al. Evaluating comparative effectiveness of psychosocial interventions adjunctive to opioid agonist therapy for opioid use disorder: A systematic review with network meta-analyses. *PLoS ONE* **2020**, *15*, e0244401. [CrossRef]
32. Cahill, K.; Lindson-Hawley, N.; Thomas, K.H.; Fanshawe, T.R.; Lancaster, T. Nicotine receptor partial agonists for smoking cessation. *Cochrane Database Syst. Rev.* **2016**. [CrossRef]
33. Schöning, C.; Flieger, M.; Pertz, H.H. Complex interaction of ergovaline with 5-HT2A, 5-HT1B/1D, and alpha1 receptors in isolated arteries of rat and guinea pig. *J. Anim. Sci.* **2001**, *79*, 2202–2209. [CrossRef] [PubMed]
34. Bigal, M.E.; Tepper, S.J. Ergotamine and dihydroergotamine: A review. *Curr. Pain Headache Rep.* **2003**, *7*, 55–62. [CrossRef]
35. Silberstein, S.D.; McCrory, D.C. Ergotamine and Dihydroergotamine: History, Pharmacology, and Efficacy. *Headache J. Head Face Pain* **2003**, *43*, 144–166. [CrossRef] [PubMed]
36. Gonzalez-Pons, R.; McRae, K.; Thompson, J.M.; Watts, S.W. 5-HT7 Receptor Restrains 5-HT–induced 5-HT2A Mediated Contraction in the Isolated Abdominal Vena Cava. *J. Cardiovasc. Pharmacol.* **2021**, *78*, 319–327. [CrossRef] [PubMed]
37. Davis, R.P.; Pattison, J.; Thompson, J.M.; Tiniakov, R.; Scrogin, K.E.; Watts, S.W. 5-hydroxytryptamine (5-HT) reduces total peripheral resistance during chronic infusion: Direct arterial mesenteric relaxation is not involved. *BMC Pharmacol.* **2012**, *12*, 4. [CrossRef]
38. Klotz, J.L.; Bush, L.P.; Smith, D.L.; Shafer, W.D.; Smith, L.L.; Vevoda, A.C.; Craig, A.M.; Arrington, B.C.; Strickland, J.R. Assessment of vasoconstrictive potential of D-lysergic acid using an isolated bovine lateral saphenous vein bioassay. *J. Anim. Sci.* **2006**, *84*, 3167–3175. [CrossRef]
39. Ji, H.; Fannin, F.; Klotz, J.; Bush, L. Tall fescue seed extraction and partial purification of ergot alkaloids. *Front. Chem.* **2014**, *2*, 110. [CrossRef]
40. Foote, A.P.; Harmon, D.L.; Brown, K.R.; Strickland, J.R.; McLeod, K.R.; Bush, L.P.; Klotz, J.L. Constriction of bovine vasculature caused by endophyte-infected tall fescue seed extract is similar to pure ergovaline. *J. Anim. Sci.* **2012**, *90*, 1603–1609. [CrossRef]
41. Egert-McLean, A.M.; Sama, M.P.; Klotz, J.L.; McLeod, K.R.; Kristensen, N.B.; Harmon, D.L. Automated system for characterizing short-term feeding behavior and real-time forestomach motility in cattle. *Comput. Electron. Agric.* **2019**, *167*, 105037. [CrossRef]

MDPI
St. Alban-Anlage 66
4052 Basel
Switzerland
Tel. +41 61 683 77 34
Fax +41 61 302 89 18
www.mdpi.com

Toxins Editorial Office
E-mail: toxins@mdpi.com
www.mdpi.com/journal/toxins

www.ingramcontent.com/pod-product-compliance
Lightning Source LLC
Chambersburg PA
CBHW040807250926
51850CB00007BA/1100
9783036537801